Rafael Bello, Rafael Falcón, Witold Pedrycz, Janusz Kacprzyk (Eds.)

# Granular Computing: At the Junction of Rough Sets and Fuzzy Sets

Studies in Fuzziness and Soft Computing, Volume 224

**Editor-in-Chief**

Prof. Janusz Kacprzyk
Systems Research Institute
Polish Academy of Sciences
ul. Newelska 6
01-447 Warsaw
Poland
E-mail: kacprzyk@ibspan.waw.pl

---

Further volumes of this series can be found on our homepage: springer.com

Vol. 210. Mike Nachtegael, Dietrich Van der Weken, Etienne E. Kerre, Wilfried Philips (Eds.)
*Soft Computing in Image Processing,* 2007
ISBN 978-3-540-38232-4

Vol. 211. Alexander Gegov
*Complexity Management in Fuzzy Systems,* 2007
ISBN 978-3-540-38883-8

Vol. 212. Elisabeth Rakus-Andersson
*Fuzzy and Rough Techniques in Medical Diagnosis and Medication,* 2007
ISBN 978-3-540-49707-3

Vol. 213. Peter Lucas, José A. Gàmez, Antonio Salmerón (Eds.)
*Advances in Probabilistic Graphical Models,* 2007
ISBN 978-3-540-68994-2

Vol. 214. Irina Georgescu
*Fuzzy Choice Functions,* 2007
ISBN 978-3-540-68997-3

Vol. 215. Paul P. Wang, Da Ruan, Etienne E. Kerre (Eds.)
*Fuzzy Logic,* 2007
ISBN 978-3-540-71257-2

Vol. 216. Rudolf Seising
*The Fuzzification of Systems,* 2007
ISBN 978-3-540-71794-2

Vol. 217. Masoud Nikravesh, Janusz Kacprzyk, Lofti A. Zadeh (Eds.)
*Forging New Frontiers: Fuzzy Pioneers I,* 2007
ISBN 978-3-540-73181-8

Vol. 218. Masoud Nikravesh, Janusz Kacprzyk, Lofti A. Zadeh (Eds.)
*Forging New Frontiers: Fuzzy Pioneers II,* 2007
ISBN 978-3-540-73184-9

Vol. 219. Roland R. Yager, Liping Liu (Eds.)
*Classic Works of the Dempster-Shafer Theory of Belief Functions,* 2007
ISBN 978-3-540-25381-5

Vol. 220. Humberto Bustince, Francisco Herrera, Javier Montero (Eds.)
*Fuzzy Sets and Their Extensions: Representation, Aggregation and Models,* 2007
ISBN 978-3-540-73722-3

Vol. 221. Gleb Beliakov, Tomasa Calvo, Ana Pradera
*Aggregation Functions: A Guide for Practitioners,* 2007
ISBN 978-3-540-73720-9

Vol. 222. James J. Buckley, Leonard J. Jowers
*Monte Carlo Methods in Fuzzy Optimization,* 2008
ISBN 978-3-540-76289-8

Vol. 223. Oscar Castillo, Patricia Melin
*Type-2 Fuzzy Logic: Theory and Applications,* 2008
ISBN 978-3-540-76283-6

Vol. 224. Rafael Bello, Rafael Falcón, Witold Pedrycz, Janusz Kacprzyk (Eds.)
*Granular Computing: At the Junction of Rough Sets and Fuzzy Sets,* 2008
ISBN 978-3-540-76972-9

Rafael Bello, Rafael Falcón, Witold Pedrycz,
Janusz Kacprzyk (Eds.)

# Granular Computing: At the Junction of Rough Sets and Fuzzy Sets

Springer

**Editors**


Prof. Rafael Bello
Univ. Central Las Villas
Depto. Ciencia Computacion
CEI
Carretera Camajuani km 5,5
54830 Santa Clara
Villa Clara
Cuba
Email: rbellop@uclv.edu.cu

Prof. Rafael Falcón
Univ. Central Las Villas
Depto. Ciencia Computacion
CEI
Carretera Camajuani km 5,5
54830 Santa Clara
Villa Clara
Cuba
Email: rfalcon@uclv.edu.cu

Prof. Dr. Witold Pedrycz
University of Alberta
Dept. Electrial & Computer
Engineering
9107 116 Street
Edmonton AB T6G 2V4
Canada
Email: pedrycz@ee.ualberta.ca

Prof. Dr. Janusz Kacprzyk
PAN Warszawa
Systems Research Instiute
Newelska 6
01-447 Warszawa
Poland
Email: kacprzyk@ibspan.waw.pl



ISBN 978-3-540-76972-9 e-ISBN 978-3-540-76973-6

Studies in Fuzziness and Soft Computing ISSN 1434-9922

Library of Congress Control Number: 2007942165





Typesetting by the authors and Scientific Publishing Services Pvt. Ltd.

Printed in acid-free paper

9 8 7 6 5 4 3 2 1

springer.com

To Elvita, strength and inspiration in tough times.
To Marilyn and my two daughters who have been vital in my life.

# Preface

Arising among an ensemble of methodologies for dealing with uncertainty in real-world problems, fuzzy and rough set theories share both a widespread use in the worldwide research community and exhibit a broad range of applications. Therefore a significant research effort being encountered today in Computational Intelligence has been devoted to these areas. A lot of innovative contributions in theoretical aspects as well as an increasing number of domains where fuzzy and rough sets have been successfully introduced are a convincing testimony to the dynamics of the area and its rapid advancements.

A recent term for grouping all theories or techniques dealing with information granules and information granulation for problem solving has been coined as Granular Computing. Information granulation can be developed in several ways and can be regarded as an important step forward when dealing with complex problem solving while overcoming many limitations present today in the traditional data-driven approach.

In this edited volume, several research papers originally submitted to the First International Symposium on Fuzzy and Rough Sets (ISFUROS 2006) held in Santa Clara, Cuba, have undergone a careful, critical review stage before becoming part of this publication. These papers clearly demonstrate the feasibility and usefulness of the methodology and algorithms of fuzzy sets and rough sets when applied to truly diversified domains such as e.g., language processing, video deinterlacing, image retrieval, evolutionary computation, bioinformatics and text mining.

We would like to express our thanks to the continuous support of the Program Committee of ISFUROS 2006 both at the initial stage of the Symposium and afterwards during the post-publication process. We do hope that the reader will greatly benefit from the potential of these methodologies.

Santa Clara, Cuba
October 2007

Rafael Bello
Rafael Falcon
Witold Pedrycz
Janusz Kacprzyk

# Contents

## Part II: Fuzzy and Rough Sets in Machine Learning and Data Mining



## Part III: Fuzzy and Rough Sets in Decision-Making

# List of Contributors

**Adnan Yazici**
Middle East Technical University
Ankara, Turkey
yazici@ceng.metu.edu.tr

**Akebo Yamakami**
Universidade Estadual de Campinas
13083-970, Campinas-SP, Brazil, 6101
akebo@dt.fee.unicamp.br

**Alain Varela**
Central University of Las Villas (UCLV)
Carretera Camajuaní km 5.5
Santa Clara, Cuba, 54830

**Andrés Soto**
Universidad Autónoma del Carmen
Campeche, Mexico, 24160
soto_andres@yahoo.com

**Boris Martínez**
Central University of Las Villas (UCLV)
Carretera Camajuaní km 5.5
Santa Clara, Cuba, 54830
boris@uclv.edu.cu

**Cristina Coppola**
Università degli Studi di Salerno
Via Ponte don Melillo
Fisciano (SA), Italy, 84084
ccoppola@unisa.it

**Daniela Bianucci**
Università di Milano–Bicocca
Via Bicocca degli Arcimboldi 8
Milano, Italy, I–20126

**Davide Ciucci**
Università di Milano–Bicocca
Via Bicocca degli Arcimboldi 8
Milano, Italy, I–20126
ciucci@disco.unimib.it

**Erick Marichal**
University of the Informatics Sciences (UCI)
Carretera San Antonio de los Baños km 2.5
Havana, Cuba
emarichal@uci.cu

**Fábio Hernandes**
Universidade Estadual do Centro-Oeste
85015-430, Guarapuava-PR, Brazil, 3010
hernandes@unicentro.br

**Fernando Gomide**
State University of Campinas
13083-970 Campinas, SP, Brazil
gomide@fee.unicamp.br

**Francisco Herrera**
Central University of Las Villas (UCLV)
Carretera Camajuaní km 5.5
Santa Clara, Cuba, 54830
herrera@uclv.edu.cu

**Genevieve Johnson**
Grant MacEwan College
Edmonton (AB), Canada, T5J 4S2

**Giangiacomo Gerla**
Università degli Studi di Salerno
Via Ponte don Melillo
Fisciano (SA), Italy, 84084
gerla@unisa.it

**Gianpiero Cattaneo**
Università di Milano–Bicocca
Via Bicocca degli Arcimboldi 8
Milano, Italy, I–20126
cattang@disco.unimib.it

**Gladys Casas**
Central University of Las Villas (UCLV)
Carretera Camajuaní km 5.5
Santa Clara, Cuba, 54830
gladita@uclv.edu.cu

**Gwanggil Jeon**
Hanyang University
17 Haengdang-dong, Seongdong-gu
Seoul, Korea
windcap315@ece.hanyang.ac.kr

**Hideo Tanaka**
Hiroshima International University
Hiroshima, Japan
h-tanaka@he.hirokoku-u.ac.jp

**Jan Komorowski**
The Linnaeus Centre for Bioinformatics, Uppsala University
Box 598 Husargatan 3
Uppsala, Sweden, SE-751 24
jan.komorowski@lcb.uu.se

**Jechang Jeong**
Hanyang University
17 Haengdang-dong, Seongdong-gu
Seoul, Korea
jjeong@ece.hanyang.ac.kr

**Jesús Fernández**
Central University of Las Villas (UCLV)
Carretera Camajuaní km 5.5
Santa Clara, Cuba, 54830

**José A. Olivas**
Universidad de Castilla La Mancha
Paseo de la Universidad 4
Ciudad Real, Spain, 13071
JoseAngel.Olivas@uclm.es

**José Luis Verdegay**
ETS de Ingeniería Informática
Universidad de Granada
Granada, Spain, E-18071
verdegay@decsai.ugr.es

**Julia Johnson**
Laurentian University
Sudbury (ON), Canada, P3E 2C6
jjohnson@cs.laurentian.ca

**Kazutomi Sugihara**
Fukui University of Technology
Fukui, Japan
sugihara@fukui-ut.ac.jp

**Kenneth Revett**
University of Westminster, Harrow
School of Computer Science
London, England, HA1 3TP
revettk@westminster.ac.uk

**Leticia Arco**
Central University of Las Villas (UCLV)
Carretera Camajuaní km 5.5
Santa Clara, Cuba, 54830
leticiaa@uclv.edu.cu

**Manuel Prieto**
Universidad de Castilla La Mancha
Paseo de la Universidad 4
Ciudad Real, Spain, 13071
Manuel.Prieto@uclm.es

**Marcin Kierczak**
The Linnaeus Centre for Bioinformatics, Uppsala University
Box 598 Husargatan 3
Uppsala, Sweden, SE-751 24

**María Matilde García**
Central University of Las Villas (UCLV)
Carretera Camajuaní km 5.5
Santa Clara, Cuba, 54830
mmgarcia@uclv.edu.cu

**Maria Teresa Lamata**
ETS de Ingeniería Informática
Universidad de Granada
Granada, Spain, E-18071
mtl@decsai.ugr.es

**Matthias Steinbrecher**
Otto-von-Guericke University of Magdeburg
Universitätsplatz 2, 39106 Magdeburg, Germany

**Mercedes Ramírez**
Universidad de Oriente, Cuba
mramirez@fie.uo.edu.cu

**Paolo Radaelli**
Università di Milano–Bicocca
Via Bicocca degli Arcimboldi 8
Milano, Italy, I–20126
radaelli@disco.unimib.it

**Pedro Albertos**
Universidad Politécnica de Valencia, Spain
pedro@aii.upv.es

**Pedro León**
University of Camagüey
Circunlavación Norte km 5.5
Camagüey, Cuba

**Piotr Grochowalski**
Rzeszów University, Poland
piotrg@univ.rzeszow.pl

**Rafael Bello**
Central University of Las Villas (UCLV)
Carretera Camajuaní km 5.5
Santa Clara, Cuba, 54830
rbellop@uclv.edu.cu

**Rafael Falcón**
Central University of Las Villas (UCLV)
Carretera Camajuaní km 5.5
Santa Clara, Cuba, 54830
rfalcon@uclv.edu.cu

**Rudolf Kruse**
Otto-von-Guericke University of Magdeburg
Universitätsplatz 2, 39106 Magdeburg, Germany
kruse@iws.cs.uni-magdeburg.de

**Serdar Arslan**
Middle East Technical University
Ankara, Turkey
aserdar@metu.edu.tr

**Silvia Calegari**
Università degli Studi di Milano–Bicocca
Via Bicocca degli Arcimboldi 8
Milano, Italy, I–20126
calegari@disco.unimib.it

**Tetsuya Murai**
Hokkaido University
Kita 14, Nishi 9, Kita-ku
Sapporo 060-0814, Japan
murahiko@main.ist.hokudai.ac.jp

**Tiziana Pacelli**
Università degli Studi di Salerno
Via Ponte don Melillo
Fisciano (SA), Italy, 84084
tpacelli@unisa.it

**Wanessa Amaral**
State University of Campinas
13083-970 Campinas, SP, Brazil
wmamaral@gmail.com

**Witold Rudnicki**
Warsaw University
Pawinskiego 5a, 02-106, Warsaw, Poland

**Yailé Caballero**
University of Camagüey
Circunlavación Norte km 5.5
Camagüey, Cuba
yailec@yahoo.com

**Yanet Rodríguez**
Central University of Las Villas (UCLV)
Carretera Camajuaní km 5.5
Santa Clara, Cuba, 54830
yrsarabia@uclv.edu.cu

**Yasuo Kudo**
Muroran Institute of Technology
Mizumoto 27-1, Muroran 050-8585, Japan
kudo@csse.muroran-it.ac.jp

**Yennely Márquez**
University of Camagüey
Circunlavación Norte km 5.5
Camagüey, Cuba

**Zbigniew Suraj**
Rzeszów University, Poland
zsuraj@univ.rzeszow.pl

# Part I: Fuzzy and Rough Sets' Theoretical and Practical Aspects

# Missing Value Semantics and Absent Value Semantics for Incomplete Information in Object-Oriented Rough Set Models

Yasuo Kudo[1] and Tetsuya Murai[2]

[1] Dept. of Computer Science and Systems Eng., Muroran Institute of Technology
Mizumoto 27-1, Muroran 050-8585, Japan
kudo@csse.muroran-it.ac.jp

[2] Graduate School of Information Science and Technology, Hokkaido University
Kita 14, Nishi 9, Kita-ku, Sapporo 060-0814, Japan
murahiko@main.ist.hokudai.ac.jp

**Summary.** We consider the "missing value" semantics and the "absent value" semantics in object–oriented rough set models proposed by Kudo and Murai. The object–oriented rough set model (OORS) is an extension of rough set theory by introducing object–oriented paradigm, and treats semi-structured objects and hierarchies among objects based on is-a and has-a relationships. In this chapter, we propose null value objects of OORS and revise Kryszkiewicz's tolerance relations, which characterize "missing value" semantics in OORS as incompleteness by "lack of parts". Moreover, we discuss connections between "absent value" semantics and is-a relationship in OORS, and revise similarity relations proposed by Stefanowski and Tsoukiàs, which characterize "absent value" semantics in OORS as incompleteness by difference of architecture of objects.

## 1 Introduction

Rough set theory was proposed by the late Professor Z. Pawlak as a mathematical basis of set-theoretical approximation of concepts and reasoning about data [8, 9]. There are many studies about treatment of incomplete information and semantics of unknown values in the framework of rough sets (for example, [2, 11, 12]). According to Stefanowski and Tsoukiàs [12], interpretations of unknown values are mainly distinguished in the following two semantics:

- the "missing value" semantics (unknown values allow any comparison)
- the "absent value" semantics (unknown values do not allow any comparison)

Kryszkiewicz has proposed tolerance relations to interpret null values in the given incomplete information table by "missing value" semantics [2]. On the other hand, Stefanowski and Tsoukiàs have proposed non-symmetric similarity relations to interpret null values by "absent value" semantics [12].

Introducing object–oriented paradigm (for detail, see [1] for example) used in computer science to rough set theory, Kudo and Murai have proposed object–oriented rough set models (for short, OORS) [3]. OORS is an extension of

R. Bello et al. (Eds.): Granular Computing, STUDFUZZ 224, pp. 3–21, 2008.
springerlink.com 

the "traditional" rough set theory which illustrates approximations of semi-structured complex objects and its hierarchical structures based on the concept of class, name, object and is-a and has-a relationships. Kudo and Murai have also proposed decision rules for OORS to illustrate characteristic combinations of semi-structured objects [5, 7].

In this chapter, as one important extension of OORS, we propose frameworks to treat incomplete information in OORS, in particular, we consider the "missing value" semantics and the "absent value" semantics in the framework of OORS. The main idea of this chapter is to characterize the concept of "missing value" and "absent value" as follows:

- the "missing value" semantics in OORS: incompleteness by lack of parts.
- the "absent value" semantics in OORS: incompleteness by difference of architecture of objects.

Moreover, we introduce tolerance relations and similarity relations in OORS. Both relations are natural extensions of the ones for incomplete information tables to the framework of OORS based on the above characterization. Note that this chapter is a revised and extended version of the authors' previous two papers [4, 6].

## 2 Backgrounds

### 2.1 Tolerance Relations and Non-symmetric Similarity Relations

Kryszkiewicz has proposed "missing value" semantics of null values in a given incomplete information table by *tolerance relations* [2]. Suppose $(U, A)$ be the given incomplete information table with the set of objects $U$ and the set of attributes $A$. Each $\mathsf{a} \in A$ is a mapping $\mathsf{a} : U \longrightarrow V_{\mathsf{a}}$, where $V_{\mathsf{a}}$ is the set of values of the attribute $\mathsf{a}$ containing a ***null value*** $*$. According to Stefanowski and Tsoukiàs [12], in the "missing value" semantics, we interpret a null value of the attribute $\mathsf{a} \in A$ in the given information table $(U, A)$ as similar to all other possible values of $\mathsf{a}$. This corresponds to the case that correct values of the object at the attributes exist, but just "missing", therefore we can compare null values of the attribute with all other possible values of the attribute.

For any set $B \subseteq A$ of attributes, a tolerance relation $T_B$ on $U$ is defined as follows:

$$xT_By \Longleftrightarrow \forall \mathsf{a} \in B, \text{ either } \mathsf{a}(x) = \mathsf{a}(y) \text{ or } \mathsf{a}(x) = * \text{ or } \mathsf{a}(y) = *, \tag{1}$$

where $\mathsf{a}(x) = *$ means that the value of the object $x$ at the attribute $\mathsf{a}$ is the null value. It is clear that $T_B$ is reflexive and symmetric, but not transitive.

Using the set $T_B(x) = \{y \in U \mid xT_By\}$ of objects $y$ which satisfy $xT_By$, for any set $X \subseteq U$ of objects, the lower approximation $\underline{T_B}(X)$ and the upper approximation $\overline{T_B}(X)$ by the tolerance relation $T_B$ are defined as follows:

$$\underline{T_B}(X) = \{x \in U \mid T_B(x) \subseteq X\}, \tag{2}$$

$$\overline{T_B}(X) = \{x \in U \mid T_B(x) \cap X \neq \emptyset\} = \bigcup\{T_B(x) \mid x \in X\}. \tag{3}$$

On the other hand, Stefanowski and Tsoukiàs have proposed "absent value" semantics of null values by *non-symmetric similarity relations* [12]. In the "absent value" semantics, we interpret null values in the given information table $(U, A)$ as "non-existing", and null values do not allow any comparison with other values. Based on this intuition, for any set $B \subseteq A$ of attributes, a non-symmetric similarity relation $S_B$ on $U$ is defined as follows:

$$xS_By \Longleftrightarrow \forall \mathsf{a} \in B \text{ such that } \mathsf{a}(x) \neq *, \ \ \mathsf{a}(x) = \mathsf{a}(y). \tag{4}$$

This definition requires that $x$ is similar to $y$ if and only if, for any attribute $\mathsf{a} \in B$, if $\mathsf{a}(x)$ is comparable, then $\mathsf{a}(x)$ is equal to $\mathsf{a}(y)$. It is easy to check that $S_B$ is reflexive and transitive, but not symmetric in general.

Next, the following two sets of objects are introduced. In general, $S_B(x)$ and $S_B^{-1}(x)$ are different sets because $S_B$ is not symmetric:

$$S_B(x) = \{y \in U \mid yS_Bx\}: \ \text{the set of objects similar to } x, \tag{5}$$

$$S_B^{-1}(x) = \{y \in U \mid xS_By\}: \ \text{the set of objects to which } x \text{ is similar.} \tag{6}$$

Using these sets, for any set $X \subseteq U$ of objects, the lower approximation $\underline{S_B}(X)$ and the upper approximation $\overline{S_B}(X)$ by the non-symmetric similarity relation $S_B$ are defined as follows:

$$\underline{S_B}(X) = \{x \in U \mid S_B^{-1}(x) \subseteq X\}, \tag{7}$$

$$\overline{S_B}(X) = \bigcup \{S_B(x) \mid x \in X\}. \tag{8}$$

Stefanowski and Tsoukiàs have shown that, for any given information table $(U, A)$ and a set $X \subseteq U$, the lower and upper approximations obtained by the non-symmetric similarity relation are a refinement of the ones by the tolerance relation [12].

*Example 1.* Suppose the following incomplete information table illustrated by Table 1 is given, where $o_1, \cdots, o_6$ are objects, $\mathsf{a}_1, \mathsf{a}_2$ and $\mathsf{a}_3$ are attributes with discrete values either 0 or 1. The symbol "$*$" illustrates the null value.

Let $B = \{\mathsf{a}_1, \mathsf{a}_2, \mathsf{a}_3\}$ be a set of attributes, and $X = \{o_1, o_3, o_5\}$ be a set of objects. We consider a tolerance relation $T_B$ and a non-symmetric similarity relation $S_B$ based on the set $B$, and lower and upper approximation of $X$ by $T_B$ and $S_B$, respectively.

First, for each $o_i$ $(1 \leq i \leq 6)$, we calculate $T_B(o_i)$ as follows:

$$T_B(o_1) = \{o_1, o_5\},\ T_B(o_2) = \{o_2, o_5\},\ T_B(o_3) = \{o_3, o_4\},$$
$$T_B(o_4) = \{o_3, o_4\},\ T_B(o_5) = \{o_1, o_2, o_4, o_5, o_6\},\ T_B(o_6) = \{o_4, o_5, o_6\}.$$

Thus, we get the following lower and upper approximations by the tolerance relation $T_B$, respectively:

$$\underline{T_B(X)} = \{o_1\},\ \ \overline{T_B(X)} = \{o_1, o_2, o_3, o_4, o_5, o_6\}.$$

**Table 1.** An incomplete information table

| | a1 | a2 | a3 |
|---|---|---|---|
| $o_1$ | 1 | 0 | 1 |
| $o_2$ | 0 | 0 | 1 |
| $o_3$ | * | 1 | 0 |
| $o_4$ | * | * | 0 |
| $o_5$ | * | 0 | * |
| $o_6$ | 1 | 0 | 0 |

Next, we calculate $S_B(o_i)$ and $S_B^{-1}(o_i)$ defined by (5) and (6), respectively:

$$\begin{array}{ll} S_B(o_1) = \{o_1, o_5\}, & S_B^{-1}(o_1) = \{o_1\}, \\ S_B(o_2) = \{o_2, o_5\}, & S_B^{-1}(o_2) = \{o_2\}, \\ S_B(o_3) = \{o_3, o_4\}, & S_B^{-1}(o_3) = \{o_3\}, \\ S_B(o_4) = \{o_4\}, & S_B^{-1}(o_4) = \{o_3, o_4, o_6\}, \\ S_B(o_5) = \{o_5\}, & S_B^{-1}(o_5) = \{o_1, o_2, o_5, o_6\}, \\ S_B(o_6) = \{o_4, o_5, o_6\}, & S_B^{-1}(o_6) = \{o_6\}. \end{array}$$

Thus, we get the following lower and upper approximations of $X$ by the non-symmetric similarity relation $S_B$, respectively:

$$\underline{S_B(X)} = \{o_1, o_3\}, \ \overline{S_B(X)} = \{o_1, o_3, o_4, o_5\}.$$

## 2.2 Object-Oriented Rough Sets

In this subsection, we review OORS briefly. Note that contents of this review are based on our previous papers [3, 5, 7].

### Class, Name, Object

We define a *class structure* $\mathcal{C}$, a *name structure* $\mathcal{N}$ and an *object structure* $\mathcal{O}$ by the following triples, respectively:

$$\mathcal{C} = (C, \ni_C, \sqsupseteq_C), \ \mathcal{N} = (N, \ni_N, \sqsupseteq_N), \ \mathcal{O} = (O, \ni_O, \sqsupseteq_O),$$

where $C$, $N$ and $O$ are finite and disjoint non-empty sets such that $|C| \leq |N|$ ($|X|$ is the cardinality of $X$). Each element $c \in C$ is called a *class*. Similarly, each $n \in N$ is called a *name*, and each $o \in O$ is called an *object*. The relation $\ni_X$ ($X \in \{C, N, O\}$) is an acyclic binary relation on $X$, and the relation $\sqsupseteq_X$ is a reflexive, transitive, and asymmetric binary relation on $X$. Moreover, $\ni_X$ and $\sqsupseteq_X$ satisfy the following property:

$$\forall x_i, x_j, x_k \in X, \ x_i \sqsupseteq_X x_j, \ x_j \ni_X x_k \ \Rightarrow \ x_i \ni_X x_k. \tag{9}$$

Two relations $\ni_X$ and $\sqsupseteq_X$ on $X \in \{C, N, O\}$ illustrate hierarchical structures among elements in $X$. The relation $\ni_X$ is called a *has-a relation*, which illustrates part / whole relationship. $x_i \ni_X x_j$ means "$x_i$ has-a $x_j$", or "$x_j$ is a part of $x_i$". For example, $c_i \ni_C c_j$ means that "the class $c_i$ has a class $c_j$", or "$c_j$ is a part of $c_i$". On the other hand, the relation $\sqsupseteq_X$ is called an *is-a relation*, which illustrates specialized / generalized relationship. $x_i \sqsupseteq_X x_j$ means that "$x_i$ is-a $x_j$". For example, $\sqsupseteq_C$ illustrates relationship between superclass and subclass, and $c_i \sqsupseteq_C c_j$ means "$c_i$ is a $c_j$", or "$c_i$ is a subclass of $c_j$".

Characteristics of class, name and object structures are as follows:

- The class structure illustrates abstract data forms and those hierarchical structures based on part / whole relationship (has-a relation) and specialized / generalized relationship (is-a relation).
- The name structure introduces numerical constraint of objects and those identification, which provides concrete design of objects.
- The object structure illustrates actual combination of objects.

**Well-Defined Structures**

Each object $x \in O$ is defined as an instance of some class $c \in C$, and the class of $x$ is identified by the *class identifier* function. The class identifier $id_C$ is a *p-morphism* between $\mathcal{O}$ and $\mathcal{C}$ (cf. [10], p.142), that is, the function $id_C : O \longrightarrow C$ satisfies the following conditions:

1. $\forall x_i, x_j \in O, x_i \ni_O x_j \Rightarrow id_C(x_i) \ni_C id_C(x_j)$.
2. $\forall x_i \in O,\ \forall c_j \in C,\ id_C(x_i) \ni_C c_j \Rightarrow \exists x_j \in O$ s.t. $x_i \ni_O x_j$ and $id_C(x_j) = c_j$,

and the same conditions are also satisfied for $\sqsupseteq_O$ and $\sqsupseteq_C$. $id_C(x) = c$ means that the object $x$ is an instance of the class $c$.

The object structure $\mathcal{O}$ and the class structure $\mathcal{C}$ are also connected through the name structure $\mathcal{N}$ by the *naming function* $nf : N \longrightarrow C$ and the *name assignment* $na : O \longrightarrow N$. The naming function provides names to each class, which enable us to use plural instances of the same class as parts of some object. On the other hand, the name assignment provides names to every objects, which enable us to identify objects by names.

Formally, the naming function $nf : N \longrightarrow C$ is a surjective p-morphism between $\mathcal{N}$ and $\mathcal{C}$, and satisfies the following *name preservation constraint*:

- For any $n_i, n_j \in N$, if $nf(n_i) = nf(n_j)$, then $H_N(c|n_i) = H_N(c|n_j)$ is satisfied for all $c \in C$,

where $H_N(c|n) = \{n_j \in N \mid n \ni_N n_j, f(n_j) = c\}$ is the set of names of $c$ that $n$ has. The requirement that $nf$ is a surjective p-morphism means that there is at least one name for each class, and structures between names reflect all structural characteristics between classes. The name preservation constraint requires that, for any classes $c_i, c_j \in C$ such that $c_i \ni_C c_j$ and any name $n \in N$ with $nf(n) = c_i$, all names of the parts of $c$ are uniquely determined. Thus, the number of names of $c_j$ is fixed as $m = |H_N(c_j|n)|$, and we can simply say that "the class $c_i$ has $m$ objects of the class $c_j$".

On the other hand, the name assignment $na : O \longrightarrow N$ is a p-morphism between $\mathcal{O}$ and $\mathcal{N}$, and satisfies the following *uniqueness condition*:

- For any $x \in O$, if $H_O(x) \neq \emptyset$, the restriction of $na$ into $H_O(x)$: $na|_{H_O(x)} : H_O(x) \longrightarrow N$ is injective,

where $H_O(x) = \{y \in O \mid x \ni_O y\}$ is the set of objects that $x$ has. $na(x) = n$ means that the name of the object $x$ is $n$. The uniqueness condition requires that all distinct parts $y \in H_O(x)$ have different names.

We say that $\mathcal{C}$, $\mathcal{N}$ and $\mathcal{O}$ are *well-defined* if and only if there exist a naming function $nf : N \longrightarrow C$ and a name assignment $na : O \longrightarrow N$ such that

$$id_C = nf \circ na, \tag{10}$$

that is, $id_C(x) = nf(na(x))$ for all $x \in O$.

In this chapter, we concentrate well-defined class, name and object structures. In well-defined structures, if a class $c_i$ has $m$ objects of a class $c_j$, then any instance $x_i$ of the class $c_i$ has exactly $m$ instances $x_{j1}, \cdots, x_{jm}$ of the class $c_j$ [3]. This good property enables us the following description for clear representation of objects. Suppose we have $x_1, x_2 \in O$, $n_1, n_2 \in N$, and $c_1, c_2 \in C$ such that $x_1 \ni_O x_2$, and $na(x_i) = n_i$, $nf(n_i) = c_i$ for $i \in \{1, 2\}$. We denote $x_1.n_2$ instead of $x_2$ by means of "the instance of $c_2$ named $n_2$ as a part of $x_1$".

Note that information tables used in "traditional" rough set theory are characterized as special cases of OORS which have the following characteristics [3]:

1. All objects $x \in U$ are instances of a unique class that represents the "schema" of the information table.
2. There is no inheritance hierarchy between classes, and there is no part / whole relationship except for objects and its values.

### Indiscernibility Relations in the Object-Oriented Rough Set Model

All indiscernibility relations in OORS are based on the concept of equivalence as instances. In [3], to evaluate equivalence of instances, an *indiscernibility relation* $\sim$ on $O$ are recursively defined as follows:

$$x \sim y \iff \begin{array}{l} x \text{ and } y \text{ satisfy the following two conditions:} \\ 1.\ id_C(x) = id_C(y), \text{ and,} \\ 2.\ \begin{cases} x.n \sim y.n,\ \forall n \in H_N(na(x)) & \text{if } H_N(na(x)) \neq \emptyset, \\ Val(x) = Val(y) & \text{otherwise,} \end{cases} \end{array} \tag{11}$$

where $H_N(na(x))$ is the set of names that $na(x)$ has. $Val(x)$ is the "value" of the "value object" $x$. Because $C$ is a finite non-empty set and $\ni_C$ is acyclic, there is at least one class $c$ such that $c$ has no other class $c'$, that is, $c \not\ni_C c'$ for any $c' \in C$. We call such class $c$ an *attribute*, and if $id_C(x) = a$ for some attribute $a$, we call such object $x$ a *value object* of the attribute $a$. The value object $x$ as an instance of the attribute $a$ represents a "value" of the attribute.

The relationship $x \sim y$ means that the object $x$ is equivalent to the object $y$ as an instance of the class $id_C(x)$. It is easy to check that $\sim$ is an equivalence relation on $O$.

To treat structural characteristics among objects by indiscernibility relations, the concept of *consistent sequence of names* is introduced as follows [7]: Let

$\mathcal{C}$, $\mathcal{N}$ and $\mathcal{O}$ be well-defined class, name and object structures, respectively. A sequence of names $n_1.\cdots.n_k$ with length $k$ $(k \geq 1)$ such that $n_i \in N$ $(1 \leq i \leq k)$ is called a consistent sequence of names if and only if either (1) $k = 1$, or (2) $k \geq 2$ and $n_{j+1} \in H_N(n_j)$ for each name $n_j$ $(1 \leq j \leq k-1)$. We denote the set of all consistent sequences of names in $\mathcal{N}$ by $N^+$. Consistent sequences describe hierarchical structures among objects correctly, and have the following good property: For any object $x$ and any consistent sequence $n_1.\cdots.n_k$, if $n_1 \in H_N(na(x))$, then the sequence $n_1.\cdots.n_k$ "connects" to the object $x$, and we can find the object $y(= x.n_1.\cdots.n_k)$ by tracing the has-a relation $\ni_O$ such that $x \ni_O \cdots \ni_O y$. Thus, we call that a consistent sequence $n_1.\cdots.n_k$ connects to an object $x$ if and only if we have $n_1 \in H_N(na(x))$.

Using consistent sequences of names, for any non-empty set of sequences $D \subseteq N^+$, an *indiscernibility relation* $\approx_D$ on $O$ to treat hierarchical structures is defined as follows [7]:

$$
\begin{aligned}
&x \approx_D y \iff \\
&\text{For each } n_1.\cdots.n_k \in D, x \text{ and } y \text{ satisfy the following conditions:} \\
&1.\ n_1.\cdots.n_k \text{ connects to } x \iff n_1.\cdots.n_k \text{ connects to } y, \text{ and} \\
&2.\ x.n_1.\cdots.n_k \sim y.n_1.\cdots.n_k.
\end{aligned} \tag{12}
$$

The condition 1 in Eq. 12 requires that the object $x$ and $y$ concern the same sequences in $D$, which means that $x$ and $y$ have the same architecture at the parts illustrated by such sequences. The condition 2 requires that, for all sequences $n_1.\cdots.n_k \in D$ that connects both $x$ and $y$, $x.n_1.\cdots.n_k$ as a part of $x$ is equivalent to $y.n_1.\cdots.n_k$ as a part of $y$. It is easy to check that the relation $\approx_D$ defined by Eq. 12 is an equivalence relation on $O$.

For any subset $X \subseteq O$ of objects, the lower approximation $\underline{\approx_D}(X)$ and the upper approximation $\overline{\approx_D}(X)$ of $X$ by $\approx_D$, and the rough set of $X$ by $\approx_D$ are defined by the same manner with "traditional" rough set theory [8, 9]:

$$\underline{\approx_D}(X) = \{x \in O \mid [x]_{\approx_D} \subseteq X\}, \tag{13}$$

$$\overline{\approx_D}(X) = \{x \in O \mid [x]_{\approx_D} \cap X \neq \emptyset\}, \tag{14}$$

$$(\underline{\approx_D}(X), \overline{\approx_D}(X)), \tag{15}$$

where $[x]_{\approx_D}$ is the equivalence class of $x$ usually defined by $\approx_D$.

*Example 2.* We consider the following object-oriented rough set model about personal computers. Let $\mathcal{C} = (C, \ni_C, \sqsupseteq_C)$ be a class structure with $C = \{$ PC, 2HDD-PC, CPU, Memory, HDD, Clock, Maker, Size$\}$ with the following is-a relationship and has-a relationship, where Maker and Size are attributes:

| Is-a relation: | Has-a relation: |
|---|---|
| $c \sqsupseteq_C c$, $\forall c \in C$, | PC $\ni_C$ CPU, PC $\ni_C$ Memory, PC $\ni_C$ HDD, |
| 2HDD-PC $\sqsupseteq_C$ PC. | CPU $\ni_C$ Maker, CPU $\ni_C$ Clock, |
| | Memory $\ni_C$ Maker, Memory $\ni_C$ Size, |
| | HDD $\ni_C$ Maker, HDD $\ni_C$ Size. |

Similarly, let $\mathcal{N} = (N, \ni_N, \sqsupseteq_N)$ is a name structure with $N = \{$ pc, 2hdd-pc, cpu, memory, hdd, hdd2, clock, maker, size$\}$ with the following relationships:

| Is-a relation: | Has-a relation: |
|---|---|
| $n \sqsupseteq_N n$, $\forall n \in N$, | pc $\ni_N$ cpu, pc $\ni_N$ memory, pc $\ni_N$ hdd, |
| 2hdd-pc $\sqsupseteq_N$ pc. | 2hdd-pc $\ni_N$ cpu, 2hdd-pc $\ni_N$ memory, |
| | 2hdd-pc $\ni_N$ hdd, 2hdd-pc $\ni_N$ hdd2, |
| | cpu $\ni_N$ Maker, cpu $\ni_N$ clock, |
| | memory $\ni_N$ maker, memory $\ni_N$ size, |
| | hdd $\ni_N$ maker, hdd $\ni_N$ size, |
| | hdd2 $\ni_N$ maker, hdd2 $\ni_N$ size. |

Moreover, suppose we have a naming function $nf : N \longrightarrow C$ such that

$$\begin{aligned} &nf(\text{pc}) = \mathsf{PC},\ nf(\text{2hdd-pc}) = \mathsf{2HDD\text{-}PC},\ nf(\text{cpu}) = \mathsf{CPU}, \\ &nf(\text{memory}) = \mathsf{Memory},\ nf(\text{hdd}) = nf(\text{hdd2}) = \mathsf{HDD}, \\ &nf(\text{maker}) = \mathsf{Maker},\ nf(\text{clock}) = \mathsf{Clock},\ nf(\text{size}) = \mathsf{Size}. \end{aligned}$$

We illustrate connections between classes and names by class diagrams of UML [13] as in Fig. 1. For example, the class diagram 2HDD-PC illustrates that the 2HDD-PC class has one object of the CPU class named "cpu", one object of the Memory class named "memory", and two objects of the HDD class named "hdd" and "hdd2", respectively.

Finally, let $\mathcal{O} = (O, \ni_O, \sqsupseteq_O)$ be an object structure with the following has-a relationship illustrated in Fig. 2, and $na : O \longrightarrow N$ be the following name assignment:

$$\begin{aligned} &na(\text{pc1}) = na(\text{pc2}) = \text{pc},\ \ na(\text{pc3}) = na(\text{pc4}) = \text{2hdd-pc}, \\ &na(\text{c}i) = \text{cpu},\ na(\text{m}i) = \text{memory},\ na(\text{h}i) = \text{hdd}\ (1 \le i \le 4), \\ &na(\text{h32}) = na(\text{h42}) = \text{hdd2},\ \ na(\text{2.2GHz}) = na(\text{2.4GHz}) = \text{clock}, \\ &na(\text{A}) = na(\text{F}) = na(\text{I}) = na(\text{S}) = na(\text{T}) = \text{maker}, \\ &na(\text{256MB}) = na(\text{512MB}) = na(\text{40GB}) = na(\text{80GB}) = \text{size}. \end{aligned}$$

We define the class identifier $id_C : O \longrightarrow C$ by $id_C = nf \circ na$. It is not hard to check that these class, name and object structures are well-defined.

This object structure $\mathcal{O}$ illustrates the following situation: There are four personal computers pc$i$ $(1 \le i \le 4)$, and, for example, the personal computer pc1 as an instance of the PC class consists of an object c1(denoted by pc1.cpu) of the CPU class, an object m1(=pc1.memory) of the Memory class, and an object h1(=pc1.hdd) of the HDD class, respectively. Moreover, the CPU c1 consists of a value object A of the attribute Maker and a value object 2.2GHz of the attribute Clock, which means that the CPU c1 is made by A company and its clock is 2.2GHz. Similarly, the memory m1 is made by S company and its size is 512MB, and the HDD h1 is made by T company and its size is 80GB, respectively.

Let $D = \{$ memory.size, hdd $\}(\subseteq N^+)$ be a set of consistent sequences of names, and $\approx_D$ be the equivalence relation based on $D$ defined by Eq. 12. Equivalence classes by $\approx_D$ are constructed as follows:

$$\begin{aligned} &[\text{pc1}]_{\approx_D} = \{\text{pc1}, \text{pc4}\},\ [\text{pc2}]_{\approx_D} = \{\text{pc2}\},\ [\text{pc3}]_{\approx_D} = \{\text{pc3}\}, \\ &[\text{c1}]_{\approx_D} = O - \bigcup [\text{pc}i]_{\approx_D}. \end{aligned}$$

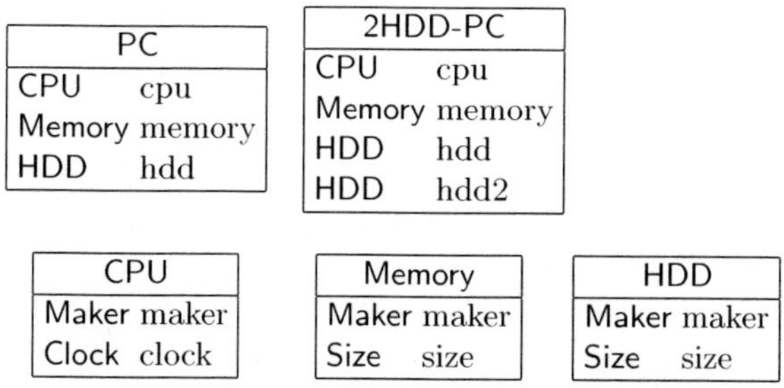


**Fig. 1.** Class diagrams in example 2

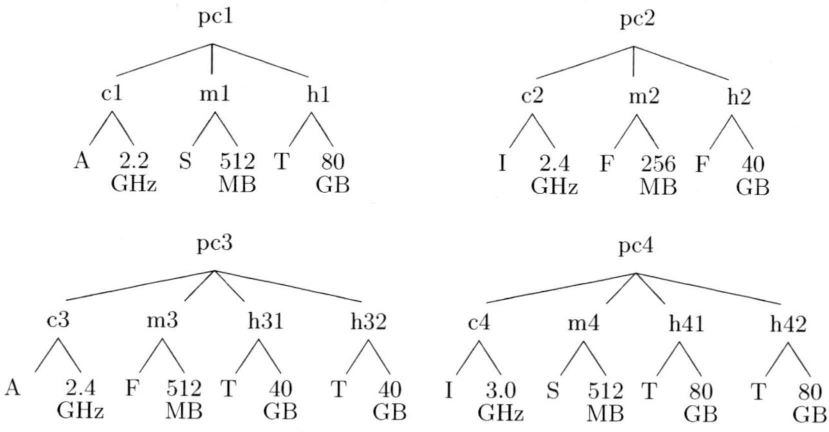


**Fig. 2.** Has-a relation $\ni_O$ in example 2

For example, the equivalence class $[\text{pc1}]_{\approx_S}$ is the set of PCs with a 512MB memory and an 80GB HDD made by T company.

Let $X = \{ \text{pc1}, \text{pc3} \}$ be the set of PCs which have the CPUs made by A company. Using the constructed equivalence classes based on $\approx_D$, we have the following lower and upper approximations, respectively:

$$\underline{\approx_D}(X) = \{\text{pc3}\}, \quad \overline{\approx_D}(X) = \{\text{pc1, pc3, pc4}\}.$$

## 3 Missing Value Semantics in the Object–Oriented Rough Sets

In this section, we extend OORS to characterize "missing value" semantics as incompleteness of information about objects that comes from "*lack of parts*". Informally, lack of parts illustrates the following situation: Suppose we have two classes $c_i$ and $c_j$ such that $c_j$ is a part of $c_i$, and an instance $x_i$ of the class $c_i$, however, there is no "actual" instance $x_j \in O$ such that $x_j$ is an instance of $c_j$ and also is a part of $x_i$. For example, "a personal computer that its CPU was taken away" has no instance of CPU class, even though any instance of PC class should have one instance of CPU class. To illustrate incompleteness we mentioned the above, we introduce *null value objects* into OORS. Note that contents of this section are based on the authors' previous paper [4].

### 3.1 Null Value Objects

We introduce null value objects and an incomplete object structure to illustrate "lack of parts" as follows.

**Definition 1.** *Let $NO$ be a finite non-empty set. An incomplete object structure $\mathcal{IO}$ is the following triple:*

$$\mathcal{IO} = (O \cup NO, \ni_I, \sqsupseteq_I), \tag{16}$$

*where $O$ is the (finite and non-empty) set of objects, $NO$ is a finite set such that $O \cap NO = \emptyset$, the relation $\ni_I$ is an acyclic binary relation on $O \cup NO$, and the relation $\sqsupseteq_I$ is a reflexive, transitive, and asymmetric binary relation on $O \cup NO$. Moreover, $\ni_I$ and $\sqsupseteq_I$ satisfy Eq. 9 and the following condition:*

$$\forall x \in NO, \forall y \in O \cup NO, x \not\ni_I y. \tag{17}$$

We call each object $x \in NO$ a *null value object.* On the other hand, each object $y \in O$ is called an *actual object.* We intend that null value objects have the following characteristics:

1. All null value objects have no objects.
2. Each null value object is an instance of some class.

The characteristic 1. means that each null value object is a special case of value objects, and it corresponds to "null value". The characteristic 2. means that each null value object is also an object of some class. This intends that we can compare null value objects and any other (null value) objects if and only if these objects are instances of the same class.

### 3.2 Well-Defined Structures with Null Value Objects

To illustrate the above characteristics of null value objects, we refine the definition of the class identifier. However, we can not directly extend the domain of the class identifier $id_C$ to $O \cup NO$ with keeping $id_C$ a p-morphism, and therefore we need to weaken the definition of p-morphism.

**Definition 2.** *Let $\mathcal{IO} = (O \cup NO, \ni_I, \sqsupseteq_I)$ and $\mathcal{C} = (C, \ni_C, \sqsupseteq_C)$ be an incomplete object structure and a class structure, respectively. We call a function $id_C : O \cup NO \longrightarrow C$ a class identifier of incomplete objects if $id_C$ satisfies the following conditions:*

1. $\forall x_i, x_j \in O \cup NO, x_i \ni_I x_j \Rightarrow id_C(x_i) \ni_C id_C(x_j)$.
2. $\forall x_i \in O,\ \forall c_j \in C,\ id_C(x_i) \ni_C c_j \Rightarrow \exists x_j \in O \cup NO$ s.t. $x_i \ni_I x_j$ and $id_C(x_j) = c_j$,

*and the same conditions are also satisfied for $\sqsupseteq_I$ and $\sqsupseteq_C$.*

$id_C(x) = c$ means that the (null value) object $x \in O \cup NO$ is an instance of the class $c$. Note that the condition 2 is weakened from the condition of p-morphism to agree with the characteristic of null value objects by Eq. 17.

We also need to extend the domain of the name assignment $na$ to $O \cup NO$ as follows.

**Definition 3.** *Let $\mathcal{IO} = (O \cup NO, \ni_I, \sqsupseteq_I)$ and $\mathcal{N} = (N, \ni_N, \sqsupseteq_N)$ be an incomplete object structure and a name structure, respectively. We call a function $na : O \cup NO \longrightarrow N$ a name assignment if $na$ satisfies the following conditions:*

1. *$na$ satisfies the condition 1 and 2 appeared in Definition 2.*
2. *$na$ satisfies the following uniqueness condition:*
   - *For any $x \in O$, if $H_I(x) \neq \emptyset$, the restriction of $na$ into $H_I(x)$: $na|_{H_I(x)} : H_I(x) \longrightarrow N$ is injective,*

   *where $H_I(x) = \{y \in O \cup NO \mid x \ni_I y\}$ is the set of "actual" and "null value" objects that $x$ has.*

$na(x) = n$ means that the name of the object $x$ is $n$.

Similar to the case of "complete" OORS, we introduce a naming function $nf : N \longrightarrow C$ as a p-morphism between $\mathcal{N}$ and $\mathcal{C}$ that satisfies the name preservation constraint. Moreover, we say that $\mathcal{C}$, $\mathcal{N}$ and $\mathcal{IO}$ are *well-defined* if and only if there exist a naming function $nf : N \longrightarrow C$ and a name assignment $na : O \cup NO \longrightarrow N$ such that $id_C = nf \circ na$, that is, $id_C(x) = nf(na(x))$ for all $x \in O \cup NO$. Hereafter, we concentrate well-defined incomplete object, name and class structures.

Now, we can explain incompleteness by "lack of parts" correctly. Suppose that any instance $x$ of a class $c_i$ should have $m$ objects of a class $c_j$, that is, there are $m$ names $n_1, \cdots, n_m$ for the class $c_j$ and $m$ instances $x.n_1, \cdots, x.n_m$ of the class $c_j$ such that $x \ni_O x.n_j$ $(j = 1, \ldots, m)$. Here, if we have $x.n_k \in NO$ for some name $n_k$, the notion $x \ni_O x.n_k$ illustrates that, even though any instance of $c$ should have $m$ "actual" objects of $c_j$, there are just $m-1$ objects of $c_j$ as parts of $x$, and there is no "actual" object that corresponds to $x.n_k$. This situation illustrates "incompleteness" of the object $x$ as an instance of $c_i$, which is triggered by "lack of parts" of $x$. Note that there are exactly $m$ (actual or null value) objects of $c_j$ as parts of $x$, and therefore, constraints about design of objects introduced by the name structure are satisfied.

### 3.3 Tolerance Relations in Object–Oriented Rough Sets

We apply the tolerance relation proposed by Kryszkiewicz [2] to interpret null value objects in well-defined incomplete object structure. According to Stefanowski and Tsoukiàs [12], the tolerance relation corresponds to "missing value" semantics that unknown values allow any comparison. Thus, we think that extended tolerance relations for OORS are suitable for treating incompleteness by "lack of parts".

**Definition 4.** *Let $\mathcal{C}$, $\mathcal{N}$ and $\mathcal{IO}$ be the well-defined class, name and incomplete object structures, respectively, and $N^+$ be the set of consistent sequences of names* *in $\mathcal{N}$. Moreover, let $D \subseteq N^+$ be a non-empty subset of consistent sequences of names, and $\approx_D$ is an indiscernibility relation on $O$ defined by Eq. 12. A tolerance relation $\tau_D$ on $O$ by $D$ for the well-defined structures is a binary relation defined as follows:*

$$x \tau_D y \iff \begin{array}{l} \text{For each } n_1.\cdots.n_k \in D \text{ that connects to both } x \text{ and } y, \\ \text{either } x \approx_{n_1.\cdots.n_k} y, \text{ or} \\ \text{there exists } n_i \ (1 \leq i \leq k) \text{ in } n_1.\cdots.n_k \text{ such that} \\ x.n_1.\cdots.n_i \in NO \text{ or } y.n_1.\cdots.n_i \in NO, \end{array} \tag{18}$$

*where $x \approx_{n_1.\cdots.n_k} y$ is the abbreviation of $x \approx_{\{n_1.\cdots.n_k\}} y$.*

It is not hard to check that the relation $\tau_D$ defined by Eq. 18 is reflexive and symmetric, however, $\tau_D$ is not transitive in general.

We intend that the definition of the relation $\tau_D$ by Eq. 18 captures incompleteness by "lack of parts", and becomes a natural extension of Kryszkiewicz's tolerance relations in the framework of OORS. Equation 18 requires that, for all sequences $n_1.\cdots.n_k \in D$ that connect both $x$ and $y$, either both the "actual objects" $x.n_1.\cdots.n_k$ and $y.n_1.\cdots.n_k$ exist and are equivalent each other, or there is some name $n_i$ $(1 \leq i \leq k)$ in the sequence $n_1.\cdots.n_k$ such that $x.n_1.\cdots.n_i$ (or $y.n_1.\cdots.n_i$ ) is a null value object. Because of Eq. 17 about null values objects, if $x.n_1.\cdots.n_i$ $(i \leq k)$ is a null value object, then an "actual object" $x.n_1.\cdots.n_k$ does not exist and it corresponds to incompleteness by "lack of parts". Thus, the relationship $x\tau_D y$ means that, for all sequences $n_1.\cdots.n_k \in D$ that connect both $x$ and $y$, we can make $x.n_1.\cdots.n_k$ and $y.n_1.\cdots.n_k$ be equivalent by replacing null values objects with relevant "actual" objects if we need.

For any subset $X \subseteq O$ of "actual" objects, we define the lower approximation $\underline{\tau_D}(X)$ and upper approximation $\overline{\tau_D}(X)$ as the same manner with [2], respectively:

$$\underline{\tau_D}(X) = \{x \in O \mid \tau_D(x) \subseteq X\}, \tag{19}$$

$$\overline{\tau_D}(X) = \{x \in O \mid \tau_D(x) \cap X \neq \emptyset\}, \tag{20}$$

where $\tau_D(X) = \{y \in O \mid x\tau_D y\}$. We call the set $\tau_D(x)$ the tolerance class of $x$. The lower approximation $\underline{\tau_D}(X)$ is the set of objects $y \in O$ such that we can make $y$ be equivalent to all objects $x \in X$. On the other hand, the upper approximation $\overline{\tau_D}(X)$ is the set of objects $y$ such that there is at least one object $x \in X$ such that we can make $x$ and $y$ be equivalent.

*Example 3.* This example is continuation of Example 2, thus we use all of the same setting in Example 2. Now, we introduce an incomplete object structure $\mathcal{IO} = (O \cup NO, \ni_I, \sqsupseteq_I)$ based on $\mathcal{O}$ in Example 2 as follows. Let $NO = \{\text{nc5}, \text{nh61}, \text{nh62}\}$ be a set of null value objects with the following $id_C$ and $na$:

$$id_C(\text{nc5}) = \mathsf{CPU},\ id_C(\text{nh61}) = id_C(\text{nh62}) = \mathsf{HDD}.$$
$$na(\text{nc5}) = \text{cpu},\ na(\text{nh61}) = \text{hdd},\ na(\text{nh62}) = \text{hdd2}.$$

Figure 3 illustrates the has-a relation $\ni_I$ about newly added actual objects and null value objects. The incomplete object structure $\mathcal{IO}$ illustrates the following

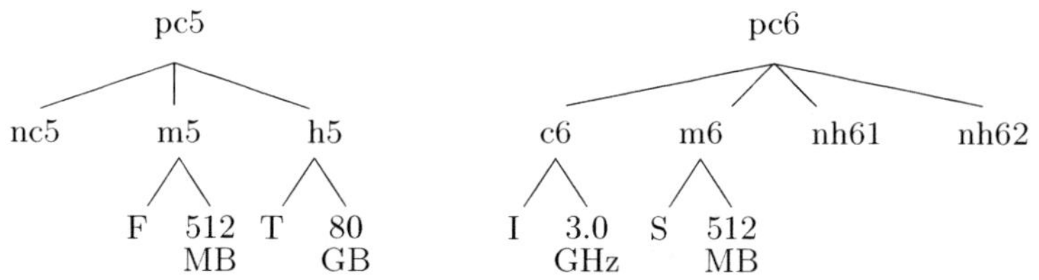


**Fig. 3.** Has-a relation $\ni_I$ between actual and null value objects

situations about newly added actual objects and null value objects: There are two personal computers pc5 and pc6 with some lack of parts, respectively. pc5 is an instance of the PC class, thus pc5 should have one CPU as a part, however, pc5 has no CPU. Similarly, pc6 is an instance of the 2HDD-DTPC class, thus pc6 should have two HDDs as parts, however, pc6 has no HDD.

Here, we consider the same problem in Example 2, that is, approximation of the set $X = \{ \text{pc1}, \text{pc3} \}$ of PCs which have the CPUs made by A company with respect to the set $D = \{ \text{memory.size}, \text{hdd} \} (\subseteq N^+)$ of consistent sequences. We construct the tolerance relation $\tau_D$ by Eq. 18, and the obtained tolerance classes are as follows:

$$\begin{array}{ll} \tau_D(\text{pc1}) = \{\text{pc1}, \text{pc4}, \text{pc5}, \text{pc6}\}, & \tau_D(\text{pc2}) = \{\text{pc2}\}, \\ \tau_D(\text{pc3}) = \{\text{pc3}, \text{pc6}\}, & \tau_D(\text{pc4}) = \{\text{pc1}, \text{pc4}, \text{pc5}, \text{pc6}\}, \\ \tau_D(\text{pc5}) = \{\text{pc1}, \text{pc4}, \text{pc5}, \text{pc6}\}, & \tau_D(\text{pc6}) = \{\text{pc1}, \text{pc3}, \text{pc4}, \text{pc5}, \text{pc6}\}, \\ \tau_D(\text{c1}) = O - \{\text{pc}i \mid 1 \le i \le 6\}. & \end{array}$$

Thus, using these tolerance classes, we have the lower approximation by Eq. 19 and the upper approximations by Eq. 20 based on the tolerance relation $\tau_D$ respectively:

$$\underline{\tau_D}(X) = \emptyset, \quad \overline{\tau_D}(X) = \{\text{pc1}, \text{pc3}, \text{pc4}, \text{pc5}, \text{pc6}\}.$$

# 4 Absent Value Semantics in the Object–Oriented Rough Sets

In this section, we illustrate that "absent value" semantics is characterized by is-a relationship in the framework of OORS. Note that contents of this section are based on the authors' previous paper [6].

## 4.1 Characterization of "Absence of Values" Based on IS-A Relationship

As we mentioned in Sect. 2.1, in the "absent value" semantics, we interpret null values in the given information table $(U, A)$ as "non-existing of objects", and null values do not allow any comparison with other values. From the viewpoint of object-orientation, we can regard all attributes $\mathsf{a} \in A$ as "classes", and all objects $x \in U$ and values of objects at any attributes as "instances of some classes".

Moreover, if there is some attribute $\mathsf{a} \in A$ such that $\mathsf{a}(x) = *$ and $\mathsf{a}(y) \neq *$, it is natural to interpret that the object $x$ does not (more precisely, can not) have any "instance" of the "class" $\mathsf{a}$, and $x$ and $y$ have different architectures. This interpretation means that $x$ and $y$ are instances of *different* classes in OORS. In particular, if we have the following property:

$$\{\mathsf{a} \mid \mathsf{a}(x) \neq *\} \subseteq \{\mathsf{a} \mid \mathsf{a}(y) \neq *\}, \qquad (21)$$

we consider that the class of $y$ is a *subclass* of the class of $x$. These indicate that "absent value" semantics is characterized by is-a relationship in the framework of OORS.

Actually, for any given incomplete information table $(U, A)$ with null values, we can construct the following well-defined structures that illustrate "absence of values" by is-a relationship. First, we define the set of classes $C$, names $N$ and objects $O$ as follows:

$$\begin{aligned} C &= 2^A \cup A, \\ N &= \{n_B \mid B \subseteq A\} \cup \{n_\mathsf{a} \mid \mathsf{a} \in A\}, \\ O &= U \cup \{v_\mathsf{a}^x \mid v \in V_\mathsf{a} \setminus \{*\} \text{ and } \mathsf{a}(x) = v\}, \end{aligned}$$

where $2^A$ is the power set of $A$, symbols $n_B$ and $n_\mathsf{a}$ are new symbols that correspond to the name of each class $B \subseteq A$ and $\mathsf{a} \in A$, respectively, $V_\mathsf{a} \setminus \{*\}$ is the set of values of $\mathsf{a}$ without the null value $*$, and $v_\mathsf{a}^x$ is a new symbol that corresponds to the value $v$ of $x$ at $\mathsf{a}$.

Next, we define is-a relations $\sqsupseteq_X$ ($X \in \{C, N, O\}$) and has-a relations $\ni_X$ as follows:

$$\begin{aligned} c_i \sqsupseteq_C c_j &\iff \text{either } c_i \subseteq A, c_j \subseteq A \text{ and } c_j \subseteq c_i, \text{ or } c_i = c_j, \\ c_i \ni_C c_j &\iff c_i \subseteq A, c_j \in A \text{ and } c_j \in c_i, \\ n_i \sqsupseteq_N n_j &\iff \text{either } n_i = n_{B_i}, n_j = n_{B_j}, \text{ and } B_j \subseteq B_i, \text{ or } n_i = n_j, \\ n_i \ni_N n_j &\iff n = n_{B_i},\ n_j = n_\mathsf{a} \text{ and } \mathsf{a} \in B_i, \\ x_i \sqsupseteq_O x_j &\iff \text{either } \forall \mathsf{a} \in A, \mathsf{a}(x_j) \neq * \text{ implies } \mathsf{a}(x_i) \neq *, \text{ or } x_i = x_j, \\ x_i \ni_O x_j &\iff \exists \mathsf{a} \in A, \exists v \in V_\mathsf{a} \text{ such that } x_j = v_\mathsf{a}^{x_i}. \end{aligned}$$

Moreover, we define the class identifier $id_C : O \longrightarrow C$, the naming function $nf : N \longrightarrow C$ and the name assignment $na : O \longrightarrow N$ as follows:

$$id_C(x) = \begin{cases} \{\mathsf{a} \mid \mathsf{a}(x) \neq *\} & \text{if } x \in U, \\ \mathsf{a} & \exists \mathsf{a} \in A, o \in U, \text{ such that } x = v_\mathsf{a}^o. \end{cases}$$

$$nf(n) = \begin{cases} B & \exists B \subseteq A \text{ such that } n = n_B, \\ \mathsf{a} & \exists \mathsf{a} \in A \text{ such that } n = n_\mathsf{a}. \end{cases}$$

$$na(x) = \begin{cases} n_B & \exists B \subseteq A \text{ such that } B = \{\mathsf{a} \mid \mathsf{a}(x) \neq *\}, \\ n_\mathsf{a} & \exists \mathsf{a} \in A, o \in U, \text{ such that } x = v_\mathsf{a}^o. \end{cases}$$

Combining these components, we construct a class structure $\mathcal{C}_S = (C, \ni_C, \sqsupseteq_C)$, a name structure $\mathcal{N}_S = (N, \ni_N, \sqsupseteq_N)$ and an object structure $\mathcal{O}_S = (O, \ni_O, \sqsupseteq_O)$, respectively. It is not hard to show that these structures are well-defined.

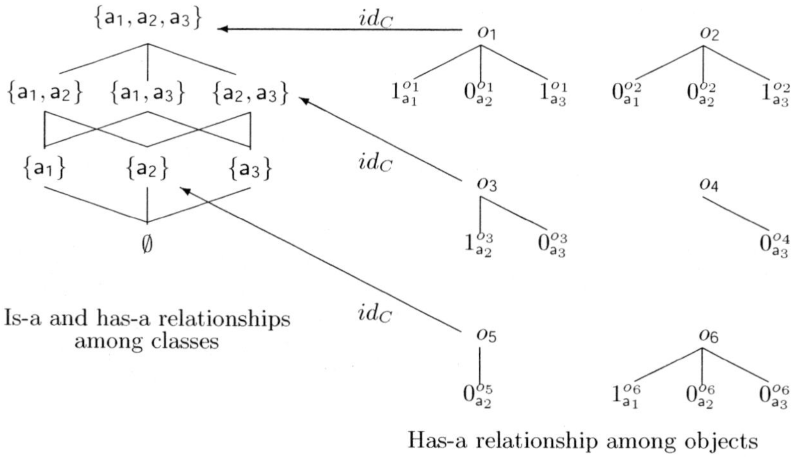

**Fig. 4.** Is-a and has-a relationship in Example 4

These constructed structures have the following good properties.

**Proposition 1.** *Let $(U, A)$ be an incomplete information table, and $\mathcal{C}_S$, $\mathcal{N}_S$ and $\mathcal{O}_S$ be the class, name and object structures constructed from $(U, A)$, respectively. These structures satisfy the following properties:*

1. *For any $x, y \in U$, the class of $y$ is a subclass of the class of $x$, that is, $id_C(y) \sqsupseteq_C id_C(x)$, if and only if $\{\mathsf{a} \mid \mathsf{a}(x) \neq *\} \subseteq \{\mathsf{a} \mid \mathsf{a}(y) \neq *\}$.*
2. *For any $x \in U$ and any $\mathsf{a} \in A$, $\mathsf{a}(x) = *$ if and only if there is no object $y \in O$ such that $id_C(y) = \mathsf{a}$ and $x \ni_O y$.*
3. *For any $x, y \in U$ and any set of names $B \subseteq N$, $B \cap H_N(na(x)) \subseteq B \cap H_N(na(y))$ if and only if $B \cap \{n_\mathsf{a} \mid \mathsf{a}(x) \neq *\} \subseteq B \cap \{n_\mathsf{a} \mid \mathsf{a}(y) \neq *\}$.*

These properties in Proposition 1 indicates that "absent value" semantics in the given incomplete information table as "non-existing of objects" are characterized by is-a relationships in the framework of OORS. The property 1 illustrates the situation that the value of the object $x$ at the attribute $\mathsf{a}$ is a null value in the given incomplete information table corresponds to the situation that $x$ has no instance of the class $\mathsf{a}$ in the constructed well-defined structures. On the other hand, the property 2 (3) illustrates that set inclusion between the set of attributes (between the set of names by attributes) provides the is-a relation $\sqsupseteq_C$ between classes (the is-a relation $\sqsupseteq_N$ between names).

*Example 4.* We construct the well-defined class structure $\mathcal{C}_S$, name structure $\mathcal{N}_S$ and object structure $\mathcal{O}_S$ from the incomplete information table illustrated in Table 1.

Figure 4 illustrates is-a and has-a relationships of the constructed structures. The Hasse's diagram in the left side of Fig. 4 illustrates "non-flat" is-a relationship and has-a relationship among classes. For example, the class $\{\mathsf{a}_1, \mathsf{a}_2, \mathsf{a}_3\}$ is a subclass of $\{\mathsf{a}_1, \mathsf{a}_2\}$ which has the class $\mathsf{a}_1$ and $\mathsf{a}_2$. On the other hand, trees in the right side illustrate actual has-a relationship among object. For example, the object $o_3$ as an instance of $\{\mathsf{a}_2, \mathsf{a}_3\}$ has an instance of $\mathsf{a}_2$ (we denote $1^{o_3}_{\mathsf{a}_2}$) and an instance of $\mathsf{a}_3$ (we denote $0^{o_3}_{\mathsf{a}_3}$), but does not have any instance of $\mathsf{a}_1$.

### 4.2 Non-symmetric Similarity Relations in Object-Oriented Rough Sets

We generalize the characteristics of "absent value" semantics based on is-a relationship illustrated in Proposition 1 to arbitrary well-defined class, name and object structures, and define non-symmetric similarity relations for OORS.

**Definition 5.** *Let $\mathcal{C} = (C, \ni_C, \sqsupseteq_C)$, $\mathcal{N} = (N, \ni_N, \sqsupseteq_N)$ and $\mathcal{O} = (O, \ni_O, \sqsupseteq_O)$ be well-defined class, name and object structures, respectively, $N^+$ be the set of consistent sequences of names in $\mathcal{N}$, and $\sim$ be the equivalence relation defined by Eq. 11. Moreover, let $D \subseteq N^+$ be a non-empty subset of consistent sequences of names. A similarity relation $\sigma_D$ on $O$ for the well-defined structures is a binary relation defined as follows:*

$$\begin{aligned} &x \sigma_D y \iff \\ &\text{For each } n_1.\cdots.n_k \in D, x \text{ and } y \text{ satisfy the following conditions:} \\ &1.\ n_1.\cdots.n_k \text{ connects to } x \Longrightarrow n_1.\cdots.n_k \text{ connects to } y, \text{ and} \\ &2.\ x.n_1.\cdots.n_k \sim y.n_1.\cdots.n_k. \end{aligned} \tag{22}$$

The difference between definitions of the indiscernibility relation $\approx_D$ by Eq. 12 and the similarity relation $\sigma_D$ by Eq. 22 is the condition 1, which is an extension of the property 3 in Proposition 1 to consistent sequences of names. Because any set of names $B \subseteq N$ is a subset of consistent sequences with length 1, that is, $B \subseteq N^+$, we can regard the property $B \cap H_N(na(x)) \subseteq B \cap H_N(na(y))$ as, for any sequence $n_1.\cdots.n_k \in D$, $n_1 \in H_N(na(x))$ implies $n_1 \in H_N(na(y))$. This condition requires that all consistent sequences of names in $D \subseteq N^+$ which connect to $x$ also connects to $y$.

We define the sets $\sigma_D(x)$ and $\sigma_D^{-1}(x)$ by the same manner of Eqs. 5 and 6 as follows, respectively:

$$\sigma_D(x) = \{y \in U \mid y\sigma_D x\}, \tag{23}$$

$$\sigma_D^{-1}(x) = \{y \in U \mid x\sigma_D y\}. \tag{24}$$

Moreover, for any set $X \subseteq O$ of objects, we define lower and upper approximations by Eqs. 7 and 8 as follows, respectively:

$$\underline{\sigma_D}(X) = \{x \in U \mid \sigma_D^{-1}(x) \subseteq X\}, \tag{25}$$

$$\overline{\sigma_D}(X) = \bigcup\{\sigma_D(x) \mid x \in X\}. \tag{26}$$

*Example 5.* We consider the "absent value" semantics in OORS by using the same setting of Example 2.

Let $\mathcal{C}$, $\mathcal{N}$ and $\mathcal{O}$ be the well-defined class, name and object structures in Example 2, respectively, and $X = \{$ pc1, pc3 $\}$ be the set of PCs which have the CPUs made by A company. Moreover, Let $D = \{$ hdd.size, hdd2.size $\}$ be a set of consistent sequences of names we consider. Here, using the non-symmetric similarity relation $\sigma_D$ defined by Eq. 22, we approximate the set $X$ by the lower approximation $\underline{\sigma_D}(X)$ defined by Eq. 25 and the upper approximation $\overline{\sigma_D}(X)$ defined by Eq. 26.

First, for each pc$i$ $(1 \leq i \leq 4)$, we calculate the sets $S_B^O(\text{pc}i)$ and $(S_B^O)^{-1}(\text{pc}i)$ as follows:

$$\begin{aligned}
\sigma_D(\text{pc1}) &= \{\text{pc1}\},\ \sigma_D^{-1}(\text{pc1}) = \{\text{pc1}, \text{pc3}, \text{pc4}\},\\
\sigma_D(\text{pc2}) &= \{\text{pc2}\},\ \sigma_D^{-1}(\text{pc2}) = \{\text{pc2}, \text{pc3}\},\\
\sigma_D(\text{pc3}) &= \{\text{pc3}\},\ \sigma_D^{-1}(\text{pc3}) = \{\text{pc3}\},\\
\sigma_D(\text{pc4}) &= \{\text{pc4}\},\ \sigma_D^{-1}(\text{pc4}) = \{\text{pc4}\}.
\end{aligned}$$

Note that pc1 is similar to pc4, however, pc4 is not similar to pc1. This is because pc1 is an instance of the PC class and there is no second HDD, then the sequence "hdd2.size" is ignored and pc1 is comparable with pc4 by the sequence "hdd.size". Consequently, we have the equivalence relationship

$$\text{pc1.hdd.size}(=80\text{GB}) \sim \text{pc4.hdd.size},$$

and therefore pc1 is similar to pc4. On the other hand, pc4 is an instance of the 2HDD-PC class and pc4 has the second HDD object h42(=pc4.hdd2), however, the object h42 is not comparable with any parts of pc1 (that is, pc4 does not satisfy the condition 1. in Eq. 22), therefore pc4 is not similar to pc1. By the same reason, pc2 is similar to pc3, but the converse is not.

Thus, we get the following lower and upper approximations of $X$ by the non-symmetric similarity relation $\sigma_D$, respectively:

$$\underline{\sigma_D(X)} = \{\text{pc3}\},\ \overline{\sigma_D(X)} = \{\text{pc1}, \text{pc3}\}.$$

## 5 Conclusion

In this chapter, we have characterized the "missing value" semantics and the "absent value" semantics in the framework of OORS as follows:

- the "missing value" semantics in OORS: incompleteness by lack of parts.
- the "absent value" semantics in OORS: incompleteness by difference of architecture of objects.

We have characterized incompleteness by lack of parts by introducing null value objects. As we discussed in [4], null value objects enable us to treat incompleteness of objects and constraints about design of objects simultaneously. Moreover, null value objects also provides flexibility of representation in a sense that name structures also illustrate "possibility" of the numbers of actual parts. As mentioned in Sect. 3.2, constraints about design of objects are satisfied in well-defined class, name and incomplete object structures. Thus, if the name structure describes that any instance $x$ of a class $c_i$ has exactly $m$ instances of a class $c_j$, there are exactly $m$ instances of $c_j$ within $k$ actual objects and $l$ null value objects as parts of $x$, where $k + l = m$ and $0 \leq k, l \leq m$. Therefore, we consider that the name structure also determines the maximum number of "actual" parts that an object *can* have, instead of determining the number of objects that an object *should* have in well-defined class, name and "complete" object structures.

The tolerance relation defined by Eq. 18 based on consistent sequences of names also illustrates incompleteness by lack of parts. Consistent sequences of names describe architecture of objects. If a sequence $n_1. \cdots .n_k$ connects to an object $x$, we expect that there is a hierarchy of has-a relationship $x \ni_O o_1 \ni_O \cdots \ni_O o_k$. Thus, if an object $x.n_1. \cdots .n_i$ $(1 \leq i \leq k-1)$ is a null value object, then the null value object "terminates" the hierarchy, and indicates that all of parts after $n_i$ do not exist. The tolerance relation detects such terminator, and illustrates indiscernibility which is tolerant to lack of parts.

On the other hand, we have characterized the "absent value" semantics as incompleteness by difference of architecture base on is-a relationship in OORS. As we have discussed in Sect. 4.1, the starting point is to regard occurrences of null values in the given incomplete information table as difference of architecture between objects, and such differences generate "non-flat" is-a relationships among classes, names, and objects in OORS. This starting point is interesting and quite different from the information table without null values, because, as we have mentioned in Sect. 2.2, well-defined structures constructed from a given "complete" information table have flat is-a relationships. Thus, generalizing this staring point to arbitrary OORS, we can treat "absent value" semantics in OORS by the non-symmetric similarity relation defined by Eq. 22 as an weakened version of the indiscernibility relation defined by Eq. 12 of OORS.

The results in this chapter indicate that we need to strictly distinguish "missing value" semantics and "absent value" semantics in OORS, and there is possibility that we use the concept of "missing value" and "absent value" in OORS simultaneously. Thus, hybridization of tolerance relations and non-symmetric similarity relations in OORS and rule generation based on these relations are interesting future works.

## Acknowledgment

We would like to express appreciation to reviewers for their helpful comments. This research was partially supported by the Grant-in-Aid for Young Scientists (B) (No.17700222), The Ministry of Education, Culture, Sports, Science and Technology, Japan.

## References

1. Budd, T.A.: An introduction of object-oriented programming, 2nd edn. Addison Wesley Longman, Reading (1997)
2. Kryszkiewicz, M.: Rough set approach to incomplete information systems. Information science 112, 39–49 (1998)

3. Kudo, Y., Murai, T.: A theoretical formulation of object-oriented rough set models. Journal of advanced computational intelligence and intelligent informatics 10(5), 612–620 (2006)
4. Kudo, Y., Murai, T.: A note on treatment of incomplete information in object–oriented rough sets. In: Proc. of the joint 3rd international conference on soft computing and intelligent systems and 7th international symposium on advanced intelligent systems, pp. 2238–2243 (2006)
5. Kudo, Y., Murai, T.: A method of generating decision rules in object-oriented rough set models. In: Greco, S., Hata, Y., Hirano, S., Inuiguchi, M., Miyamoto, S., Nguyen, H.S., Słowiński, R. (eds.) RSCTC 2006. LNCS (LNAI), vol. 4259, pp. 338–347. Springer, Heidelberg (2006)
6. Kudo, Y., Murai, T.: Absent value semantics as IS-A relationship in object-oriented rough set models. In: Proc. of the international symposium on fuzzy and rough sets (2006)
7. Kudo, Y., Murai, T.: Semi-structured decision rules in object-oriented rough set models for Kansei engineering. In: Yao, J.T., et al. (eds.) Rough sets and knowledge technology. LNCS (LNAI), vol. 4481, pp. 219–227. Springer, Heidelberg (2007)
8. Pawlak, Z.: Rough sets. International journal of computer and information science 11, 341–356 (1982)
9. Pawlak, Z.: Rough sets: Theoretical aspects of reasoning about data. Kluwer Academic Publishers, Dordrecht (1991)
10. Popkorn, S.: First steps in modal logic. Cambridge University Press, Cambridge (1994)
11. Słowiński, R., Stefanowski, J.: Rough classification in incomplete information systems. Mathematical computing modeling 12(10–11), 1347–1357 (1989)
12. Stefanowski, J., Tsoukiàs, A.: Incomplete information tables and rough classification. Computational intelligence 17(3), 545–565 (2001)
13. UML resource page, http://www.uml.org/

# Similarities for Crisp and Fuzzy Probabilistic Expert Systems

Cristina Coppola, Giangiacomo Gerla, and Tiziana Pacelli

Dipartimento di Matematica e Informatica, Università degli Studi di Salerno
Via Ponte don Melillo, 84084 Fisciano (SA), Italy
{ccoppola,gerla,tpacelli}@unisa.it

**Summary.** As stressed in [1] and [12] an interesting question on philosophy of probability is to assign probabilistic valuations to individual phenomenon. In [10] such a question was discussed and a solution was proposed. In this chapter we start from the ideas in [10] to sketch a method to design expert systems, probabilistic in nature. Indeed, we assume that the probability an individual satisfies a property is the percentage of similar individuals satisfying such a property. In turn, we call "similar" two individuals sharing the same observable properties. Such an approach is extended to the case of vague properties. We adopt a formalism arising from formal concept analysis.

**Keywords:** Fuzzy Formal Context, Fuzzy Similarity, State, Probabilistic Expert Systems, Foundation of Probability.

## 1 Introduction

An interesting question in philosophy of probability is to assign a probabilistic valuation to an individual phenomenon. As an example, imagine we claim that

(i) the probability of the statement "*a bird is able to fly*" is 0.9,
and compare such a claim with the following one
(ii) the probability of the statement "*Tweety is able to fly*" is 0.9.

Then, as emphasized by F.Bacchus in [1] and J.Y.Halpern in [12] the justification of these probabilistic assignations looks to be very different. In fact, (i) expresses a statistical information about the proportion of fliers among the set of birds. Such information, related to the whole class of birds, is statistical in nature. Instead, it seems very hard to justify (ii) from a statistical point of view, since the statement (ii) refers to a particular bird (Tweety) and not to a class of elements. As a matter of fact, either Tweety is able to fly or not, and the probabilistic valuation in (ii) is a *degree of belief* depending on the level of our knowledge about the capabilities of Tweety. In [10] it is proposed the idea that in such a case we can refer to the class of birds "similar" to Tweety. More precisely, the belief expressed in (ii) is based on the past experience about the percentage of birds similar to Tweety and able to fly. Obviously, the valuation of the similarity depends on the information on Tweety we have. So, both the probabilistic assignments in (i) and in (ii) are statistical in nature.

R. Bello et al. (Eds.): Granular Computing, STUDFUZZ 224, pp. 23–42, 2008.
springerlink.com 

**Table 1.** Notation

| symbols | meaning |
|---|---|
| $F$ | set of formulas |
| $B$ | Boolean algebra |
| $v$ | Boolean valuation |
| $\mu$ | probability valuation |
| $p$ | probability on $B$ |
| $(B, v, p)$ | $B$-probability valuation |
| $Ob$ | set of objects |
| $AT$ | set of attributes |
| $tr$ | information function |
| $(Ob, AT, tr)$ | formal context |
| $PC$ | set of past cases |
| $OBS$ | set of observable attributes |
| $an$ | (characteristic function of) the set of analogous past cases |
| $w$ | weight function |
| $SIB$ | statistical inferential basis |
| $v(\alpha)$ | (characteristic function of) the set of past cases satisfying $\alpha$ |
| $e$ | indiscernibility relation |
| $a_c$ | actual case |
| $T$ | piece of information on $a_c$ |
| $e_T$ | indiscernibility relation given $T$ |
| $SIB(T)$ | statistical inferential basis induced by $T$ in $SIB$ |
| $an_T$ | (characteristic function of) the set of analogous past cases given $T$ |
| $v_T(\alpha)$ | (char. funct. of) the set of past cases indiscernible from $a_c$ satisfying $\alpha$, given $T$ |
| $p_T$ | probability in $SIB(T)$ |
| $\mu_T$ | probability valuation in $SIB(T)$ |
| $A$ | $MV$-algebra |
| $\oplus$ | $t$-conorm, in particular Łukasiewicz disjunction |
| $\otimes$ | $t$-norm, in particular Łukasiewicz conjunction |
| $\leftrightarrow_\otimes$ | associated biresiduation to $\otimes$ |
| $E$ | $\otimes$-fuzzy similarity |
| $v_f$ | $MV$-valuation |
| $p_f$ | state on $A$ |
| $\mu$ | $MV$-probability valuation |
| $(A, v_f, p_f)$ | $A$-probability valuation |
| $tr_f$ | fuzzy information function |
| $sim$ | fuzzy set of similar past cases |
| $SIB_f$ | fuzzy statistical inferential basis |
| $v_f(\alpha)$ | fuzzy set of past cases satisfying $\alpha$ |
| $E_T$ | similarity given $T$ |
| $SIB_f(T)$ | fuzzy statistical inferential basis induced by $T$ in $SIB_f$ |
| $sim_T$ | fuzzy set of past cases similar to $a_c$ given $T$ |
| $v_f^T(\alpha)$ | fuzzy set of past cases similar to $a_c$ satisfying $\alpha$ given $T$ |
| $p_f^T$ | state in $SIB_f(T)$ |
| $\mu_T$ | $MV$-probability valuation in $SIB_f(T)$ |

On the basis of such an idea, in [2] a method to design probabilistic expert systems was proposed, by means of the crucial notion of *analogous*. In accordance with Leibniz' principle, two individuals are called analogous provided that they share the same *observable* properties.

In this chapter we reformulate the approach sketched in [2] and we extend it in order to admit vague properties. In particular we show how the notion of *fuzzy similarity* [23] can be utilized in order to design such probabilistic expert systems. Moreover, in doing this, we adopt a new formalism which is very close to *formal concept analysis* (see [9], [19], [21]) and which is adequate for a suitable extension to the fuzzy framework. This leads also to consider the crucial notion of *state* [14], [24], when we have to evaluate the probability that an individual satisfies an eventually vague property.

## 2 Probabilistic Valuations of the Formulas in Classical Logic

In this section we recall some basic notions of probabilistic logic. In the following we denote by $F$ the set of formulas of a classical zero-order language.

**Definition 1.** *Let $B = (B, \vee, \wedge, -, 0, 1)$ be a Boolean algebra. A Boolean valuation of $F$ (briefly $B$-valuation) is any map $v : F \to B$ satisfying the following properties, for any $\alpha$ and $\beta \in F$:*

- $v(\alpha \vee \beta) = v(\alpha) \vee v(\beta)$,
- $v(\alpha \wedge \beta) = v(\alpha) \wedge v(\beta)$,
- $v(\neg\alpha) = 1 - v(\alpha)$.

If $B$ is $(\{0, 1\}, \vee, \wedge, -, 0, 1)$, then the *$B$-valuation* coincides with the usual truth assignment of the formulas in classical logic. A *$B$-valuation* is truth-functional by definition, i.e. the truth value of a compound formula depends on the truth values of its components, unambiguously. A formula $\alpha$ is called *tautology* if $v(\alpha) = 1$ and *contradiction* if $v(\alpha) = 0$ for any $v$. Moreover, two formulas $\alpha$ and $\beta$ are called *logically equivalent* if $v(\alpha) = v(\beta)$ for any $v$.

**Definition 2.** *A probability valuation of $F$ is any map $\mu : F \to [0, 1]$ such that:*

1. $\mu(\alpha) = 1$, *for every tautology* $\alpha$,
2. $\mu(\alpha \vee \beta) = \mu(\alpha) + \mu(\beta)$, *if $\alpha \wedge \beta$ is a contradiction,*
3. $\mu(\alpha) = \mu(\beta)$, *if $\alpha$ is logically equivalent to $\beta$.*

Let us observe that if $\mu$ is a probability valuation, then $\mu(\alpha) = 0$ for every contradiction $\alpha$.

As it is well known, probability valuations are not truth-functional. Nevertheless, the truth-functionality can be obtained by means of the notion of $B$-valuation.

**Definition 3.** *A $B$-probability valuation of $F$ is a structure $(B, v, p)$ where*

- *$B$ is a Boolean algebra,*
- *$v : F \to B$ is a $B$-valuation (truth-functional),*
- *$p : B \to [0, 1]$ is a finitely additive probability on $B$.*

The notion of $B$-probability valuation and that one of probability valuation are strictly related as it is asserted in the following proposition [3].

**Proposition 1.** *Let $(B, v, p)$ be a $B$-probability valuation and let us define $\mu : F \to [0,1]$ by setting $\mu(\alpha) = p(v(\alpha))$ for every $\alpha \in F$. Then $\mu$ is a probability valuation. Conversely, let $\mu : F \to [0,1]$ be any probability valuation in $F$. Then a Boolean algebra $B$ and a $B$-probability valuation $(B, v, p)$ exist such that $\mu(\alpha) = p(v(\alpha))$.*

Due to the Representation Theorem of Boolean algebras [2], [15], it is not restrictive to assume that $B$ is an algebra of subsets of a set $S$. Moreover, we prefer identifying the subsets of a set with the related characteristic functions. So we refer to Boolean algebras as $\{0,1\}^S$ instead of $P(S)$ as we will see later on.

## 3 Formal Contexts, Statistical Inferential Bases and Indiscernibility

The first important step to design a probabilistic expert system is to create a database storing information about past cases we consider related to the actual one, according to the idea in [3], [10]. The notion of *formal context* ([9], [21]) seems suitable to represent this kind of collected information. It is usually used to identify patterns in data and it recognizes similarities between sets of objects based on their attributes.

**Definition 4.** *A formal context is a structure $(Ob, AT, tr)$ where:*

- *$Ob$ is a finite set whose elements we call objects,*
- *$AT$ is a finite set whose elements we call attributes,*
- *$tr : Ob \times AT \to \{0,1\}$ is a binary relation from $Ob$ to $AT$.*

Given an object $o$ and an attribute $\alpha$, $tr(o, \alpha) = 1$ means that the object $o$ possesses the attribute $\alpha$, while $tr(o, \alpha) = 0$ means that $o$ doesn't satisfy $\alpha$. It is easy to represent a formal context by a table, where the rows are the objects, the columns are the attributes and in the cells there are 0 or 1.

We consider as set of objects a set of "past cases" and we distinguish two types of attributes: we call *observable* the properties for which it is possible to discover directly whether they are satisfied or not by the examined case. Otherwise, a property is called *non observable*. As an example, an event that will happen in the future is a non observable property. The "actual case", i.e. the new examined case different from past cases, is considered *analogous* to a class of past cases if it satisfies their same observable properties.

**Definition 5.** *A (complete) statistical inferential basis is a structure $SIB = (PC, AT, OBS, an, tr, w)$ such that*

- *$(PC, AT, tr)$ is a formal context,*
- *$OBS$ is a subset of $AT$,*

- *$an : PC \rightarrow \{0,1\}$ is a map from $PC$ to $\{0,1\}$,*
- *$w : PC \rightarrow \mathbf{N}$ is a function called weight function.*

We call the elements of $PC$ *past cases* and the map $tr : PC \times AT \rightarrow \{0,1\}$ *information function*. The set $OBS$ is the subset of the *observable attributes* and the map $an$ is regarded as the (characteristic function of the) set of past cases analogous to the actual one. The meaning of the number $w(c) = n$ is that the past case $c$ is the representative of $n$ analogous cases. Then, we set the *total weight* of a statistical inferential basis $SIB$ as

$$w(SIB) = \sum\{w(c)an(c)|c \in PC\}. \tag{1}$$

It corresponds to the number of the past cases analogous to the actual case represented globally by $SIB$. If $w(SIB) \neq 0$ then we say that the statistical inferential basis is *consistent*. We denote by $F$ (by $F_{obs}$) the set of formulas of the propositional calculus whose set of propositional variables is $AT$ (is $OBS$, respectively). As usual, the function $tr$ can be extended to the whole set $F$ of formulas by setting, for every formula $\alpha$ and $\beta$,

- $tr(c, \alpha \wedge \beta) = min\{tr(c, \alpha), tr(c, \beta)\}$,
- $tr(c, \alpha \vee \beta) = max\{tr(c, \alpha), tr(c, \beta)\}$,
- $tr(c, \neg\alpha) = 1 - tr(c, \alpha)$.

In this way, any past case is associated by $tr$ with a classical valuation of the formulas in $F$.

In accordance with the basic notions of probabilistic logic, exposed in the previous section, now we provide some definitions of valuations associated to a statistical inferential basis $SIB$.

**Proposition 2.** *Every consistent statistical inferential basis $SIB = (PC, AT, OBS, an, tr, w)$ defines a B-probability valuation $(B, v, p)$ in $F$ such that:*

- *$B$ is the Boolean algebra $(\{0,1\}^{PC}, \cup, \cap, \neg, c_{\emptyset}, c_{PC})$,*
- *$v(\alpha) : PC \rightarrow \{0,1\}$ is (the characteristic function of) the set of past cases satisfying $\alpha$, i.e.*

$$v(\alpha)(c) = tr(c, \alpha), \tag{2}$$

- *$p : B \rightarrow [0,1]$ is the probability in $B$ defined by setting, for any $s \in \{0,1\}^{PC}$,*

$$p(s) = \frac{\sum\{w(c)an(c)s(c)|c \in PC\}}{w(SIB)} \tag{3}$$

As a consequence of Proposition 1 and Proposition 2 any statistical inferential basis $SIB$ can be associated with a probability valuation $\mu$ of the formulas. So we have, for every formula $\alpha$,

$$\mu(\alpha) = p(v(\alpha)) = \frac{\sum\{w(c)an(c)tr(c, \alpha)|c \in PC\}}{w(SIB)}. \tag{4}$$

In other words, $\mu(\alpha)$ represents the percentage of past cases (analogous to the actual case) in which $\alpha$ is true according to the stored dates.

According to the main idea we refer, it is important to specify which relation we take into account in order to consider "analogous" two cases. In the following, we introduce a formalism very close to Pawlak's one [17], based on Leibniz's indiscernibility principle, saying that two individuals are indiscernible if they share the same properties.

**Definition 6.** *Let $A$ be a subset of $AT$. Let $\leftrightarrow$ be the operation corresponding to the equivalence in the classical zero-order language and $e : PC \times PC \to \{0,1\}$ be a relation on $PC$ defined by setting*

$$e(c_1, c_2) = inf_{\alpha \in A} tr(c_1, \alpha) \leftrightarrow tr(c_2, \alpha). \tag{5}$$

*If $e(c_1, c_2) = 1$ we call the two cases $c_1$ and $c_2$ $A$-indiscernible.*

Let us observe that two cases are $A$-indiscernible if $tr(c_1, \alpha) = tr(c_2, \alpha)$ for every $\alpha \in A$, i.e. if they satisfy the same properties in $A$. It is immediate that $e$ is (the characteristic function of) an equivalence relation on $PC$. Then, for every case $c$, we can consider the corresponding equivalence class $[c]_A$.

In particular, we are interested to identify the past cases satisfying the same observable properties of the actual case. Let us recall that by "actual case" we intend a case different from past cases in which the only available information is represented by "observable" formulas. To our aim it is important to give an adequate definition of actual case.

**Definition 7.** *We call actual case any map $a_c : OBS \to \{0,1\}$ from the set of the observable $OBS$ to $\{0,1\}$. We call piece of information about $a_c$ any subset of $a_c$, i.e. any partial map $T : OBS \to \{0,1\}$ such that $a_c$ is an extension of $T$. We say that $T$ is complete if $T = a_c$.*

So, we identify the actual case with the "complete information" about its observable properties. As we will see in the next sections, we can collect pieces of information about $a_c$ by a query process. In the following we denote the actual case by $a_c$ or by the family $\{(\alpha, a_c(\alpha)\}_{\alpha \in OBS}$, indifferently. We extend the information function $tr$ to the actual case by setting $tr(a_c, \alpha) = a_c(\alpha)$ for every $\alpha \in OBS$ and then to the whole set $F_{obs}$ of observable formulas in the usual way. We also extend the relation $e$ by considering pieces of information on the actual case $a_c$. Indeed, given a piece of information $T$, we set

$$e_T(c, a_c) = inf_{\alpha \in Dom(T)} tr(c, \alpha) \leftrightarrow tr(a_c, \alpha). \tag{6}$$

If $e_T(c, a_c) = 1$, then $c$ is a past case which is *$OBS$-indiscernible from the actual case $a_c$ given the information $T$*. If $T$ is complete then we write $e(c, a_c)$ instead of $e_T(c, a_c)$.

**Definition 8.** *Let $SIB$ be a statistical inferential basis. We say that a piece of information $T$ is consistent with $SIB$ if there exists a past case $c \in PC$ such that $an(c) \neq 0$ and $e_T(c, a_c) \neq 0$.*

Let us observe that if $T$ is consistent with $SIB$ there is a past case $c$ analogous to the actual case according to the available information $T$, i.e. a past case $c$ exists such that it satisfies the same observable property of $a_c$ with respect to $T$.

Given a statistical inferential basis $SIB$, representing the basic information, and a piece of information $T = \{(\alpha_1, T(\alpha_1)), ..., (\alpha_n, T(\alpha_n))\}$ on the actual case, we obtain a new statistical inferential basis $SIB(T)$ from $SIB$.

**Definition 9.** *Let $SIB$ be a statistical inferential basis and $T$ a piece of information on $a_c$ consistent with $SIB$. We call statistical inferential basis induced by $T$ in $SIB$ the structure $SIB(T) = (PC, AT, OBS, an_T, tr, w)$, where $an_T$ is defined by setting $an_T(c) = an(c)e_T(c, a_c)$.*

In accordance with Proposition 2 and also considering the $B$-probability valuation $(B, v, p)$ associated to $SIB$, the statistical inferential basis $SIB(T)$ defines a $B$-probability valuation $(B, v_T, p_T)$ where:

- $B$ is the Boolean algebra $(\{0,1\}^{PC}, \cup, \cap, \neg, c_\emptyset, c_{PC})$,
- $v_T : F \to B$ is a $B$-valuation of the formulas in $F$ defined by

$$v_T(\alpha)(c) = an_T(c)v(\alpha)(c) = an(c)e_T(c, a_c)tr(c, \alpha), \tag{7}$$

  i.e. $v_T(\alpha)$ is (the characteristic function of) the set of past cases which are indiscernible from $a_c$ (given the available information $T$) and verifying $\alpha$,
- $p_T : B \to [0,1]$ is the probability on $B$ defined by setting, for any $s \in \{0,1\}^{PC}$,

$$p_T(s) = \frac{\sum\{w(c)an_T(c)s(c) | c \in PC\}}{w(SIB(T))}. \tag{8}$$

As usual, we have a probability valuation $\mu_T$ of the formulas defined, for every formula $\alpha$, as $\mu_T(\alpha) = p_T(v_T(\alpha))$, i.e.

$$\mu_T(\alpha) = \frac{\sum\{w(c)an_T(c)tr(c, \alpha) | c \in PC\}}{w(SIB(T))}. \tag{9}$$

The number $\mu_T(\alpha)$ is the percentage of the past cases verifying $\alpha$ among the cases in $SIB$ considered analogous to $a_c$ taking into account the available information $T$.

Let us observe that the probability $p_T$, defined in 8, can be regarded as the *conditioned probability* $p(\_/m_T)$, where $m_T$ denotes the set of past cases indiscernible from $a_c$ given $T$. Indeed, for any $s \in \{0,1\}^{PC}$, we have

$$p_T(s) = \frac{p(s \cap m_T)}{p(m_T)} = p(s/m_T).$$

Consequently, for every formula $\alpha$, also the probability valuation $\mu_T$ can be regarded as the *conditioned probability* $\mu_T(\alpha) = \mu(\alpha/m_T)$.

## 4 A Step-by-Step Inferential Process

In this section we describe how the step-by-step inferential process works. We imagine an expert system whose inferential engine contains an *initial* statistical inferential basis $SIB$, i.e. a statistical inferential basis such that $an$ is constantly equal to 1. This means that initially and in absence of information on $a_c$ we assume that all the past cases are analogous to the actual case. Successively, we can obtain information on $a_c$ by a sequence $\alpha_1, ..., \alpha_n$ of queries about observable properties. So, we set $T_0 = \emptyset$ and, given a new query $\alpha_i$, we set $T_i = T_{i-1} \cup \{(\alpha_i, \lambda_i)\}$ where $\lambda_i = 1$ if the answer is positive (the actual case verifies $\alpha_i$) and $\lambda_i = 0$ otherwise. As a consequence, we obtain a sequence of corresponding inferential statistical bases $\{SIB(T_i)\}_{i=1,...,n}$. At every step we can evaluate the probability that $a_c$ *satisfies* $\beta$ *given the available information.* Obviously, we are interested to a non observable property $\beta$.

**Definition 10.** *Let $SIB$ be an initial statistical inferential basis and $\beta$ be a formula in $F$. Let $T_n$ be the available information on $a_c$ obtained by a sequence of $n$ queries. Then we call probability that $a_c$ satisfies $\beta$ given the information $T_n$, the probability of $\beta$ in the statistical inferential basis $SIB(T_n)$ induced by $T_n$ in $SIB$.*

More precisely, we have the following step-by-step process:

1. Set $T_0 = \emptyset$ and $SIB_0 = SIB(\emptyset) = SIB$.
2. Given $T_k$ and $SIB_k = SIB(T_k)$, after the query $\alpha_{k+1}$ and the answer $\lambda_{k+1}$, put $T_{k+1} = T_k \cup \{(\alpha_{k+1}, \lambda_{k+1})\})$ and $SIB_{k+1} = SIB(T_{k+1})$.
3. If the information is *sufficient* or *complete* goto 4, otherwise goto 2.
4. Set $\mu(\beta) = \mu_{T_{k+1}}(\beta)$ as defined in (9).
5. If $T_{k+1}$ is inconsistent with $SIB_{k+1}$ then the process is *failed.*

Let us observe that we have different processes depending on the choice of the queries and on the *stop*-criterion expressed by the term *sufficient.* As an example, the query $\alpha_i$ can be selected in order to minimize the expected value of the entropy. This is achieved by minimizing the value $|\mu(\alpha_i) - \mu(\neg\alpha_i)|$ where $\mu$ is the valuation related to $SIB_i$. Also, let us notice that once a complete information on $a_c$ is obtained (in the language of the observable properties), $T_n = a_c$ and the inferential process necessarily terminates.

In other words: "*The probability that the actual case $a_c$ satisfies the property $\beta$ is given by the percentage of the cases $OBS$-indiscernible from $a_c$ that in the past verified $\beta$*". Such a point of view gives an answer to the question about the probabilities related to single cases [10].

## 5 Vague Properties and Similarities

In the previous sections we have considered only the presence of crisp attributes. An object satisfies or does not satisfy a property. But the real world has a fuzzy nature. In the most real situations an object verifies a property with a "degree".

So, if we consider the presence of eventually "vague" properties, it is necessary to extend the notions we have considered so far. Firstly, we give some basic notions in multi-valued logic. In many-valued logics (see [4], [5], [11]], [15]) truth degrees are not two yet, but three or more and many different algebraic structures are used for the evaluation of formulas.

In this section we present a class of these structures, the class of $MV$-algebras, devised by C.C.Chang [4], and then we introduce some other notion concerning multi-valued logic, such as *fuzzy set* and *fuzzy similarity.*

**Definition 11.** *An MV-algebra [5] is a structure* $A = (A, \oplus, \neg, 0)$ *such that* $(A, \oplus, 0)$ *is a commutative monoid satisfying the following additional properties:*

1. $\neg\neg a = a$;
2. $a \oplus \neg 0 = \neg 0$;
3. $\neg(\neg a \oplus b) \oplus b = \neg(\neg b \oplus a) \oplus a$.

On each $MV$-algebra $A$ we define the element 1 and the operation $\otimes$ as follows:

$$1 = \neg 0 \text{ and } a \otimes b = \neg(\neg a \oplus \neg b).$$

A well known example of $MV$-algebra is given by the *Łukasiewicz algebra* $([0,1], \oplus, \neg, 0)$, where $\oplus$ is the *Łukasiewicz disjunction* defined by

$$a \oplus b = min(1, a + b)$$

and $\neg a = 1 - a$. As a consequence the operation $\otimes$ is the *Łukasiewicz conjunction* defined by

$$a \otimes b = max(0, a + b - 1). \tag{10}$$

Łukasiewicz conjuction and disjunction are, respectively, examples of *t-norm* and *t-conorm* (see [11], [15]).

**Definition 12.** *A triangular norm (briefly t-norm) is a binary operation* $\otimes$ *on* $[0,1]$ *such that,* $\otimes$ *is commutative, associative, isotone in both arguments, i.e.,* $x_1 \leq x_2 \Rightarrow x_1 \otimes y \leq x_2 \otimes y$ *and* $y_1 \leq y_2 \Rightarrow x \otimes y_1 \leq x \otimes y_2$*, and* $\otimes$ *verifies the boundary conditions, i.e.* $1 \otimes x = x = x \otimes 1$ *and* $0 \otimes x = 0 = x \otimes 0$*, for all* $x, y, z, x_1, x_2, y_1, y_2 \in [0,1]$.

**Definition 13.** *A t-conorm is a binary operation* $\oplus : [0,1]^2 \to [0,1]$ *such that* $\oplus$ *is commutative, associative, isotone in both arguments and such that* $0 \oplus x = 0 = x \oplus 0$ *and* $1 \oplus x = x = x \oplus 1$.

Moreover, the t-conorm $\oplus$ is dual to a given t-norm $\otimes$ if, for every $x, y \in [0,1]$,

$$x \oplus y = 1 - ((1 - x) \otimes (1 - y)).$$

For each t-norm, we can consider the *associated biresiduation*, suitable to represent the truth function of equivalence. In the case of Łukasiewicz conjunction, it is defined by

$$a \leftrightarrow_{\otimes} b = 1 - |a - b|, \tag{11}$$

and some its properties are listed in the following:

- $x \leftrightarrow_{\otimes} x = 1$,
- $x \leftrightarrow_{\otimes} y = 1 \Leftrightarrow x = y$,
- $(x \leftrightarrow_{\otimes} y) \otimes (y \leftrightarrow_{\otimes} z) \leq x \leftrightarrow_{\otimes} z$,
- $x \leftrightarrow_{\otimes} y = y \leftrightarrow_{\otimes} x$.

Fuzzy set theory [22] can be regarded as an extension of the classical one, where an element either belongs or does not belong to a set. It permits the gradual assessment of the membership of elements to a set, by a generalized characteristic function.

**Definition 14.** *Let $S$ be a set and let us consider the complete lattice $[0,1]$. We call fuzzy subset of $S$ any map $s : S \to [0,1]$ and we denote by $[0,1]^S$ the class of all the fuzzy subsets of $S$.*

Given any $x$ in $S$, the value $s(x)$ is the *degree of membership* of $x$ to $s$. In particular, $s(x) = 0$ means that $x$ is not included in $s$, whereas 1 is assigned to the elements fully belonging to $s$. Any fuzzy subset $s$ such that $s(x) \in \{0,1\}$, for any $x \in S$, is called *crisp set*. Given $\lambda \in [0,1]$, we denote by $s^\lambda$ the fuzzy set constantly equal to $\lambda$.

**Definition 15.** *Let $\otimes$ be the Lukasiewicz conjunction and $\oplus$ be the Lukasiewicz disjunction. We define the union, the intersection and the complement by setting, respectively, for any $s, s' \in [0,1]^S$ and for every $x \in S$,*

- $(s \cup_{\oplus} s')(x) = s(x) \oplus s'(x)$
- $(s \cap_{\otimes} s')(x) = s(x) \otimes s'(x)$
- $(\neg s)(x) = -s(x)$.

**Proposition 3.** *The structure $([0,1]^S, \cup_{\oplus}, \cap_{\otimes}, \neg, s^0, s^1)$ is an $MV$-algebra extending the Boolean algebra $(\{0,1\}^S, \cup, \cap, \neg, \emptyset, S)$ of the subsets of $S$.*

In the following we denote this $MV$-algebra also by $([0,1]^S, \oplus, \neg, s^0)$.

A special class of fuzzy sets is given by the concept of *similarity* [22], which is essentially a generalization of an equivalence relation.

**Definition 16.** *Let $\otimes$ be the Łukasiewicz conjunction. A $\otimes$-fuzzy similarity on a set $S$ is a fuzzy relation on $S$, i.e. a fuzzy subset of $S \times S$, $E : S \times S \to [0,1]$, satisfying the following properties:*

1. *$E(x,x) = 1$ (reflexivity)*
2. *$E(x,y) = E(y,x)$ (symmetry)*
3. *$E(x,y) \otimes E(y,z) \leq E(x,z)$ ($\otimes$-transitivity).*

The logical meaning of the $\otimes$-transitivity is that "if $x$ is similar to $y$ with a degree $E(x,y)$ and $y$ is similar to $z$ with a degree $E(y,z)$ then $x$ is similar to $z$ with a degree $E(x,z)$ greater or equal to $E(x,y) \otimes E(y,z)$".

Let us recall that for any t-norm we can have a corresponding notion of fuzzy similarity but we give the definition directly by the Łukasiewicz conjunction because we will use it in the proposed inferential process. In the following sections we refer to the following basic theorem ([20]) enabling to extend the Definition 6 to vague properties. In a sense, it is also related to Leibniz's indiscernibility principle.

**Proposition 4.** *Let us consider a finite family $(s_i)_{i\in I}$ of fuzzy subsets of a set $S$. Let $\otimes$ be the Łukasiewicz conjunction and $\leftrightarrow_{\otimes}$ be its associated biresiduation. Let us define the fuzzy relation $E : S \times S \to [0,1]$ by*

$$E(x,y) = \otimes_{i\in I} s_i(x) \leftrightarrow_{\otimes} s_i(y).$$

*Then $E$ is a $\otimes$-similarity on $S$.*

## 6 Probabilistic Logic in Fuzzy Framework

In this section we extend the basic notions of probabilistic logic, exposed in Section 2. Since we will admit the presence of eventually "vague" properties in the inferential process, we have to consider probabilistic valuation of fuzzy subsets ([6]). In particular, we refer to the concept of *state* ([14]), which is a generalization on $MV$-algebras of the classical notion of (finitely additive) probability measure on Boolean algebras. In the following, we denote by $F$ the set of formulas in the language of a many-valued logic. More precisely, we refer to a logic whose propositional calculus assumes truth values in an $MV$-algebra.

**Definition 17.** *Let $(A, \oplus, \neg, 0)$ be an $MV$-algebra. An $MV$-valuation is any map $v_f : F \to A$ satisfying the following properties:*

- $v_f(\alpha \vee \beta) = v_f(\alpha) \oplus v_f(\beta)$,
- $v_f(\alpha \wedge \beta) = v_f(\alpha) \otimes v_f(\beta)$,
- $v_f(\neg\alpha) = \neg v_f(\alpha)$.

Trivially, $v_f$ is a truth-functional map by definition. Moreover, a formula $\alpha$ is called *tautology* if $v_f(\alpha) = 1$ and it is called *contradiction* if $v_f(\alpha) = 0$, for any $MV$-valuation $v_f$. Two formulas $\alpha$ and $\beta$ are *logically equivalent* if $v_f(\alpha) = v_f(\beta)$ for any $v_f$.

**Definition 18.** *A state of an $MV$-algebra $A$ is a map $p_f : A \to [0,1]$ satisfying the following conditions:*

1. $p_f(0) = 0$,
2. $p_f(1) = 1$,
3. $p_f(a \oplus b) = p_f(a) + p_f(b)$, *for every $a, b \in A$ such that $a \otimes b = 0$.*

A natural example of state in the $MV$-algebra $([0,1]^X, \oplus, \neg, s^0)$, where we have Łukasiewicz disjunction, is given by [24]:

**Proposition 5.** *Let $X$ be a finite set and $p : \{0,1\}^X \to [0,1]$ an arbitrary probability measure on $\{0,1\}^X$. Let the map $p_f : [0,1]^X \to [0,1]$ be defined, for every $s \in [0,1]^X$, by*

$$p_f(s) = \sum \{s(x)p(x) | x \in X\}.$$

*Then $p_f$ is a state of the $MV$-algebra $([0,1]^X, \oplus, \neg, s^0)$.*

We introduce the notions of *MV-probability valuation* of formulas and, then, of *A-probability valuation* which enables us to obtain the truth-functionality of the first one.

**Definition 19.** *An MV-probability valuation of $F$ is any map $\mu : F \to [0,1]$ such that:*

- $\mu(\alpha) = 1$ *for every tautology* $\alpha$,
- $\mu(\alpha \vee \beta) = \mu(\alpha) + \mu(\beta)$ *if* $\alpha \wedge \beta$ *is a contradiction,*
- $\mu(\alpha) = \mu(\beta)$ *if* $\alpha$ *is logically equivalent to* $\beta$.

Let us observe that the only difference with Definition 2 is that the notions of "tautology", "contradiction" and "logically equivalent" are intended in the sense of Definition 17.

**Definition 20.** *An A-probability valuation is a structure $(A, v_f, p_f)$ where*

- $A$ *is an MV-algebra,*
- $v_f : F \to A$ *is a truth-functional MV-valuation of formulas,*
- $p_f : A \to [0,1]$ *is a state on* $A$.

The notion of $A$-probability valuation is connected to that one of $MV$-probability valuation [16].

**Proposition 6.** *Let $(A, v_f, p_f)$ be an A-probability valuation and let us define $\mu : F \to [0,1]$ by setting $\mu(\alpha) = p_f(v_f(\alpha))$ for every $\alpha \in F$. Then $\mu$ is an MV-probability valuation. Conversely, let $\mu : F \to [0,1]$ be any MV-probability valuation in $F$. Then an MV-algebra $A$ and an A-probability valuation $(A, v_f, p_f)$ exist such that $\mu(\alpha) = p_f(v_f(\alpha))$.*

## 7 Fuzzy Statistical Inferential Bases

In order to create a database of past cases verifying eventually vague properties in this section we extend the definitions presented in Section 3. We refer to a generalization of the basic notion of *formal concept analysis* ([9], [21]).

**Definition 21.** *A fuzzy formal context ([19]) is a structure $(Ob, AT, tr_f)$ where:*

- $Ob$ *is a finite set whose elements we call objects,*
- $AT$ *is a finite set whose elements we call attributes,*
- $tr_f : Ob \times AT \to [0,1]$ *is a fuzzy binary relation from $Ob$ to $AT$.*

The fuzzy relation $tr_f$ connects any object with any attribute, i.e. the value $tr_f(o, \alpha)$ is the truth degree of the claim "the object $o$ satisfies the property $\alpha$". As in Section 3, we consider as set of objects the set of "past cases" and we distinguish two types of attributes: we call *observable* the attributes for which it is possible to discover directly whether they are satisfied by the examined case, *non observable* the others. The observable properties are considered to yield the similarity between past cases and the "actual case", different from past cases. We want to evaluate the probability with which the actual case verifies a non observable property.

**Definition 22.** *A (complete) fuzzy statistical inferential basis is a structure* $SIB_f = (PC, AT, OBS, sim, tr_f, w)$ *such that*

- $(PC, AT, tr_f)$ *is a fuzzy formal context,*
- $OBS$ *is a subset of* $AT$*,*
- $sim : PC \to [0, 1]$ *is a fuzzy subset of* $PC$*,*
- $w : PC \to \mathbf{N}$ *is a function called weight function.*

The set $PC$ is that one of *past cases* and the map $tr_f$ is called *fuzzy information function*. It provides the degree with which a past case satisfies an attribute. The set $OBS$ is the (classical) subset of the observable attributes and the map $sim$ is interpreted as the fuzzy set of past cases "similar" to the actual one. The value $w(c)$ gives the number of past cases whose representative is $c$. Then, we set the *total weight* of a fuzzy statistical inferential basis $SIB_f$ as

$$w(SIB_f) = \sum\{w(c)sim(c)|c \in PC\}. \tag{12}$$

If $w(SIB_f) \neq 0$ then we say that $SIB_f$ is *consistent*.

As in the previous section, we denote by $F$ (by $F_{obs}$) the set of formulas of a multivalued propositional calculus whose set of propositional variables is $AT$ (or $OBS$, respectively). So we extend $tr_f$ to the whole set $F$ of formulas by setting

- $tr_f(c, \alpha \wedge \beta) = tr_f(c, \alpha) \otimes trf(c, \beta)$,
- $tr_f(c, \alpha \vee \beta) = tr_f(c, \alpha) \oplus tr_f(c, \beta)$,
- $tr_f(c, \neg\alpha) = 1 - tr_f(c, \alpha)$.

By referring to the notions introduced in Section 6, we provide definitions of valuations associated to a fuzzy statistical inferential basis.

**Proposition 7.** *Every consistent fuzzy statistical inferential basis* $SIB_f$ *defines an A-probability valuation* $(A, v_f, p_f)$ *in* $F$ *such that:*

- $A$ *is the* $MV$*-algebra* $([0, 1]^{PC}, \oplus, \neg, s^0)$*,*
- $v_f(\alpha) : PC \to [0, 1]$ *is the fuzzy subset of the past cases satisfying the formula* $\alpha$*, i.e.*

$$v_f(\alpha)(c) = tr_f(c, \alpha),$$

- $p_f : A \to [0, 1]$ *is the state on* $A$ *defined, for any* $s \in [0, 1]^{PC}$*, as in Proposition 5, i.e.*

$$p_f(s) = \sum\{s(c)p(c)|c \in PC\},$$

*where* $p$ *is the probability on* $\{0, 1\}^{PC}$*, defined by*

$$p(c) = \frac{w(c)sim(c)}{w(SIB_f)}.$$

By Proposition 6, any fuzzy statistical inferential basis $SIB_f$ can be associated with an $MV$-probability valuation $\mu$ of the formulas, defined, for every $\alpha$, by

$$\mu(\alpha) = p_f(v_f(\alpha)) = \sum\{v_f(\alpha)(c)p(c) | c \in PC\}. \tag{13}$$

In other words we have

$$\mu(\alpha) = \frac{\sum\{w(c)sim(c)v_f(\alpha)(c) | c \in PC\}}{w(SIB_f)}, \tag{14}$$

and this value represents the percentage of past cases similar to the actual case in which $\alpha$ is verified.

## 8 The Actual Case and Its Similar Past Cases

The indiscernibility relation in Definition 6, used for "crisp" properties, is not sufficient anymore for "vague" properties. Indeed, in a classification process, given a set of (eventually vague) properties $B$, and a property $\beta \in B$, if $tr(c_1, \alpha) = tr(c_2, \alpha)$ for every $\alpha \in B - \{\beta\}$ and $tr(c_1, \beta) = 0,8$ and $tr(c_2, \beta) = 0,9$, it is not reasonable to consider the two case $c_1$ and $c_2$ not "analogous". Therefore, it is necessary to take into account an extension of the relation, such that it results appropriate to a classification handling "vague" properties and in order to consider "similar" two cases with respect to these properties. As an immediate consequence of Proposition 4, we obtain the following one, where $\oplus$ and $\otimes$ denote the Łukasiewicz conjunction and disjunction, respectively.

**Proposition 8.** *Let $SIB_f$ be a fuzzy statistical inferential basis and let $(A, v_f, p_f)$ be the A-probability valuation associated to it. Then, for any subset $B$ of $AT$, the fuzzy relation $E : PC \times PC \to [0,1]$, defined by setting*

$$E(c_1, c_2) = \otimes_{\alpha \in B}(v_f(\alpha)(c_1) \leftrightarrow_{\otimes} v_f(\alpha)(c_2)), \tag{15}$$

*is a $\otimes$-fuzzy similarity.*

Since the fuzzy set $v_f(\alpha) : PC \to [0,1]$ of past cases satisfying the property $\alpha$ is defined by $v_f(\alpha)(c) = tr_f(c, \alpha)$, we can rewrite 15 as

$$E(c_1, c_2) = \otimes_{\alpha \in B}(tr_f(c_1, \alpha) \leftrightarrow_{\otimes} tr_f(c_2, \alpha)). \tag{16}$$

The value $E(c_1, c_2)$ yields the "degree of similarity" between the two past cases $c_1$ and $c_2$ in $SIB_f$. From the logical point of view, it is the valuation of the claim "every property satisfied by $c_1$ is satisfied by $c_2$ and vice-versa".

As usual, a similarity can be interpreted in terms of *fuzzy similarity classes*, one for each element of the universe. In our situation, for every case $c_j$, we can consider a fuzzy subset $sim_{c_j} : PC \to [0,1]$ as the fuzzy class of the past cases "similar" to $c_j$, by setting $sim_{c_j}(c) = E(c, c_j)$. In particular, we have to identify the past cases similar to the actual one. Let us recall that by "actual case" we intend a case different from past cases in which the only available information is that one expressed by the set $F_{obs}$ in the language of "observable" properties. The definition of "actual case" in the fuzzy situation is a generalization of that one in the crisp case (Definition 7).

**Definition 23.** *We call (fuzzy) actual case any map $a_c : OBS \to [0,1]$ from the set of observable properties to the interval $[0,1]$. We call piece of information about $a_c$ any subset of $a_c$, i.e. any partial map $T : OBS \to [0,1]$ such that $a_c$ is an extension of $T$. We say that $T$ is complete if $T = a_c$.*

We denote the actual case by $a_c$ or by the family $\{a_c, a_c(\alpha)\}_{\alpha \in OBS}$, indifferently. The last notation is more useful in describing the inferential process, where we identify the actual case with the "information" about its observable properties, collected by a query process. We extend the fuzzy information function $tr_f$ and the similarity $E$, given in 16 to the actual case by setting $tr_f(a_c, \alpha) = a_c(\alpha)$ for every $\alpha \in OBS$. Given a piece of information $T$ about $a_c$, we set, for every $\alpha \in Dom(T)$, $E_\alpha(c, a_c) = tr_f(c, \alpha) \leftrightarrow_\otimes tr_f(a_c, \alpha)$ and

$$E_T(c, a_c) = \otimes_{\alpha \in Dom(T)} E_\alpha(c, a_c). \tag{17}$$

$E_T(c, a_c)$ yields the *similarity between $c$ and the actual case $a_c$, given the information $T$*. If $T$ is complete then we write $E(c, a_c)$ instead of $E_T(c, a_c)$.

**Definition 24.** *Let $SIB_f$ be a fuzzy statistical inferential basis. We say that a piece of information $T$ is consistent with $SIB_f$ if there exists $c \in PC$ such that $sim(c) \neq 0$ and $E_T(c, a_c) \neq 0$.*

If $T$ is consistent with $SIB_f$, then in our database there is at least a past case $c$ similar to $a_c$ according to the available information $T$.

## 9 Fuzzy Statistical Inferential Bases Induced by a Piece of Information and the Step-by-Step Inferential Process

Given a fuzzy statistical inferential basis $SIB_f$ and a piece of information $T$ on the actual case $a_c$, we obtain a new fuzzy statistical inferential basis $SIB_f(T)$ from $SIB_f$.

**Definition 25.** *Let $SIB_f$ be a consistent fuzzy statistical inferential basis and $T$ be a piece of information on $a_c$. We call fuzzy statistical inferential basis induced by $T$ in $SIB_f$ the structure $SIB_f(T) = (PC, AT, OBS, sim_T, tr_f, w)$, where $sim_T$ is defined by setting $sim_T(c) = E_T(c, a_c)$, where $E_T$ is defined in 17.*

Let us observe that $sim_T$ can be regarded as the fuzzy class of the past cases "similar" to '$a_c$ given the information $T$. Then, in accordance with Proposition 7 and given the $A$-probability valuation $(A, v_f, p_f)$ associated to $SIB_f$, the induced fuzzy statistical inferential basis $SIB_f(T)$ defines an $A$-probability valuation $(A, v_f^T, p_f^T)$ where:

- $A$ is the $MV$-algebra $([0,1]^{PC}, \oplus, \neg, s^0)$,
- $v_f^T : F \to A$ is an $MV$-valuation of the formulas and $v_f^T(\alpha)$ is the fuzzy set of the past cases similar to $a_c$ (given the information $T$) and verifying the formula $\alpha$, i.e.

$$v_f^T(\alpha)(c) = sim_T(c) \otimes v_f(\alpha)(c),$$

- $p_f^T : A \to [0,1]$ is the state on $A$ defined by setting, for any $s \in [0,1]^{PC}$,

$$p_f^T(s) = \sum\{s(c)p_T(c)|c \in PC\},$$

where $p_T$ is the probability on $\{0,1\}^{PC}$ given by

$$p_T(c) = \frac{w(c)sim_T(c)}{w(SIB_f(T))}.$$

So, given a fuzzy statistical inferential basis $SIB_f$ and a piece of information $T$, we obtain an $MV$-probability valuation $\mu_T$ of the formulas, defined, for every formula $\alpha$, by

$$\mu_T(\alpha) = p_f^T(v_f^T(\alpha)) = \sum\{v_f^T(\alpha)(c)p_T(c)|c \in PC\}. \tag{18}$$

Let us observe that we obtain

$$\mu_T(\alpha) = \frac{\sum\{w(c)sim_T(c)v_f^T(\alpha)(c)|c \in PC\}}{\sum\{w(c)sim_T(c)|c \in PC\}}, \tag{19}$$

and it represents the percentage of the past cases verifying $\alpha$ among the cases in $SIB_f$ considered similar to $a_c$ according to the available information $T$.

Now, let us imagine the expert system has to evaluate the probability that an actual case $a_c$ verifies a non observable formula $\beta$ . Let us suppose that in the *initial* fuzzy statistical inferential basis $SIB_f$, the map *sim* is constantly equal to 1, i.e. we are considering all the past cases "similar" to the actual one. The information on $a_c$ can be obtained by a query-strategy. Let us denote by $\alpha_1, ..., \alpha_n$ a sequence of appropriate queries about observable properties of $a_c$. Then, we set $T_0 = \emptyset$ and, given a new query $\alpha_i$, we set $T_i = T_{i-1} \cup \{(\alpha_i, \lambda_i)\}$, where $\lambda_i \in [0,1]$ is the degree with which the actual case verifies the property $\alpha_i$. Consequently, we obtain a sequence of corresponding fuzzy inferential statistical basis $\{SIB_f(T_i)\}_{i=1,...,n}$. At every step we have the probability that $a_c$ *satisfies $\beta$ given the available information.*

**Definition 26.** *Let $SIB_f$ be an initial fuzzy statistical inferential basis and $\beta$ be a formula in $F$. Let $T_n$ be the available information on $a_c$ obtained by a sequence of $n$ queries. Then we call probability that $a_c$ satisfies $\beta$ given the information $T_n$, the probability of $\beta$ in the fuzzy statistical inferential basis $SIB_f(T_n)$ induced by $T_n$ in $SIB_f$.*

**The step-by-step inferential process:**

1. Set $T_0 = \emptyset$ and $SIB_0 = SIB_f(\emptyset) = SIB_f$.
2. Given $T_k$ and $SIB_k = SIB_f(T_k)$, after the query $\alpha_{k+1}$ and the answer $\lambda_{k+1}$, set $T_{k+1} = T_k \cup \{(\alpha_{k+1}, \lambda_{k+1})\}$ and $SIB_{k+1} = SIB_f(T_{k+1})$, in which

$$sim_{T_{k+1}}(c) = sim_{T_k} \otimes E_{\alpha_{k+1}}(c, a_c) = sim_{T_k} \otimes (v_f(\alpha_{k+1})(c) \leftrightarrow_\otimes \lambda_{k+1}). \tag{20}$$

3. If the information is *sufficient* or *complete* goto 4, otherwise goto 2.

4. Set $\mu(\beta) = \mu_{T_{k+1}}(\beta)$ as defined in 19.
5. If $T_{k+1}$ is inconsistent with $SIB_{k+1}$ then the process is *failed.*

Let us notice that if the information on $a_c$ is complete, then $T_n = a_c$ and the inferential process terminates. Let us remark that in 20, $\otimes$ is calculated as in 10 and $\leftrightarrow_\otimes$ is calculated as in 11.

We give a representation by tables of the $SIB_f$ induced by information we obtain at every step. This example is not based on real data, but its only aim is to show how the inferential process works. Let $c_1$, $c_2$ and $c_3$ be the past cases and $\alpha_1$, $\alpha_2$, $\alpha_3$ and $\beta$ the attributes. Let us suppose that $\alpha_1$, $\alpha_2$ and $\alpha_3$ are the observable attributes for the actual case and $\beta$ is the non-observable attribute. So we want to evaluate the probabilty that $a_c$ verifies $\beta$. We are supposing that we have no information on $a_c$ i.e. $T_0 = \emptyset$ and that $sim_{T_0} = 1$. The initial statistical inferential basis $SIB_0$ is represented by the following table:

Let $\alpha_1$ be the first query on $a_c$ and $\lambda_1 = 0.854$, i.e. $T_1 = \{(\alpha_1, 0.854)\}$. For every case $c_i$, $sim_{T_1}(c_i) = sim_{T_0}(c_i) \otimes E_{\alpha_1}(c_i, a_c)$ (see 20). The representation of $SIB_1$ is:
At this step, the probability that $a_c$ verifies $\beta$ is $\mu_{T_1}(\beta) = 0.899$ (see 19).

Let $\alpha_2$ be the second query on $a_c$ and $\lambda_2 = 0.973$, i.e. $T_2 = T_1 \cup \{(\alpha_2, 0.973)\}$. As in the previous step, it is possible to evaluate $sim_{T_2}$. The representation of $SIB_2$ is:

At this step, the probability that $a_c$ verifies $\beta$ is $\mu_{T_2}(\beta) = 0.829$ (see 19).

Let $\alpha_3$ be the last query on $a_c$ and $\lambda_3 = 1.000$, i.e. $T_3 = T_2 \cup \{(\alpha_3, 1.000)\}$. It is possible to evaluate $sim_{T_3}$ and the representation of $SIB_3$ is:
The final probability that $a_c$ verifies $\beta$ is $\mu_{T_3}(\beta) = 0.824$ (see 19).

Let us observe that the proposed process is also interpretable in the framework of *case-based reasoning* (see for example [18]). Indeed, if we interpret the set of non observable attributes as the set of probable "solutions" and the value $tr_f(c, \alpha)$, with $\alpha \in AT - OBS$, as the "validity degree" of the solution $\alpha$ for the

**Table 2.** Representation of $SIB$

| case | $v_f(\alpha_1)$ | $v_f(\alpha_2)$ | $v_f(\alpha_3)$ | $v_f(\beta)$ | $w$ | $sim_{T_0}$ |
|---|---|---|---|---|---|---|
| $c_1$ | 0.954 | 0.853 | 0.974 | 1.000 | 20 | 1.000 |
| $c_2$ | 0.873 | 0.921 | 1.000 | 0.977 | 40 | 1.000 |
| $c_3$ | 0.737 | 0.897 | 1.000 | 0.892 | 20 | 1.000 |

**Table 3.** Representation of $SIB_1$

| case | $v_f(\alpha_1)$ | $v_f(\alpha_2)$ | $v_f(\alpha_3)$ | $v_f(\beta)$ | $w$ | $sim_{T_1}$ |
|---|---|---|---|---|---|---|
| $c_1$ | 0.954 | 0.853 | 0.974 | 1.000 | 20 | 0.900 |
| $c_2$ | 0.873 | 0.921 | 1.000 | 0.977 | 40 | 0.954 |
| $c_3$ | 0.737 | 0.897 | 1.000 | 0.892 | 20 | 0.775 |

**Table 4.** Representation of $SIB_2$

| case | $v_f(\alpha_1)$ | $v_f(\alpha_2)$ | $v_f(\alpha_3)$ | $v_f(\beta)$ | $w$ | $sim_{T_2}$ |
|---|---|---|---|---|---|---|
| $c_1$ | 0.954 | 0.853 | 0.974 | 1.000 | 20 | 0.780 |
| $c_2$ | 0.873 | 0.921 | 1.000 | 0.977 | 40 | 0.929 |
| $c_3$ | 0.737 | 0.897 | 1.000 | 0.892 | 20 | 0.807 |

**Table 5.** Representation of $SIB_3$

| case | $v_f(\alpha_1)$ | $v_f(\alpha_2)$ | $v_f(\alpha_3)$ | $v_f(\beta)$ | $w$ | $sim_{T_3}$ |
|---|---|---|---|---|---|---|
| $c_1$ | 0.954 | 0.853 | 0.974 | 1.000 | 20 | 0.754 |
| $c_2$ | 0.873 | 0.921 | 1.000 | 0.977 | 40 | 0.929 |
| $c_3$ | 0.737 | 0.897 | 1.000 | 0.892 | 20 | 0.807 |

case $c$ collected in the database, the final value $\mu_T(\alpha)$ represents a "validity degree" of the solution $\alpha$ for the actual case. In such a case $\mu_T(\alpha)$ is the percentage of the past cases similar to $a_c$ for which $\alpha$ was a "good solution". Our approach is close to case-based reasoning systems, since we make a prediction on a new case by observing precedent cases. On the other hand, the prediction, probabilistic in nature, is obviously different from that one used in other approaches, generally possibilistic in nature [8].

## 10 Conclusions and Future Work

In this chapter we sketch a method to define an expert system probabilistic in nature. The implementation of such a method has no difficulty since it is sufficient to refer to a suitable relational database management system. In the crisp case this was done in [3]. An interesting feature of the step-by-step inferential process we propose is that in each step we obtain a reliable probabilistic valuation of the question we are interested. Another feature is the evolutionary character of the system. In fact the initial database storing the past cases can be continuously enriched by adding new cases. Also, the proposed formalism enables us to define suitable querying strategies in which the choice of the successive query is addressed to minimize the expected value of the entropy, the costs, the times and so on.

However, our researches are at an initial state and there are several open questions. The main one is to test again such an idea in some concrete situations and, in particular, in the cases in which vague properties are involved. Moreover, we proposed the inferential process by the Łukasiewicz t-norm, but it should be interesting to examine which t-norm is the most suitable with respect to the data to manage.

We intend also to test the fuzzy inferential process in the early development effort/cost estimation in planning a software project. This topic is a very critical management activity, heavily affecting the competitiveness of a software company. In the context of software engineering, numerous methods to estimate software development cost have been proposed, conventionally divided in *model* and *non-model based methods*. While the latter mainly take into account expert judgments (thus with highly-subjective factors), model-based methods rely on a formal approach, involving the application of an algorithm that, based on a number of inputs, produce an effort estimate. Several techniques for constructing effort estimation have been proposed. The inputs for these algorithms are factors that influence the development effort of software projects. Among the techniques, the case-based reasoning methods have been investigated and utilized in empirical studies, such as in [7], [13]. Nevertheless, it could be interesting comparing these methods with the idea sketched in this chapter.

An open question is related to the difficulties of interpreting the probabilistic valuation of a formula in $SIB_f(T)$ as a conditioned probability in $SIB_f$ (as we have made for the crisp framework in Section 3). In fact, we can define the *conditioned state*, as in the classical probability theory, by setting $p_f(s/t) = p_f(s \otimes t)/p_f(t)$ and, due to the associativity of $\otimes$, $p_f$ satisfies the *iteration rule* of the classical conditioning for a probability, i.e. $p_f(s/t \cap_\otimes v) = p_f(s \cap_\otimes t/v)/p_f(t/v)$. This is a basic property which is useful in the inferential process and for a possible implementation of the expert system. Unfortunately, $p_f$ does not result a state, since $\otimes$ is not distributive with respect to $\oplus$. So, we might look for an adequate definition of state, such that the corresponding conditioned state verifies the *iteration rule.*

Finally, another question is related to the kind of the available information on the actual case. Indeed, it should be natural to admit that this information can be expressed by intervals, i.e. $T = \{a_c, I(\alpha)\}$ where $I(\alpha)$ is a closed interval in $[0, 1]$. In fact, it is possible to admit that the truth value of $\alpha$ cannot be given in a precise way and that it is approximated by an interval $I(\alpha)$. The intended meaning is that the precise truth value with which $a_c$ verifies $\alpha$ is in $I(\alpha)$. In other words we can refer to interval-valued fuzzy subsets to represent the extension of a vague predicate. If we admit such a possibility, then it is necessary to find an analogue of Proposition 4 enabling us to define an *interval-valued similarity* by considering interval-valued fuzzy subsets.

## References

1. Bacchus, F.: Lp, a Logic for Representing and Reasoning with Statistical Knowledge. Computational Intelligence 6, 209–231 (1990)
2. Burris, S., Sankappanavar, H.P.: A Course in Universal Algebra. Springer, Heidelberg (1982)
3. Calabrò, D., Gerla, G., Scarpati, L.: Extension principle and probabilistic inferential process. Lectures on Soft Computing and Fuzzy Logic, pp. 113–127. Springer, Heidelberg (2001)
4. Chang, C.C.: Algebraic analysis of many valued logics. Trans. AMS 93, 74–80 (1958)
5. Cignoli, R., D'Ottaviano, I.M.L., Mundici, D.: Algebraic Foundations of many-valued reasoning. Trends in Logic, vol. 7. Kluwer, Dordrecht (2000)

6. Coppola, C., Gerla, G., Pacelli, T.: Fuzzy Formal Context, Similarity and Probabilistic Expert System. In: ISFUROS 2006. Proceedings of International Symposium on Fuzzy and Rough Sets, Santa Clara, Cuba (2006)
7. Costagliola, G., Di Martino, S., Ferrucci, F., Gravino, C., Tortora, G., Vitiello, G.: Effort Estimation Modeling Techniques: A Case Study for Web Applications. In: ICWE 2006. ACM Proceedings of the 6th International Conference on Web Engineering, Palo Alto, CA, USA, pp. 9–16 (2006)
8. Dubois, D., Hüllermeier, E., Prade, H.: Fuzzy set-based methods in instance-based reasoning. IEEE Transactions on Fuzzy Systems 10(3), 322–332 (2002)
9. Ganter, B., Wille, R.: Formal Concept Analysis: Mathematical Foundations. Springer, Heidelberg (1999)
10. Gerla, G.: The probability that Tweety is able to fly. International Journal of Intelligent Systems 9, 403–409 (1994)
11. Hájek, P.: Metamathematics of Fuzzy Logic. Kluwer Academic Publishers, Dordrecht (1998)
12. Halpern, J.Y.: An analysis of first-order logic of probability. Artificial Intelligence 46, 331–350 (1990)
13. Mendes, E., Di Martino, S., Ferrucci, F., Gravino, C.: Effort Estimation: How Valuable is it for a Web company to Use a Cross-company Data Set, Compared to Using Its Own Single-company Data Set? In: WWW 2007. ACM Proceedings of the 6th International World Wide Web Conference, Banff, Canada (2007)
14. Mundici, D.: Averaging the truth-value in Łukasiewicz logic. Studia Logica 55(1), 113–127 (1995)
15. Novak, V., Perfilieva, I., Mockor, J.: Mathematical Principles of Fuzzy Logic. Kluwer Academic Publishers, London (1999)
16. Pacelli, T.: Similarities, distances and incomplete information. Ph.D Thesis, Università degli Studi di Salerno, Italy (2006)
17. Pawlak, Z.: Rough sets. International Journal of Information and Computer Science 11, 341–356 (1982)
18. Plaza, E., Esteva, F., Garcia, P., Godo, L., Lopez de Mantaras, R.: A logical approach to case-based reasoning using fuzzy similarity relations. Information Sciences 106, 105–122 (1998)
19. Quan, T.T., Hui, S.C., Cao, T.H.: FOGA: A Fuzzy Ontology Generation Framework for Scholarly Semantic Web. In: ECML/PKDD-2004 KDO Workshop (2004)
20. Valverde, L.: On the Structure of F-Indistinguishability Operators. Fuzzy Sets and Systems 17, 313–328 (1985)
21. Wille, R.: Restructuring lattice theory: an approach based on hierarchies of concepts. In: Ivan Rival, R. (ed.) Ordered Sets, pp. 445–470. Reidel, Dordecht, Boston (1982)
22. Zadeh, L.A.: Fuzzy Sets. Information and Control 8, 338–353 (1965)
23. Zadeh, L.A.: Similarity relations and fuzzy orderings. Inf. Sci. 3, 177–200 (1971)
24. Zadeh, L.A.: Probability measures of fuzzy events. Journal of Mathematical Analysis and Applications 23, 421–427 (1968)

# An Efficient Image Retrieval System Using Ordered Weighted Aggregation

Serdar Arslan[1] and Adnan Yazici[2]

[1] Dept. of Computer Engineering, Middle East Technical University, Ankara, Turkey
`aserdar@metu.edu.tr`

[2] Dept. of Computer Engineering, Middle East Technical University, Ankara, Turkey
`yazici@ceng.metu.edu.tr`

**Summary.** In this study, an XML based content-based image retrieval system that combines three visual descriptors of MPEG-7 (Dominant Color (DC), Color Layout (CL) and Edge Histogram (EH)) is introduced. The system is extended to support high dimensional indexing for efficient search and retrieval from a native XML-based DBMS. To do this, an index structure, called M-Tree, which uses Euclidean distance function for each feature, is used. In addition the Ordered Weighted Aggregation (OWA) operators are adapted for aggregating the distance functions of these features. The system supports nearest neighbor queries and various types of fuzzy queries; feature-based, image-based and color-based queries. The experimental results show that our system is effective in terms of retrieval efficiency.

**Keywords:** Content-Based Image Retrieval, MPEG-7, M-Tree, Fuzzy Query, OWA, XML Database.

## 1 Introduction

The tremendous growth in the amount of multimedia is driving the need for more effective methods for storing, searching and retrieving digital images, video and audio data. In content-based image retrieval (CBIR) systems [1] [2] [3] [4], images are indexed on the basis of low-level features, such as color, texture, and shape. A typical content-based image retrieval system is depicted in Figure 1.

In general, most CBIR systems suffer from several drawbacks [5]: First of all, feature extraction is a very expensive process. Since low-level features are very complicated for extraction, CBIR systems need improve efficiency of this process. Secondly, the quality of results tends to be low. Thirdly, performances of querying are often unsatisfactory. Finally, user interfaces are much too complicated for average users.

The CBIR system described in this chapter has the following features: (1) The system has an efficient extraction of low-level color and texture features: Dominant Color (DC), Color Layout (CL) and Edge Histogram (EH). These features need very complex extraction process, so in this study we use MPEG-7 Descriptors [6]. (2) In this study we improve the performance of various types of flexible queries by adapting an indexing technique, namely, M-tree [7], which is

R. Bello et al. (Eds.): Granular Computing, STUDFUZZ 224, pp. 43–54, 2008.
springerlink.com 

a high-dimensional and distance-based index structure. (3) We use combination of multi-features to improve the query performance and quality of the results. Most of the CBIR systems combine these features by associating weights to individual features. One of the main problems with them is that the same weights are associated with the same features for all images in database and the sum of these weighted features are used to build an index structure. However, when comparing two specific images, one feature may be more distinctive than others; therefore, such a feature must be associated with a higher weight. Also, for comparing other images, the same feature may be less distinctive than the other features and for this reason the same feature must be associated with a lower weight. Our proposed solution to this problem is to adapt the OWA [14] operator for aggregation of distance functions for various features. (4) Our system supports flexible queries. Using fuzzy queries in the retrieval system gives a flexibility that is more appropriate for human vision. Fuzzy evaluation of queries is mainly dependent on similarity measures and there are three types of fuzzy queries supported in this study; image-based, feature-based and color-based. For simplicity, in image-based and feature-based fuzzy queries, we restrict the user to express queries using only *Almost Same, Very Similar, Similar* and *Not Similar.* In color-based fuzzy queries, the user expresses queries using only *Mostly, Many, Normally, Few* and *Very Few.* By using conjunction and disjunction rules [8], we also support aggregation of multiple queries. (5) In this chapter, we include a number of performance tests for various query types. With these tests, a number of distance computations for various query types are measured and the retrieval efficiency of the system is evaluated by using Average Normalized Modified Retrieval Rank (ANMRR) metric [9].

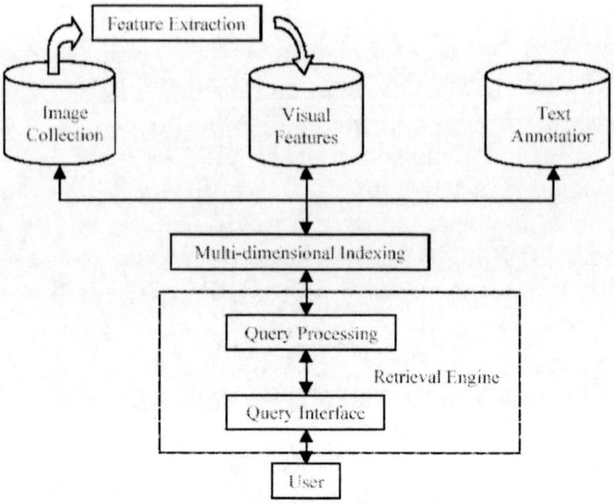


**Fig. 1.** A typical content-based image retrieval system [4]

The rest of this chapter is organized as follows: In Section 2 we present our proposed CBIR system. The performance tests of the system introduced in this chapter are given in Section 3. Finally, Section 4 concludes the chapter.

## 2 Image Retrieval System

In this section, we describe our approach to image retrieval using MPEG-7 Descriptors and the OWA operator along with the similarity measurement.

### 2.1 Feature Extraction Process

In [10], the visual content descriptors, which are extracted with MPEG-7 Descriptors [6], are analyzed from the statistical point of view and the main results show that the best descriptors for combination are Color Layout (CL), Dominant Color (DC), Edge Histogram (EH), and Texture Browsing (TB). The others are highly dependent on these. In this study we choose MPEG-7 Color Descriptors, DC, and CL as the low level features. In order to increase the efficiency of our developed system, a texture descriptor, EH, is added to these color descriptors. These descriptors are extracted by using MPEG-7 eXperimentation Model (XM) [11] [12] Software. After creating each feature's XML document separately, we insert them into our XML database. We use the randomly selected image collection included in Corel Database [13], which has 1000 images.

MPEG-7 XM Software is also used in the process of querying the database. Since this CBIR system uses Query By Example (QBE), the same steps in creating XML documents of each feature for an image collection are applied to the query image. The query image is given to the client application of MPEG-7 XM Software as a parameter and three features are extracted from that image and stored in a document for further processing. In the standard client application of MPEG-7, XM Software has a searching module for querying but we excluded this module from the client application.

### 2.2 Multi-dimensional Index Structure

For indexing multimedia data we use M-Tree [7] known as a dynamic and balanced access structure. The M-tree is a dynamic paged structure that can be efficiently used to index multimedia databases, where the object is represented by means of complex features and the object proximity is defined by a distance function [18-19]. Similarity queries of the objects require the computation of time-consuming distance functions. The details of the M-tree indexing structure and the algorithms of inserting, querying and bulk loading are reported in [7]. A major differentiation of M-tree from the other trees [16] [18] is that the design of M-tree gives an efficient secondary storage organization, since M-tree is a paged, balanced and dynamic structure [17]. In this study, we construct a single M-Tree for the combination of three features, DC, CL, and EH to retrieve images from a image database efficiently. An overview of the M-tree index structure is shown in Figure 2.

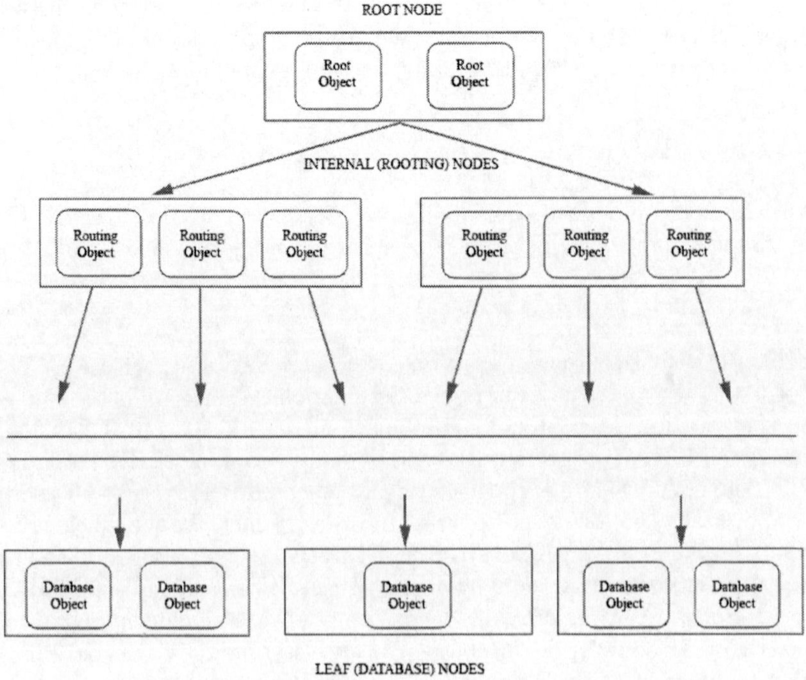


**Fig. 2.** M-Tree Overview

### 2.3 Similarity Measurement

To evaluate similarity measurement, we use Euclidean distance function with the Ordered Weighted Averaging Operator [14]. An OWA operator of dimension $n$ is a mapping:

$$\mathbf{F} : \mathbf{R^n} \rightarrow \mathbf{R} \tag{1}$$

which has an associated weighting vector W

$$\mathbf{W} = [\mathbf{w_1 w_2 ... w_n}]^\mathbf{T} \tag{2}$$

such that

$$\sum_{i=1}^{n} \mathbf{w_i} = 1 \tag{3}$$

where $\mathbf{w_i} \in [0, 1]$ and where

$$\mathbf{F(a_1, a_2, ..., a_n)} = \sum_{i=1}^{n} (\mathbf{w_i} \times \mathbf{b_i}) \tag{4}$$

where $b_i$ is the $i^{th}$ largest element of the collection of the aggregated objects $a_1, ..., a_n$. The function value $F(a_1, ..., a_n)$ determines the aggregated value of arguments, $a_1, ..., a_n$. For example, let us assume that W = [0.4 0.3 0.2 0.1]. Then,

F(0.7,1,0.3,0.6) = (0.4)(1)+(0.3)(0.7)+(0.2)(0.6)+(0.1)(0.3)= 0.76

A fundamental aspect of the OWA operator is the re-ordering step, in particular an argument $a_i$ is not associated with a particular weight $w_i$ but rather a weight $w_i$ is associated with a particular ordered position $i$ of the arguments. A known property of the OWA operator is that it includes the *max, min* and *arithmetic mean* operators.

In general, similarity evaluation of a query object with respect to an object in database is done by applying some distance function to these two objects. In this case, what is actually measured is the distance between feature values, so the distance function returns a dissimilarity value between two objects. It means that high distances correspond to low scores and low distances correspond to high scores.

Commonly used distance function is Minkowski-form distance ($Lp$) [2]:

$$\mathbf{D(x,y)} = [\sum_{i=1}^{d} \mathbf{w_i}|\mathbf{x_i} - \mathbf{y_i}|^{p}]^{\frac{1}{\mathbf{p}}} \tag{5}$$

where $x$ and $y$ feature vectors and $d$ is feature dimension. If

- $p = 1$, $L_1$ is Manhattan or city-block distance [2]
- $p = 2$, $L_2$ is Euclidean distance [2]
- $p = \infty$ , $L_\infty$ is maximum distance [2]

In this study, we implemented two versions of M-Tree, one is with weighted sum of distance functions by using equal weights and one is by utilizing OWA. In both versions distance evaluation is carried out by weighted-Euclidean distance function. Since there are three low-level features that represent the image content, the system evaluates different distance values for each feature, by using Euclidean distance function. But our system computes an overall distance of these three distance values. For this purpose we adapt the OWA operator in our system.

For the CL feature, the distance function is as follows:

$$\mathbf{D_{YCoefficient}} = \sqrt{\sum_{i=0}^{5}(\mathbf{YC[i]} - \mathbf{YC'[i]})^2} \tag{6}$$

$$\mathbf{D_{CbCoefficient}} = \sqrt{\sum_{i=0}^{2}(\mathbf{CbC[i]} - \mathbf{CbC'[i]})^2} \tag{7}$$

$$\mathbf{D_{CrCoefficient}} = \sqrt{\sum_{i=0}^{2}(\mathbf{CrC[i]} - \mathbf{CrC'[i]})^2} \tag{8}$$

$$\mathbf{D_{CL}} = \mathbf{D_{YCoefficient}} + \mathbf{D_{CbCoefficient}} + \mathbf{D_{CrCoefficient}} \tag{9}$$

and for the DC feature, the distance function is:

$$\mathbf{D_{DC}} = \sqrt{\sum_{i=0,j=0,k=0}^{n} (\mathbf{Percentage[i][j][k]} - \mathbf{Percentage'[i][j][k]})^2} \quad (10)$$

where $n = 31$ and for the EH feature, the distance function is:

$$\mathbf{D_{EH}} = \sqrt{\sum_{i=0}^{n} (\mathbf{BinCounts[i]} - \mathbf{BinCounts'[i]})^2} \quad (11)$$

where $n = 79$.

To compute the overall distance between two images, we compute CL, DC and EH distances and apply normalization to each of them separately so that the range is from '0' (similar) to '1' (dissimilar). After normalization of each feature's distances, we compute the overall distance value from these three distances by using the OWA operator. From the definition of OWA method [14], the overall distance is in [0, 1].

Suppose that $(d_1, d_2, ..., d_n)$ are $n$ distance values and order these numbers increasingly: $d_1 \leq d_2 \leq ... \leq d_n$ . The OWA operator associated to the $n$ nonnegative weights $(w_1, w_2, ..., w_n)$ with

$$\sum_{i=1}^{n} \mathbf{w_i} = 1 \quad (12)$$

where $w_i \, \epsilon [0,1]$ and $w_n \leq ... \leq w_2 \leq w_1$ corresponds to

$$\mathbf{F(d_1, d_2, ..., d_n)} = \sum_{i=1}^{n} (\mathbf{w_i} \times \mathbf{d_i}) \quad (13)$$

It should be noted that the weight $w_n$ is linked to the greatest value, $d_1$ and $w_1$ is linked to the lowest value $d_n$ to emphasize similarity between two objects. For example, for two objects $O_1$ and $O_2$, we want to compute distance between them, let's say $d(O_1, O_2)$, and assume that, for each feature, CL, DC and EH, the normalized Euclidean distance values are;

$d_{CL}(O_1, O_2) = 0.325$,
$d_{DC}(O_1, O_2) = 0.570$,
$d_{EH}(O_1, O_2) = 0.450$

and the OWA weights are;

$w_1 = 0.7, w_2 = 0.2, w_3 = 0.1$, that is, $w_1 + w_2 + w_3 = 0.7+0.2+0.1 = 1$,

then the overall distance is:

$d(O_1, O_2) = F(d_{CL}(O_1, O_2), d_{DC}(O_1, O_2), d_{EH}(O_1, O_2))$
$= w_1 * d_{CL}(O_1, O_2) + w_2 * d_{EH}(O_1, O_2) + w_3 * d_{DC}(O_1, O_2)$
$= 0.7 * 0.325 + 0.2 * 0.450 + 0.1 * 0.570$
$= 0.3745$.

## 2.4 Querying the M-Tree

M-tree is able to support processing of two main types of queries [7]: *range queries*; finding all objects that are within a specific distance from a given object and *k-nearest Neighbor Query (k-NN);* finding a specific number, $k$, of closest objects to a given query object. These queries are defined as follows:

*Range Query*

Given a query object $Q \epsilon D$ , where $D$ is the domain of feature values, and for a distance (range) $r(Q)$, the range query $range(Q, r(Q))$ selects all indexed objects $O_j$ such that

$$\mathbf{d(O_j, Q) \leq r(Q)} \tag{14}$$

For example, a range query becomes:
*"Find all images which have a distance value less than 0.2 from query image"*

*k-Nearest Neighbors Query (k-NN)*

Given a query object $Q \epsilon D$ and an integer $k \geq 1$ , the k-NN query $NN(Q, k)$ selects the $k$ indexed objects which have the shortest distance from $Q$. An example, k-NN query is:
*"Find 10 nearest images to the given query image"*

In this study, we also support three types of fuzzy queries: Image-based, Feature-based, and Color-based. Here we briefly describe each one.

*Image-Based Fuzzy Query*

If the image query is selected, the user has to select a similarity degree for a query image, which is assumed to be either *'Almost Same'*, *'Very Similar'*, *'Similar'* or *'Not Similar'* in this study. Then the system maps this similarity degree into a distance range, which is defined according to our data set, and searches the tree to retrieve matched images, which have a distance to a query image in that range. And finally, retrieved results are shown to the user with their distance value to the query image.

The general syntax of this type of query is as follows:
QUERY={{<Similarity>}}, where
Similarity={<Almost Same> | <Very Similar> | <Similar > | <Not Similar >}

For example, suppose that a user gives the following similarity degree for the query image; *'Find images which are Very Similar to the given Query Image'*. Then our query is:

*QUERY= Very Similar to Query Image*

And, for simplicity, suppose we have the similarity values mapped into the following ranges: *'Almost Same'*: [1, 0.95), *'Very Similar'*: [0.95, 0.85), *'Similar'*: [0.85, 0.5), *'Not Similar'*: [0.5, 0.0].

Since we use distance for indexing, the final distance range for this query becomes (0.05, 0.15]. Finally the system retrieves the images, which have a distance value from the query image in the range (0.05 - 0.15].

*Feature-Based Fuzzy Query*

Another type of query that our system supports is the feature-based fuzzy query. In this type, the user must supply similarity values for all three features DC, CL and EH. Again, for simplicity, these similarity values are assumed to be the same as the ones in an image-based fuzzy query. For combining these similarities, AND/OR operators are used. Then, the system applies some conjunction/disjunction rules [8] to get final similarity values and maps these values into a distance range.

$$ConjunctionRule : \mu_{A \wedge B} = \min(\mu_A(x), \mu_B(x)) \tag{15}$$

$$DisjunctionRule : \mu_{A \vee B} = \max(\mu_A(x), \mu_B(x)) \tag{16}$$

If AND operator is supplied to combine feature similarities, the system uses conjunction rule, and if OR operator is supplied to combine feature similarities, the system uses disjunction rule. The general syntax of this type of query is as follows:

QUERY={{<Similarity><Feature>}<AND | OR> {<QUERY> | <>}}, where

Similarity = {<Almost Same> | <Very Similar> | <Similar > | <Not Similar >}

Feature = {<CL> | <DC> | <EH >}

For an example, suppose that user specifies the following similarity values for the features; *'Very Similar'* for CL feature, *'Similar'* for DC feature, *'Almost Same'* for EH feature. Then our query is defined as:

*QUERY= Very Similar in CL OR Similar in DC AND Almost Same as EH*

To get the final similarity, the system combines these similarities as follows: First, AND operator between DC and EH feature is taken into account and conjunction rule is applied to this part. After that the system combines CL feature similarity with this part by applying disjunction rule. Then the final distance range for this similarity range is calculated. Finally the system retrieves the images, which have a distance value from the query image in that range.

*Color-Based Fuzzy Query*

Color-Based Fuzzy Query differs from the other fuzzy queries that we have just discussed. For this type of queries the user has to supply a degree of percentages of three colors in expected images. To support this query type, the system requires that the expected main colors be in the image. To do this, the user supplies each color's percentage vaguely, using the terms in natural language, such as *'mostly'*, *'many'*, *'normally'*, *'few'*, *'very few.'* Thus, the user can pose a composite query in terms of colors. The general syntax of this type of query is as follows:

QUERY={{<Content><Color>}<AND | OR>{<QUERY> | <>}} where

Content= {<mostly> | <many> | <normally> | <few> | <very few>}

Color = {<Red> | <Green> | <Blue>}

An example query is as follows:

*QUERY=many red AND mostly green OR very few blue.*

Mapping function of these linguistic terms into similarity values is defined according to the data set. For example, in our study, for testing on Corel Dataset, we use the following values:

'*Mostly*': [1, 0.88), '*Many*': [0.88, 0.85), '*Normally*': [0.85, 0.82), '*Few*': [0.82, 0.80), '*Very Few*': [0.80, 0.0].

After defining the query, system searches the tree for each color separately by using predefined query features in DC and CL for *red, green* and *blue* colors. The EH feature is not so important since the query is a color query, so the distance value for the EH feature is 0 (zero). Then the result sets of each color's query are combined into the final result set. If the AND operator is used in a composite query then all objects in both result sets are shown to the user with a similarity degree. If the OR operator is used, then all objects of both result sets are shown to the user with a similarity degree.

## 3 Performance Experiments

The performances of various types of queries supported by the system are tested by using a number of test cases. To test the performance of our content-based image retrieval system, we used 400 images from Corel Database [13]. For M-Tree, two different types of weighted sum of Euclidean Distance function are used: Euclidean distance with equal weights and with OWA. While using the M-tree in querying, the construction time of the tree, the retrieval efficiency of the system, the number of distance computations and query cost time are computed and evaluated.

To evaluate the retrieval effectiveness of querying the M-Tree, we use ANMRR performance metric [9]. Basically, if this value is equal to 0 (zero) then this means that the system has a perfect retrieval process, but if this value is equal to 1 (one) then the system has an inefficient retrieval process.

We have 335 queries for test over two version of M-Tree and compare each tree's ANMRR results. And we also compare the ANMRR results of our system with the ANMRR results of MPEG-7 XM Search Engine which uses these three features (CL, DC and EH) separately. The results are included in Table 1.

The results of this experiment show that our system is more efficient then MPEG-7 XM in terms of query relevancy. We use OWA operators for aggregation of distance functions for three low level features. Note that the features have different effects to the retrieval results. Among three features, the most relevant feature (or the most distinctive one) to the query image is treated as the main feature for comparing query object with database objects. This property provides better performance according to MPEG-7 XM, because MPEG-7 XM search engine uses one feature and that specific feature may not be 'the best' or

**Table 1.** ANMRR results of our System and XM Software for 335 queries (DB size= 100 - 400 images)

| Index Structure | Distance Function | ANMRR Value |
|---|---|---|
| M-Tree | with OWA | 0.342271 |
| M-Tree | with equal weights | 0.394931 |
| MPEG-7 XM | CL Feature | 0.338113 |
| MPEG-7 XM | DC Feature | 0.407258 |
| MPEG-7 XM | EH Feature | 0.423513 |

'the most distinctive' feature for comparing query object with database objects. Our system shows nearly the same performance as MPEG-7 XM search engine with CL feature, with the image dataset used in our experiments. Moreover, we achieve a significant improvement when compared with M-Tree using Euclidean distances with equal weights.

The number of distance computations is another performance improvement of our system. For k-NN queries, this number is important for performance of the CBIR system. Based on the test results that we have done, we observe that the approach using Euclidean distance function with OWA operator has less distance computations than when Euclidean distance function with equal weights is used. Since the distance function is a complex one, the evaluation time of this distance function becomes more important for query response time. By adapting the OWA operator, we use the best feature's effect on the query results; thus, the system prunes more branches and it becomes more effective. Note that pruning directly effects the query response time.

To evaluate the effectiveness of a k-NN query, we use 400 queries to retrieve top 10 images (k=10) from the XML database, which has 400 images. And as in previous tests, we test two versions of M-Tree. In this test, fill factor (minimum utilization value) is 0.1 and the page size is 16K and we use the hyper plane split function and the random promote function [7].The results of the number of distance computations are shown in Table 2.

For building M-Tree, number of distance computations and cost time are the key values for evaluating efficiency of the system. For this purpose, tests for building the tree include the number of distance computations and construction time for both version of M-Tree. To evaluate construction time and number of distance computations for building M-Tree, we use five different minimum

**Table 2.** Minimum and Maximum Computed Distances for 400 Queries in 10-NN Queries

| Distance Function | Min. Comp. Dist. | Max. Comp. Dist. |
|---|---|---|
| M-Tree with OWA | 215 | 403 |
| M-Tree with equal weights | 383 | 406 |

utilization values and five different page sizes. Four different image groups are used in our experiments.

Tests for building the tree have been made for two different promotions, Confirmed and Random [7]. Page size parameter of the index structure varies from 8K to 32K and minimum utilization parameter is between 0.1 and 0.5.

The number of computed distances is another important value for evaluating the efficiency of the system. Tests have been made for calculating the number of computed distances with same parameters and same databases. Results of these tests show that a significant improvement ,approximately %11, for number of computed distances and also for construction time can be achieved by using OWA operators.

## 4 Conclusion

In this chapter, we present an efficient content-based image retrieval system that evaluates the similarity of each image for various features. For the distance evaluation between images, we use the weighted sum of Euclidean distance and each weight is evaluated by using the OWA operator. In this system, we use three descriptors of MPEG-7, CL, DC and EH. These features are extracted by using MPEG-7 XM Software. The system stores these features in a XML database, Berkeley XML DB [15]. The system has been tested on images of the Corel database [13].

The experiment results show a performance improvement using the OWA operator for aggregating the weights of Euclidean. The system supports flexible queries by using fuzzy logic in retrieval process. Fuzzy evaluation of queries gives a flexibility that is more appropriate for human vision and mainly depends on similarity measures.

A possible future work to be done is to enhance the effectiveness of building the M-Tree. The pruning efficiency of M-tree and the performance of building and querying the tree may be improved further. In our system, only images are used for indexing and retrieval. Another possible future study is to adapt our system for video/audio databases.

## References


1. Sikora, T.: The MPEG-7 Visual Standard for Content Description-An Overview. IEEE Transactions on Circuits and Systems for Video Technology 11(6) (2001)
2. Ying, L., Wan, X., Jay, K.C.: Introduction to Content-Based Image Retrieval-Overview of Key Techniques. In: Castelli, V., Bergman, D. (eds.) Image DBs, pp. 261–284. John Wiley & Sons, Chichester (2002)
3. Koskela, M., Laaksonen, J., Oja, E.: Comparison of Techniques for CBIR. In: Proc. of the 12th Scandinavian Conf. on Im.e Analysis, Norway, pp. 579–586 (2001)
4. Rui, Y., Hang, T.S., Chang, S.: Image retrieval: Current technique, promising directions, and open issues. J. of Visual Comm. and Image Representation 10, 39–62 (1999)

5. Breiteneder, C., Eidenberger, H.: CBIR in Digital Libraries. In: Proc. of Digital Lib's Conf., Japan, pp. 67–74 (2000)
6. Int. Org. Stanart, MPEG-7 Overview (ver. 9) (2003)
7. Ciaccia, P., Patella, M., Zezula, P.: Mtree: An efficient access method for similarity search in metric space. In: Proc. of the 23rd VLDB Int. Conf., Athens, pp. 426–435 (1997)
8. Fagin, R.: Combining Fuzzy Information from Multiple Systems. In: Proc.15th ACM Symp. On Prn. of Db. Sys., Montreal, pp. 216–226 (1996)
9. Manjunath, B.S., Salembier, P., Sikora, T.: Introduction to MPEG-7: Multimedia Content Description Interface. John Wiley & Sons, Chichester (2002)
10. Eidenberger, H.: How good are the visual MPEG-7 features. In: Proc. of the 5th ACM SIGMM Int. WS on Mm. info. retrieval, Berkeley, pp. 130–137 (2003)
11. MPEG-7 XM Homepage, `http://www.lis.ei.tum.de/research/bv/topics/mmdb.html`
12. Ojala, T., Aittola, M., Matinmikko, E.: Empirical Evaluation of MPEG-7 XM Color Descriptors in Content-Based Retrieval of Semantic Image Categories. In: Proc. 16th Int. Con. on Pattern Recognition, Canada, vol. 2, pp. 1021–1024 (2002)
13. Corel database, `http://www.corel.com`
14. Yager, R.R.: On ordered weighted averaging aggregation operators in multi-criteria decision making. IEEE Trans. Sys. Man Cyb. 18, 183–190 (1988)
15. SleepyCat Software, `www.sleepycat.com`
16. Berchtold, S., Keim, D.A., Kriegel, H.P.: The X-tree: An Index Structure for High-dimensional Data. In: Proc. of VLDB (1996)
17. Bohm, C., Berchtold, S., Keim, D.A.: Searching in High-Dimensional Spaces-Index Structures for Improving the Performance of Multimedia Databases. ACM Comp. Surv. 83, 322–373 (2001)
18. Gaede, V., Gunther, O.: Multidimensional Access Methods. ACM Comp. Surv. 30(2) (1998)
19. Chavez, E., Navarro, G., Yates, R.B., Marroquin, J.L.: Searching in Metric Spaces. ACM Comp. Surv. 33, 273–321 (2001)

# Entropy and Co–entropy of Partitions and Coverings with Applications to Roughness Theory

Gianpiero Cattaneo, Davide Ciucci, and Daniela Bianucci

Dipartimento Di Informatica, Sistemistica e Comunicazione
Università di Milano – Bicocca
Via Bicocca degli Arcimboldi 8, I–20126 Milano (Italia)
{cattang,ciucci}@disco.unimib.it

**Summary.** The abstract notion of rough approximation space is applied to the concrete cases of topological spaces with the particular situation of clopen–topologies generated by partitions, according to the Pawlak approach to rough set theory. In this partition context of a finite universe, typical of complete information systems, the probability space generated by the counting measure is analyzed, with particular regard to a local notion of rough entropy linked to the Shannon approach to these arguments. In the context of partition the notion of entropy as measure of uncertainty is distinguished from the notion of co–entropy as measure of granularity.

The above considerations are extended to the case of covering, typical situation of incomplete information systems with the associated similarity relation.

## 1 Abstract Rough Approximation Spaces

The notion of *rough approximation space* introduced in [1] with the aim of giving an abstract axiomatization to the Pawlak rough set theory [2], is defined as a set whose points represent vague, uncertain elements which can be approximated from the bottom and the top by crisp, sharp elements. Formally, an abstract *rough approximation space* (see [1, 3]) is a structure: $\mathfrak{R} := \langle \Sigma, \mathbb{L}(\Sigma), \mathbb{U}(\Sigma) \rangle$, where:

(1) $\langle \Sigma, \wedge, \vee, 0, 1 \rangle$ is a *distributive complete lattice* with respect to the partial order relation $a \leq b$ iff $a = a \wedge b$ (or, equivalently, $b = a \vee b$); bounded by the least element $0$ ($\forall a \in \Sigma$, $0 \leq a$) and the greatest element $1$ ($\forall a \in \Sigma$, $a \leq 1$). Elements from $\Sigma$ are interpreted as concepts, data, etc., and are said to be the elements which can be *approximated*;

(2) $\mathbb{L}(\Sigma)$ and $\mathbb{U}(\Sigma)$ are sublattices of $\Sigma$ whose elements are called *lower* (also, *inner*) and *upper* (also, *outer*) *definable* respectively.

The structure satisfies the following conditions.

(Ax1) For any element $a \in \Sigma$ which can be approximated, there exists (at least) one element $l(a)$, called the *lower approximation* (also *interior*) of $a$, such that: (In1) $l(a) \in \mathbb{L}(\Sigma)$; (In2) $l(a) \leq a$; (In3) $\forall \beta \in \mathbb{L}(\Sigma)$, $\beta \leq a \Rightarrow \beta \leq l(a)$.

R. Bello et al. (Eds.): Granular Computing, STUDFUZZ 224, pp. 55–77, 2008.
springerlink.com 

(Ax2) For any element $a \in \Sigma$ which can be approximated, there exists (at least) one element $u(a)$, called the *upper approximation* (also *closure*) of $a$, such that: (Up1) $u(a) \in \mathbb{U}(\Sigma)$; (Up2) $a \leq u(a)$; (Up3) $\forall \gamma \in \mathbb{U}(\Sigma)$, $a \leq \gamma \Rightarrow u(a) \leq \gamma$.

Therefore, $l(a)$ (resp., $u(a)$) is the best approximation of the "vague", "imprecise", "uncertain" element $a$ from the bottom (resp., top) by lower (resp., upper) definable elements. For any element $a \in \Sigma$ which can be approximated, the lower $l(a) \in \mathbb{L}(\Sigma)$ and the upper $u(a) \in \mathbb{U}(\Sigma)$ definable elements, whose existence is assured by (Ax1) and (Ax2), are unique. Thus, it is possible to introduce in an *equivalent way* a rough approximation space as a structure $\langle \Sigma, \mathbb{L}(\Sigma), \mathbb{U}(\Sigma), l, u \rangle$, consisting of a bounded distributive lattice and two sublattices of its, under the assumption of the *existence* of a *lower approximation mapping* $l : \Sigma \mapsto \mathbb{L}(\Sigma)$ and an *upper approximation mapping* $u : \Sigma \mapsto \mathbb{U}(\Sigma)$, given for an arbitrary $a \in \Sigma$ respectively by the laws:

$$l(a) := \max\{\beta \in \mathbb{L}(\Sigma) : \beta \leq a\} \quad \text{and} \quad u(a) := \min\{\gamma \in \mathbb{U}(\Sigma) : a \leq \gamma\} \tag{1}$$

The *rough approximation* of any element $a \in \Sigma$ is then the lower–upper pair $r(a) := \langle l(a), u(a) \rangle \in \mathbb{L}(\Sigma) \times \mathbb{U}(\Sigma)$, with $l(a) \leq a \leq u(a)$, which is the image of the element $a$ under the *rough approximation mapping* $r : \Sigma \mapsto \mathbb{L}(\Sigma) \times \mathbb{U}(\Sigma)$ described by the following diagram:

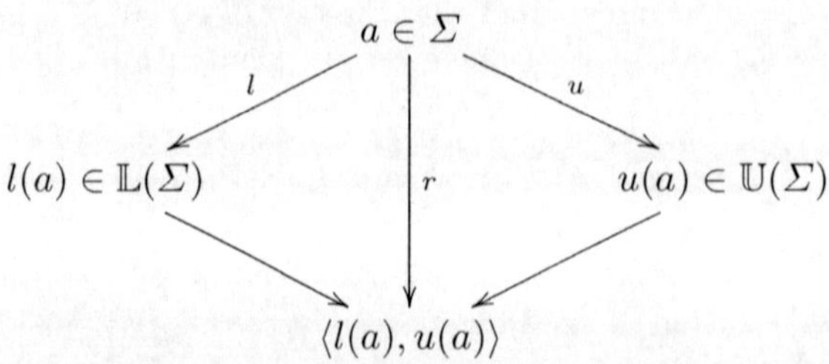


Following [1], an element $e$ of $X$ is said to be *crisp* (also *exact*, *sharp*) if and only if its lower and upper approximations coincide: $l(e) = u(e)$, equivalently, iff its rough approximation is the trivial one $r(e) = (e, e)$. Owing to (In1) and (Up1) this happens iff $e$ is simultaneously a lower and an upper definable element; therefore, $\mathbb{L}\,\mathbb{U}(\Sigma) := \mathbb{L}(\Sigma) \cap \mathbb{U}(\Sigma)$ is the collection of all crisp elements, which is not empty since $0, 1 \in \mathbb{L}(\Sigma) \cap \mathbb{U}(\Sigma)$.

A particular case of abstract approximation space is the one in which the lattice of approximable elements $\Sigma$ is equipped with an *orthocomplementation mapping* $' : \Sigma \to \Sigma$ satisfying the conditions: (oc-1) $(a')' = a$ (double negation law); (oc-2) $a \leq b$ implies $b' \leq a'$ (which is equivalent to both the de Morgan laws, $(a \wedge b)' = a \vee b'$ and $(a \vee b)' = a' \wedge b'$); (oc-3) $a \wedge a' = 0$ (non–contradiction law) and $a \vee a' = 1$ (excluded middle law). Let us note that in the context of orthocomplemented lattices $\Sigma$, given a sublattice of inner definable elements $\mathbb{L}(\Sigma)$ the set $\mathbb{U}_d(\Sigma) := \{\gamma \in \Sigma : \exists \beta \in \mathbb{L}(\Sigma) \ s.t. \ \gamma = \beta'\}$ is naturally a lattice of upper definable elements, called the *dual* of $\mathbb{L}(\Sigma)$ (i.e., it satisfies all the above conditions (In1)–(In3)). Moreover, the mapping defined for any $a \in \Sigma$

by $u_d(a) := (l(a'))'$ is an upper approximation map, *dual* of $l$. In this way the triplet $\langle \Sigma, \mathbb{U}_d(\Sigma), u_d \rangle$ is an *upper approximation space* dual of the original lower approximation space $\langle \Sigma, \mathbb{L}(\Sigma), l \rangle$.

Similarly, in the case of a lattice of upper definable elements $\mathbb{U}(\Sigma)$ the collection $\mathbb{L}_d(\Sigma) := \{\beta \in \Sigma : \exists \gamma \in \mathbb{U}(\Sigma) \ s.t. \ \ \beta = \gamma'\}$ is the *dual* lattice of lower approximable elements (i.e., it satisfies the above conditions (Up1)–(Up3)). Also in this case the mapping assigning to any $a \in \Sigma$ the element $l_d(a) := (u(a'))'$ is a lower approximation map, *dual* of $u$. Hence, the triplet $\langle \Sigma, \mathbb{L}_d(\Sigma), l_d \rangle$ is a *lower approximation space* dual of the original upper approximation space $\langle \Sigma, \mathbb{U}(\Sigma), u \rangle$.

Two new notions involving the elements of $\Sigma$ can be now introduced: given $a \in \Sigma$ its *exterior* is the lower definable element $e(a) := u(a)' \in \mathbb{L}(\Sigma)$, i.e., the complement of its upper approximation; moreover, its *boundary* is the upper definable element $b(a) := u(a) \wedge l(a)' \in \mathbb{U}(\Sigma)$, i.e., the relative complement of its "interior" with respect to the "closure." For any approximable element $a \in \Sigma$ the triplet $\{l(a), b(a), e(a)\}$ consists of mutually orthogonal elements (two elements $a, b$ of an orthocomplemented lattice $\Sigma$ are said to be *orthogonal*, written $a \perp b$, iff $a \leq b'$, or equivalently $b \leq a'$). Moreover, they *orthogonally decompose* the whole distributive lattice $\Sigma$ since $l(a) \vee b(a) \vee e(a) = 1$.

What we have outlined in the present section is the so–called approach to the *roughness theory* based on the abstract notion of a lattice $\Sigma$, which we want to distinguish from the usual *rough set theory* based on a concrete set, the universe of the discourse $X$, and its power set $\mathcal{P}(X)$ as lattice of all subsets of $X$ (i.e., in this concrete approach the role of $\Sigma$ is played by $\mathcal{P}(X)$). This latter concrete situation furnishes one of the possible models of the former abstract theory.

### 1.1 Topological Rough Approximation Spaces

A first concrete example of rough approximation space is based on the notion of topological space, in this context called *topological rough approximation space.* To this purpose, let us consider a topological space defined as a pair $(X, \mathcal{O}(X))$ consisting of a nonempty set $X$ equipped with a family of *open subsets* $\mathcal{O}(X)$, satisfying the following conditions: (O1) the empty set $\emptyset$ and the whole space $X$ are open; (O2) the family $\mathcal{O}(X)$ is closed with respect to arbitrary unions; (O3) the family $\mathcal{O}(X)$ is closed with respect to finite intersections. As it is well known, a subset of $X$ is said to be *closed* iff it is the set theoretic complement of an open set. Therefore, the collection $\mathcal{C}(X)$ of all closed subsets of $X$ satisfies the following conditions: (C1) both $\emptyset$ and $X$ are closed; (C2) the family $\mathcal{C}(X)$ is closed with respect to finite unions; (C3) the family $\mathcal{C}(X)$ is closed with respect to arbitrary intersections. In this framework one can consider the structure

$$\mathfrak{R}_T = \langle \mathcal{P}(X), \mathcal{O}(X), \mathcal{C}(X), l, u \rangle$$

where the role of the set $\Sigma$ of approximable elements is played by the power set $\mathcal{P}(X)$ of $X$, collection of all its subsets. $\Sigma = \mathcal{P}(X)$ is a distributive (complete) lattice $\langle \mathcal{P}(X), \cap, \cup, {}^c, \emptyset, X \rangle$ with respect to set theoretic intersection $\cap$, union

$\cup$, and the set theoretic complementation $^c$; this lattice is bounded by the least element $\emptyset$ and the greatest element $X$. From the above outlined properties of open and closed sets it immediately follows that $\mathcal{O}(X)$ (resp., $\mathcal{C}(X)$) plays the role of lattice of lower (resp., upper) definable elements, i.e., $\mathbb{L}(\Sigma) = \mathcal{O}(X)$ (resp., $\mathbb{U}(\Sigma) = \mathcal{C}(X)$), where the two lattice structures $\mathcal{O}(X)$ and $\mathcal{C}(X)$ are in mutual duality. Trivially, for any subset $A$ of $X$ it is possible to introduce, according to the (1), the following definitions:

$$l(A) := \bigcup\{O \in \mathcal{O}(X) : O \subseteq A\} \quad \text{and} \quad u(A) := \bigcap\{C \in \mathcal{C}(X) : A \subseteq C\}$$

In other words, $l(A) \in \mathcal{O}(X)$ (resp., $u(A) \in \mathcal{C}(X)$) is the *topological interior* (resp, *closure*), usually denoted by $A^o$ (resp., $A^*$), of the set $A$. In particular, owing to (O2) (resp., (C3)), $A^o$ (resp., $A^*$) is the open (resp., closed) set which furnishes the best approximation of the approximable subset $A$ of $X$ by open (resp., closed) subsets from the bottom (resp., the top), i.e., it is the *rough lower* (resp., *upper*) *approximation* of $A$.

The *topological rough approximation mapping* $r_T : \mathcal{P}(X) \mapsto \mathcal{O}(X) \times \mathcal{C}(X)$ is the mapping which assigns to any subset $A$ of the topological space $X$ the pair $r_T(A) = \langle A^o, A^* \rangle$, with $A^o \subseteq A \subseteq A^*$, consisting of its interior (open subset $A^o \in \mathcal{O}(X)$) and its closure (closed subset $A^* \in \mathcal{C}(X)$). Trivially, a subset $E$ of a topological space $X$ is crisp (exact, sharp) iff it is *clopen* (in particular the empty set and the whole space are clopen, and so exact).

Note that for any subset $A$, the induced partition of the universe by mutually disjoint sets is $X = l(A) \cup b(A) \cup e(A)$, where $b(A)$ is the (topological) *boundary* and $e(A)$ the (topological) *exterior* of the subset $A$.

## 2 The Partition Approach to Rough Set Theory

The usual approach to rough set theory as introduced by Pawlak [2, 4] is formally (and essentially) based on a concrete *partition space*, that is a pair $(X, \pi)$ consisting of a nonempty set $X$, the *universe* (with corresponding power set $\mathcal{P}(X)$, the collection of sets which can be *approximated*), and a partition $\pi := \{A_i \in \mathcal{P}(X) : i \in I\}$ of $X$ (indexed by the index set $I$) whose elements are the *elementary sets*. The partition $\pi$ can be characterized by the induced equivalence relation $\mathcal{R} \subseteq X \times X$, defined as

$$(x, y) \in \mathcal{R} \quad \text{iff} \quad \exists A_j \in \pi : x, y \in A_j \tag{2}$$

In this case $x, y$ are said to be *indistinguishable* with respect to $\mathcal{R}$ and the equivalence relation $\mathcal{R}$ is called an *indistinguishability* relation. In this indistinguishability context the partition $\pi$ is considered as the support of some knowledge available on the objects of the universe and so any equivalence class (i.e., elementary set) is interpreted as a *granule* (or *atom*) of knowledge contained in (or supported by) $\pi$. For any object $x \in X$ we shall denote by $gr(x)$, called the *granule generated* by $x$, the (unique) equivalence class which contains $x$ (if $x \in A_i$, then $gr(x) = A_i$).

A *definable set* is any subset of $X$ obtained as the set theoretic union of elementary subsets: $E_J = \cup\{A_j \in \pi : j \in J \subseteq I\}$. The collection of all such definable sets plus the empty set $\emptyset$ will be denoted by $\mathcal{E}_\pi(X)$ and it turns out to be a Boolean algebra $\langle \mathcal{E}_\pi(X), \cap, \cup, {}^c, \emptyset, X\rangle$ with respect to set theoretic intersection, union, and complement. This Boolean algebra is atomic whose atoms are just the elementary sets from the partition $\pi$.

From the topological point of view $\mathcal{E}_\pi(X)$ contains both the empty set and the whole space, moreover it is closed with respect to any arbitrary set theoretic union and intersection, i.e., it is a family of clopen subsets for a topology on $X$: $\mathcal{E}_\pi(X) = \mathcal{O}(X) = \mathcal{C}(X)$. In this way we can construct the concrete *rough approximation space* $\mathfrak{R}_P := \langle \mathcal{P}(X), \mathcal{E}_\pi(X), \mathcal{E}_\pi(X), l_\pi, u_\pi\rangle$ based on the partition $\pi$, simply written as $\langle \mathcal{P}(X), \mathcal{E}_\pi(X), l_\pi, u_\pi\rangle$, consisting of:

(1) the Boolean (complete) atomic lattice $\mathcal{P}(X)$ of all *approximable* subsets of the universe $X$, whose atoms are the singletons;

(2) the Boolean (complete) atomic lattice $\mathcal{E}_\pi(X)$ of all *definable* subsets of $X$, whose atoms are the equivalence classes of the partition $\pi(X)$;

(3) the *lower approximation map* $l_\pi : \mathcal{P}(X) \mapsto \mathcal{E}_\pi(X)$ associating with any subset $Y$ of $X$ its lower approximation defined by the (clopen) definable set

$$l_\pi(Y) := \cup\{E \in \mathcal{E}_\pi(X) : E \subseteq Y\} = \cup\{A \in \pi : A \subseteq Y\}$$

(4) the *upper approximation map* $u_\pi : \mathcal{P}(X) \mapsto \mathcal{E}_\pi(X)$ associating with any subset $Y$ of $X$ its upper approximation defined by the (clopen) definable set

$$u_\pi(Y) := \cap\{F \in \mathcal{E}_\pi(X) : Y \subseteq F\} = \cup\{B \in \pi : Y \cap B \neq \emptyset\}$$

The *rough approximation* of a subset $Y$ of $X$ is then the clopen pair $r_\pi(Y) := \langle l_\pi(Y), u_\pi(Y)\rangle$, with $l_\pi(Y) \subseteq Y \subseteq u_\pi(Y)$.

## 2.1 Entropy (as Measure of Average Uncertainty) and Co–entropy (as Measure of Average Granularity) of Partitions

Let us now assume that the universe is finite ($|X| < \infty$). The set $\mathcal{E}_\pi(X)$ of all definable elements induced from the (necessarily finite) partition $\pi = \{A_1, A_2, \ldots, A_N\}$ of the universe $X$ has also the structure of $\sigma$–algebra of sets for a measurable space $\langle X, \mathcal{E}_\pi(X)\rangle$ (see [5]); in this context elements from $\mathcal{E}_\pi(X)$ are also called *events* and the ones from the original partition $\pi$ *elementary events*. On this measurable space we will consider the so–called *counting measure* $m : \mathcal{E}_\pi(X) \mapsto \mathbb{R}_+$ assigning to any event $E \in \mathcal{E}_\pi(X)$ the corresponding measure $m(E) = |E|$, i.e., the *cardinality* of the measurable set (event) under examination. Thus, the following two $N$–component vectors depending from the partition $\pi$ can be constructed:

(md) the *measure distribution*

$$\mathbf{m}(\pi) = (m(A_1), m(A_2), \ldots, m(A_N)), \quad \text{with } m(A_i) = |A_i|\,.$$

The quantity $m(A_i)$ expresses the *measure of the granule* $A_i$, and the total sum of $\mathbf{m}(\pi)$ is $M(\pi) := \sum_{i=1}^{N} m(A_i) = m(X)$, which is constant with respect to the variation of the partition $\pi$;

(pd) the *probability distribution* [6]

$$\mathbf{p}(\pi) = (p(A_1), p(A_2), \ldots, p(A_N)), \quad \text{with } p(A_i) = \frac{m(A_i)}{m(X)}.$$

The quantity $p(A_i)$ describes the *probability of the event* $A_i$, and $\mathbf{p}(\pi)$ is a finite collection of non–negative real numbers ($\forall i$, $p(A_i) \geq 0$), whose sum is one ($\sum_{i=1}^{N} p(A_i) = 1$).

One must not confuse the *measure* $m(A_i)$ of the "granule" $A_i$ with the occurrence *probability* $p(A_i)$ of the "event" $A_i$. They are two semantical concepts very different between them. Of course, both these distributions depend on the choice of the partition $\pi$ and if one changes the partition $\pi$ inside the collection $\Pi(X)$ of all possible partitions of $X$, then different distributions $\mathbf{m}(\pi)$ and $\mathbf{p}(\pi)$ are obtained. Fixed the partition $\pi$, on the basis of these two distributions it is possible to introduce two really different (non–negative) discrete random variables:

(RV-G) the *granularity random variable*

$$\mathbf{G}(\pi) := (\log m(A_1), \log m(A_2), \ldots, \log m(A_N))$$

where the non–negative real quantity $G(A_i) := \log m(A_i)$ represents the *measure of the granularity* associated to the knowledge supported by the "granule" $A_i$ of the partition $\pi$;

(RV-U) the *uncertainty random variable*

$$\mathbf{I}(\pi) := (-\log p(A_1), -\log p(A_2), \ldots, -\log p(A_N))$$

where the non–negative real quantity $I(A_i) := -\log p(A_i)$ is interpreted (see [7], and also [8, 9]) as a *measure of the uncertainty* related to the probability of occurrence of the "event" $A_i$ of the partition $\pi$.

Also in the case of these two discrete random variables, their semantical/terminological confusion is a real metatheoretical disaster. Indeed, $G(A_i)$ involves the measure $m(A_i)$ (*granularity measure* of the "granule" $A_i$), contrary to $I(A_i)$ which involves the probability $p(A_i)$ of occurrence of $A_i$ (*uncertainty measure* of the "event" $A_i$). Note that these two measures generated by $A_i \in \pi$ are both non–negative since whatever be the event $A_i$ it is $m(A_i) \geq 1$ (see figure (1)).

Moreover they are mutually "complementary" with respect to the quantity $\log m(X)$, which is invariant with respect to the choice of the partition $\pi$:

$$G(A_i) + I(A_i) = \log m(X) \tag{3}$$

The granularity measure $G$ is *strictly monotonic* with respect to the set theoretic inclusion: $A \subset B$ implies $G(A) < G(B)$. On the contrary, the uncertainty measure $I$ is *strictly anti–monotonic*: $A \subset B$ implies $I(B) < I(A)$.

As it happens for any discrete random variable, it is possible to calculate its average with respect to the fixed probability distribution $\mathbf{p}(\pi)$, obtaining the two results:

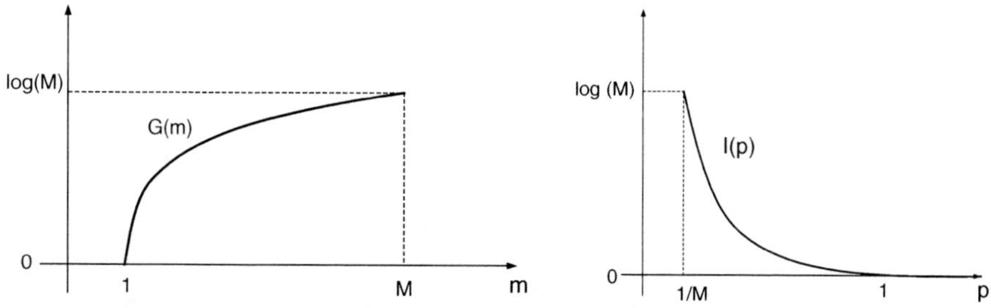


**Fig. 1.** Graphs of the granularity $G(m)$ and the uncertainty $I(p)$ measures in the "positivity" domains $m \in [1, M]$ and $p = m/M \in [1/M, 1]$ with $M = m(X)$

(GA) the *granularity average* with respect to $\mathbf{p}(\pi)$ expressed by the average

$$Av(\mathbf{G}(\pi), \mathbf{p}(\pi)) := \sum_{i=1}^{N} G(A_i) \cdot p(A_i) = \frac{1}{m(X)} \sum_{i=1}^{N} m(A_i) \cdot \log m(A_i)$$

which in the sequel will be simply denoted by $E(\pi)$;

(UA) the *uncertainty average* with respect to $\mathbf{p}(\pi)$ expressed by the average

$$Av(\mathbf{I}(\pi), \mathbf{p}(\pi)) := \sum_{i=1}^{N} I(A_i) \cdot p(A_i) = -\frac{1}{m(X)} \sum_{i=1}^{N} m(A_i) \cdot \log \frac{m(A_i)}{m(X)}$$

which is the *information entropy* $H(\pi)$ of the partition $\pi$ according to the Shannon approach to information theory [10] (and see also [6, 8] for introductory treatments).

Thus, the quantity $E(\pi)$ furnishes the (average) *measure of the granularity* carried by the partition $\pi$ as a whole, whereas the entropy $H(\pi)$ furnishes the (average) *measure of the uncertainty* associated to the same partition. In conclusion, also in this case the *granularity measure* must not be confused with the *uncertainty measure* supported by $\pi$. Analogously to the (3), related to a single event $A_i$, these averages satisfy the following identity, which holds for any *arbitrary* partition $\pi$ of the universe $X$:

$$H(\pi) + E(\pi) = \log m(X) \tag{4}$$

Also in this case the two measures *complement* each other with respect to the constant quantity $\log m(X)$, which is *invariant* with respect to the choice of the partition $\pi$ of $X$. This is the reason for the name of co–entropy given to $E(\pi)$ in a previous work of ours [11].

*Remark 1.* Let us recall that in [12] Wierman has interpreted the entropy $H(\pi)$ of the partition $\pi$ as a *granularity measure*, defined as the quantity which "measures the uncertainty (in bits) associated with the prediction of outcomes where elements of each partition sets $A_i$ are indistinguishable." This is the kind of

semantical confusion which must be avoided, preferring to distinguish the *uncertainty measure* of the partition $\pi$ given by $H(\pi)$ from the *granularity measure* of the same partition described by $E(\pi)$.
Note that in [13] it is remarked that the Wierman "*granularity measure*" coincides with the Shannon entropy $H(\pi)$, more correctly interpreted as the "*information measure of knowledge*" furnished by the partition $\pi$.

The co–entropy (average granularity measure) $E(\pi)$ ranges into the real (closed) interval $[0, \log|X|]$ with the minimum obtained by the discrete partition $\pi_d = \{\{x_1\}, \{x_2\}, \dots, \{x_{|X|}\}\})$, collection of all singletons from $X$, and the maximum obtained by the trivial partition $\pi_t = \{X\}$, consisting of the unique element $X$: that is $\forall \pi \in \Pi(X)$, $0 = E(\pi_d) \leq E(\pi) \leq E(\pi_t) = \log|X|$. Since the discrete partition is the one which generates the "best" sharpness ($\forall Y \in \mathcal{P}(X)$, $r_{\pi_d}(Y) = \langle Y, Y \rangle$), formalized by the fact that the boundary of *any* $Y$ is $b_{\pi_d}(Y) = u_{\pi_d}(Y) \setminus l_{\pi_d}(Y) = \emptyset$ (i.e., any subset is sharp), whereas the trivial partition is the one which generates the "worst" sharpness ($\forall Y \in \mathcal{P}(X) \setminus \{\emptyset, X\}$, $r_{\pi_t}(Y) = \langle \emptyset, X \rangle$; with $\emptyset$ and $X$ the unique crisp sets since $r_{\pi_t}(\emptyset) = \langle \emptyset, \emptyset \rangle$ and $r_{\pi_t}(X) = \langle X, X \rangle$), formalized by the fact that the boundary of *any nontrivial* subset $Y (\neq \emptyset, X)$ is the whole universe $b_{\pi_t}(Y) = X$. For these reasons, the interval $[0, \log|X|]$ is assumed as the *reference scale* for *measuring roughness* (or *sharpness*): the less is the value the worst is the roughness (or the best is the sharpness).

| $0$ ———————— | $\log\lvert X\rvert$ |
|---|---|
| maximum sharpness<br>minimum roughness | minimum sharpness<br>maximum roughness |

### 2.2 The Lattice of Partitions and the Monotonic Behavior of Entropy and Co–entropy

Up to now we discussed the notion of co–entropy (granularity average measure) $E(\pi)$ and of entropy (uncertainty average measure) $H(\pi)$ for a fixed partition $\pi \in \Pi(X)$ of the universe $X$. Now it is of a great importance to study what happens when the partition $\pi$ of $X$ changes in $\Pi(X)$. First of all, let us remark that on the family $\Pi(X)$ of all partitions on $X$ it is possible to introduce a partial order relation which in the context of partitions can be formulated in at least three mutually equivalent ways (where for a fixed partition $\pi$ we denote by $gr_\pi(x)$ the granule of $\pi$ which contains the point $x$):

(po-1) $\pi_1 \preceq \pi_2$ iff $\forall A \in \pi_1, \exists B \in \pi_2$: $A \subseteq B$.
(po-2) $\pi_1 \ll \pi_2$ iff $\forall B \in \pi_2, \exists \{A_{i_1}, \dots, A_{i_k}\} \subseteq \pi_1$: $B = A_{i_1} \cup \dots \cup A_{i_k}$.
(po-3) $\pi_1 \trianglelefteq \pi_2$ iff $\forall x \in X$, $gr_{\pi_1}(x) \subseteq gr_{\pi_2}(x)$.

*Remark 2.* The introduction on $\Pi(X)$ of these binary relations $\preceq$, $\ll$, and $\trianglelefteq$ defining a unique partial ordering on $\Pi(X)$ might seem a little bit redundant, but the reason of listing them in this partition context is essentially due to the fact that in the case of coverings of $X$ they give rise to different quasi–ordering relations, as we will see in the sequel.

With respect to this partial ordering, $\Pi(X)$ turns out to be a lattice lower bounded by the discrete partition ($\forall \pi$, $\pi_d \preceq \pi$) and upper bounded by the trivial partition ($\forall \pi$, $\pi \preceq \pi_t$), which are then the *least* and *greatest* element of the lattice, respectively. The strict ordering on partitions is as usual defined as $\pi_1 \prec \pi_2$ iff $\pi_1 \preceq \pi_2$ and $\pi_1 \neq \pi_2$, in this case it is said that $\pi_1$ (resp. $\pi_2$) is finer (resp., coarser) than $\pi_2$ (resp., $\pi_1$). Note that $\pi_1 \prec \pi_2$ happens if making use of **(po-2)** there is at least an elementary event $B \in \pi_2$ such that it is the union $B = A_{i_1} \cup \ldots \cup A_{i_k}$ of at least two different elements from $\pi_1$, i.e., for $k \geq 2$. Then, it is a standard result (see [11]) that the co–entropy is a *strictly monotonic* mapping with respect to the partition ordering, i.e.,

$$\pi_1 \prec \pi_2 \quad \text{implies} \quad E(\pi_1) < E(\pi_2)$$

Thus from (4) it follows the *strict anti–monotonicity* behavior of entropy:

$$\pi_1 \prec \pi_2 \quad \text{implies} \quad H(\pi_2) < H(\pi_1)$$

Let us recall that with respect to the above partial ordering on partitions, the lattice meet of $\pi_1 = (A)_{i=1,\ldots,M}$ and $\pi_2 = (B_j)_{j=1,\ldots,N}$ is the partition $\pi_1 \wedge \pi_2 = (A_i \cap B_j)_{\substack{i=1,\ldots,M \\ j=1,\ldots,N}}$, where some of the $A_i \cap B_j$ might be the empty set. The probability distribution corresponding to this partition (where some of its terms might be 0, but this does not constitute any problem in calculating the involved entropies) is then the length $M \cdot N$ vector:

$$\mathbf{p}(\pi_1 \wedge \pi_2) = \left( p(A_l \cap B_k) = \frac{m(A_l \cap B_k)}{m(X)} \right)_{\substack{k=1,\ldots,N \\ l=1,\ldots,M}}$$

Note that the following probability vector of $\pi_1$ *conditioned* by $\pi_2$ is not a probability distribution

$$\mathbf{p}(\pi_1|\pi_2) = \left( p(A_l|B_k) = \frac{m(A_l \cap B_k)}{m(B_k)} \right)_{\substack{k=1,\ldots,N \\ l=1,\ldots,M}}$$

Indeed, giving as assured the condition (1) $\forall l, k$, $p(A_l|B_k) \geq 0$, we can only state that (2) $\forall k$, $\sum_{k=1}^{N} p(A_l|B_k) = 1$, which leads to $\sum_{lk} p(A_l|B_k) = M$.

Generalizing the **(RV-U)** of section 2.1, let us consider now the following two discrete uncertainty random variables.

(RV-UM) The *uncertainty* random variable of the *meet* partition $\pi_1 \wedge \pi_2$

$$\mathbf{I}(\pi_1 \wedge \pi_2) = \Big( -\log p(A_l \cap B_k) \Big)_{\substack{k=1,\ldots,N \\ l=1,\ldots,M}}$$

(RV-UC) The *uncertainty* random variable of the partition $\pi_1$ *conditioned* by the partition $\pi_2$

$$\mathbf{I}(\pi_1|\pi_2) = \Big( -\log p(A_l|B_k) \Big)_{\substack{k=1,\ldots,N \\ l=1,\ldots,M}}$$

The *uncertainty* of the partition $\pi_1 \wedge \pi_2$, as average of the random variable **(RV-UM)** with respect to the probability distribution $\mathbf{p}(\pi_1 \wedge \pi_2)$, is so expressed by the meet entropy

$$H(\pi_1 \wedge \pi_2) = -\sum_{l,k} p(A_l \cap B_k) \log p(A_l \cap B_k)$$

Whereas the *entropy* of the partition $\pi_1$ *conditioned* by the partition $\pi_2$ is defined as the average of the discrete random variable **(RV-UC)** with respect to the probability distribution $\mathbf{p}(\pi_1 \wedge \pi_2)$, expressed by the non–negative quantity:

$$H(\pi_1|\pi_2) := -\sum_{l,k} p(A_l \cap B_k) \log p(A_l|B_k)$$

As a first result, we recall that the following relationship holds:

$$E(\pi_1 \wedge \pi_2) = \frac{1}{m(X)} \sum_{l,k} m(A_l \cap B_k) \cdot \log m(A_l \cap B_k)$$

$$E(\pi_1 \wedge \pi_2) = E(\pi_2) - H(\pi_1|\pi_2) \tag{5}$$

Moreover, introduced the *co–entropy* of the partition $\pi_1$ *conditioned* by the partition $\pi_2$ as the quantity

$$E(\pi_1|\pi_2) := \frac{1}{m(X)} \sum_{lk} m(A_l \cap B_k) \cdot \log \left[ \frac{m(X)}{m(B_k)} m(A_l \cap B_k) \right] \tag{6a}$$

$$= \frac{1}{m(X)} \sum_{lk} [m(B_k) \cdot p(A_l|B_k)] \log [m(X) \cdot p(A_l|B_k)] \tag{6b}$$

it is easy to show that $E(\pi_1|\pi_2) \geq 0$. Furthermore, the expected relationship holds:

$$H(\pi_1|\pi_2) + E(\pi_1|\pi_2) = \log m(X) \tag{7}$$

Note that from (5) it follows that

$$\pi_1 \preceq \pi_2 \quad \text{implies} \quad E(\pi_2) = E(\pi_1) + H(\pi_1|\pi_2) \tag{8}$$

### 2.3 Local Rough Granularity Measure in the Case of Partitions

From the point of view of the rough approximations of subsets $Y$ of the universe $X$ with respect to its partitions $\pi$, we shall consider now the situation in which during the time evolution $t_1 \to t_2$ one tries to relate the corresponding variation of partitions $\pi_{t_1} \to \pi_{t_2}$ with, for instance, the corresponding boundary modification $b_{t_1}(Y) \to b_{t_2}(Y)$. Let us note that

$$\text{if } \pi_1 \preceq \pi_2, \quad \text{then} \quad l_{\pi_2}(Y) \subseteq l_{\pi_1}(Y) \subseteq Y \subseteq u_{\pi_1}(Y) \subseteq u_{\pi_2}(Y)$$

i.e., the rough approximation of $Y$ with respect to the partition $\pi_1$, $r_{\pi_1}(Y) = (l_{\pi_1}(Y), u_{\pi_1}(Y))$, is *better* than the rough approximation of the same subset with respect to $\pi_2$, $r_{\pi_2}(Y) = (l_{\pi_2}(Y), u_{\pi_2}(Y))$. This fact can be denoted by the binary relation of partial ordering on subsets: $r_{\pi_1}(Y) \sqsubseteq r_{\pi_2}(Y)$. This leads to a first but only *qualitative* valuation of the *roughness* expressed by the law:

$$\pi_1 \preceq \pi_2 \quad \text{implies that} \quad \forall Y,\, b_{\pi_1}(Y) \subseteq b_{\pi_2}(Y)$$

The delicate point is that the condition of strict ordering $\pi_1 \prec \pi_2$ does not assure that $\forall Y$, $b_{\pi_1}(Y) \subset b_{\pi_2}(Y)$. It is possible to give some very simple counter–examples (see for instance example 1) in which notwithstanding $\pi_1 \prec \pi_2$ one has that $\exists Y_0$: $b_{\pi_1}(Y_0) = b_{\pi_2}(Y_0)$ [14, 11], and this is not a desirable behavior of such a qualitative valuation of roughness.

*Example 1.* In the universe $X = \{1, 2, 3, 4, 5, 6\}$, let us consider the two partitions $\pi_1 = \{\{1\}, \{2\}, \{3\}, \{4, 5, 6\}\}$ and $\pi_2 = \{\{1, 2\}, \{3\}, \{4, 5, 6\}\}$, with respect to which $\pi_1 \prec \pi_2$. The subset $Y_0 = \{1, 2, 4, 6\}$ is such that $l_{\pi_1}(Y_0) = l_{\pi_2}(Y_0) = \{1, 2\}$ and $u_{\pi_1}(Y_0) = u_{\pi_2}(Y_0) = \{1, 2, 4, 5, 6\}$. This result implies that $b_{\pi_1}(Y_0) = b_{\pi_2}(Y_0) = \{4, 5, 6\}$.

On the other hand, in many practical applications (for instance in the attribute reduction procedure), it is interesting not only to have a possible qualitative valuation of the roughness of a generic subset $Y$, but also a *quantitative* valuation formalized by a mapping $E : \Pi(X) \times 2^X \to [0, K]$ (with $K$ suitable non–negative real number) assumed to satisfy (at least) the following two minimal requirements:

(re1) The *strict monotonicity condition*: for any $Y \in 2^X$, $\pi_1 \prec \pi_2$ implies $E(\pi_1, Y) < E(\pi_2, Y)$.
(re2) The *boundary conditions*: $\forall Y \in 2^X$, $E(\pi_d, Y) = 0$ and $E(\pi_t, Y) = 1$.

In the sequel, sometimes we will use $E_\pi : 2^X \to [0, K]$ to denote the above mapping in which the partition $\pi \in \Pi(X)$ is considered fixed once for all. The interpretation of condition (re2) is possible under the assumption that a quantitative valuation of the roughness $E_\pi(Y)$ should be *directly* related to its boundary by $|b_\pi(Y)|$. From this point of view, the value 0 corresponds to the discrete partition for which the boundary of any subset $Y$ is empty, and so its rough approximation is $r_{\pi_d}(Y) = (Y, Y)$ with $|b_{\pi_d}(Y)| = 0$, i.e., a crisp situation. On the other hand, the trivial partition is such that the boundary of any nontrivial subset $Y (\neq \emptyset, X)$ is the whole universe, and so its rough approximation is $r_{\pi_t}(Y) = (\emptyset, X)$ with $|b_{\pi_t}(Y)| = |X|$. For all other partitions $\pi$ we must recall that $\pi_d \preceq \pi \prec \pi_t$ and $0 = |b_{\pi_d}(Y)| \leq |b_\pi(Y)| \leq |b_{\pi_t}(Y)| = |X|$, i.e., the maximum of roughness (or minimum of sharpness) valuation is reached by the trivial partition $\pi_t$.

This being stated, in literature one can find a lot of quantitative *measures of roughness* of $Y$ relatively to a given partition $\pi \in \Pi(X)$ formalized as mappings $\rho_\pi : 2^X \to [0, 1]$ such that:

(rm1) the *monotonicity condition* holds: $\pi_1 \preceq \pi_2$ implies that $\forall Y \in 2^X$, $\rho_{\pi_1}(Y) \leq \rho_{\pi_2}(Y)$;
(rm2) $\forall Y \in 2^X$, $\rho_{\pi_d}(Y) = 0$ and $\rho_{\pi_t}(Y) = 1$.

The *accuracy* of the set $Y$ with respect to the partition $\pi$ is then defined as $\alpha_\pi(Y) = 1 - \rho_\pi(Y)$. The interpretation of the condition (rm2) is that in general a roughness measure directly depends from a valuation of the cardinality of the boundary $b_\pi(Y)$ of $Y$ relatively to $\pi$. Two of the more interesting roughness measures are

$$\rho_\pi^{(P)}(Y) := \frac{|b_\pi(Y)|}{|u_\pi(Y)|} \quad \text{and} \quad \rho_\pi^{(C)}(Y) := \frac{|b_\pi(Y)|}{|X|}$$

with the latter (considered in [11]) producing a better description of the former (introduced by Pawlak in [15]) with respect to the absolute scale of sharpness previously introduced, since whatever be the subset $Y$ it is $\rho_\pi^{(C)}(Y) \leq \rho_\pi^{(P)}(Y)$. These roughness measures satisfy the above "boundary" condition (re2), but their drawback is that the strict condition on partitions $\pi_1 \prec \pi_2$ does not assure a corresponding strict behavior $\forall Y, b_{\pi_1}(Y) \subset b_{\pi_2}(Y)$, and so the strict correlation $\rho_{\pi_1}(Y) < \rho_{\pi_2}(Y)$ cannot be inferred. It might happen that notwithstanding the strict partition order $\pi_1 \prec \pi_2$, the two corresponding roughness measures for a certain subset $Y_0$ turn out to be equal $\rho_{\pi_1}(Y_0) = \rho_{\pi_2}(Y_0)$ as illustrated in the following example.

*Example 2.* Making reference to example 1 we have that although $\pi_1 \prec \pi_2$, for the subset $Y_0$ we get $\rho_{\pi_1}(Y_0) = \rho_{\pi_2}(Y_0)$ (for both roughness measures $\rho_\pi^{(P)}(Y_0)$ and $\rho_\pi^{(C)}(Y_0)$).

Summarizing we can only state the following monotonicity with respect to the partition ordering:

$$\pi_1 \prec \pi_2 \quad \text{implies} \quad \forall Y \subseteq X : \rho_{\pi_1}(Y) \leq \rho_{\pi_2}(Y)$$

Taking inspiration from [14] a *local co–entropy measure* of $Y$, in the sense of a "co–entropy" assigned not to the whole universe $X$ but to any possible of its subset $Y$, is then defined as the product of the above (local) roughness measure times the (global) co–entropy:

$$E_\pi(Y) := \rho_\pi(Y) \cdot E(\pi) \tag{9}$$

For a fixed partition $\pi$ of $X$ also this quantity ranges into the closed real interval $[0, \log|X|]$ whatever be the subset $Y$, with the extreme values reached for $E_{\pi_d}(Y) = 0$ and $E_{\pi_t}(Y) = \log|X|$, i.e., $\forall Y \subseteq X$ it is

$$0 = E_{\pi_d}(Y) \leq E_\pi(Y) \leq E_{\pi_t}(Y) = \log|X|$$

Moreover, for any fixed subset $Y$ this local co–entropy is strictly monotonic with respect to partitions:

$$\pi_1 \prec \pi_2 \quad \text{implies} \quad \forall Y \subseteq X : E_{\pi_1}(Y) < E_{\pi_2}(Y) \tag{10}$$

Making use of the above interpretation (see the end of the section 2.1) of the real interval $[0, \log|X|]$ as an absolute scale of sharpness, from this result we have that, according to our intuition, *the finer is the partition the best is the sharpness of the rough approximation of* $Y$, i.e., $E_\pi : Y \in \mathcal{P}(X) \to E_\pi(Y) \in [0, \log_2|X|]$ can be considered as a (local) *rough granularity* mapping.

*Example 3.* Let us consider the universe $X = \{1,2,3,4,5,6,7,8,9,10,11\}$, its subset $Y = \{2,3,5,8,9,10,11\}$, and the following three different partitions of the universe $X$ by granules:

$$\begin{aligned}\pi_1 &= \{\{2,3,5,8,9\}, \{1,4\}, \{6,7,10,11\}\},\\ \pi_2 &= \{\{2,3\}, \{5,8,9\}, \{1,4,\}, \{6,7,10,11\}\},\\ \pi_3 &= \{\{2,3\}, \{5,8,9\}, \{1,4,\}, \{7,10\}, \{6,11\}\}\end{aligned}$$

with $\pi_3 \prec \pi_2 \prec \pi_1$. The lower and upper approximations of $Y$ with respect to $\pi_1$, $\pi_2$ and $\pi_3$ are equal, and given, respectively by: $i_{\pi_k}(Y) = \{2,3,5,8,9\}$ and $o_{\pi_k}(Y) = \{2,3,5,6,7,8,9,10,11\}$, for $k = 1,2,3$. Note that necessarily $e_{\pi_1}(Y) = e_{\pi_2}(Y) = e_{\pi_3}(Y) = \{1,4\}$. Therefore, the corresponding roughness measures are exactly the same: $\rho_{\pi_1}(Y) = \rho_{\pi_2}(Y) = \rho_{\pi_3}(Y)$, even though from the point of view of the granularity knowledge we know that the lower approximations of $Y$ are obtained by different collections of granules: $gr_i^{\pi 2}(Y) = \{\{2,3\},\{5,8,9\}\} = gr_i^{\pi 3}(Y)$, as collection of two granules, are better (finer) than $gr_i^{\pi 1}(Y) = \{\{2,3,5,8,9\}\}$, a single granule, this fact formally written as $gr_i^{\pi 2}(Y) = gr_i^{\pi 3}(Y) \prec gr_i^{\pi 1}(Y)$. Similarly, always from the granule knowledge point of view, we can see that the best partitioning for the upper approximation of $Y$ is obtained with $\pi_3$ since $gr_o^{\pi 1}(Y) = \{\{2,3,5,8,9\},\{6,7,10,11\}\}, gr_o^{\pi 2}(Y) = \{\{2,3\},\{5,8,9\},\{6,7,10,11\}\}$, and $gr_o^{\pi 3}(Y) = \{\{2,3\},\{5,8,9\},\{7,10\},\{6,11\}\}$, and thus $gr_o^{\pi 3}(Y) \prec gr_o^{\pi 2}(Y) \prec gr_o^{\pi 1}(Y)$.

It is clear that the roughness measure $\rho_\pi(Y)$ is not enough when we want to catch any possible advantage in terms of granularity knowledge given by different partitioning, even when the new partitioning does not increase the cardinality of the internal and the closure approximation sets. On the contrary, this difference is measured by the local co–entropy (9) since according to (10), and recalling that $\pi_3 \prec \pi_2 \prec \pi_1$, we have the following strict monotonicity: $E_{\pi_3}(Y) \prec E_{\pi_2}(Y) \prec E_{\pi_1}(Y)$.

### 2.4 Application to Complete Information Systems

These considerations can be applied to the case of a complete Information System (IS), formalized by a triple $IS := \langle X, Att, F\rangle$ consisting of a nonempty finite set $X$ of objects, a nonempty finite set of attribute $Att$, and a mapping $F : X \times Att \to V$ which assigns to any object $x \in A$ the value $F(x,a)$ assumed by the attribute $a \in Att$ [16, 15, 17]. Indeed, in this IS case the partition generated by a set of attributes $\mathcal{A}$, denoted by $\pi_{\mathcal{A}}(IS)$, consists of equivalence classes of *indistinguishable* objects with respect to the equivalence relation $R_{\mathcal{A}}$ involving any pair of object $x, y \in X$:

(In) $(x,y) \in R_{\mathcal{A}}$ iff $\forall a \in \mathcal{A},\ F(x,a) = F(y,a)$.

The equivalence class generated by the object $x \in X$ is the *granule of knowledge* $gr_{\mathcal{A}}(x) := \{y \in X : (x,y) \in R_{\mathcal{A}}\}$. In many applications it is of a certain interest to analyze the variations occurring inside two information systems labelled with two parameters $t_1$ and $t_2$. In particular, one has to do mainly with the following two cases in both of which the set of objects remains invariant:

(1) *dynamics* (see [3]), in which $IS_{t_1} = (X, Att_1, F_1)$ and $IS_{t_2} = (X, Att_2, F_2)$ are under the conditions that $Att_1 \subset Att_2$ and $\forall x \in X$, $\forall a_1 \in Att_1$: $F_2(x, a_1) = F_1(x, a_1)$. This situation corresponds to a dynamical increase of knowledge ($t_1$ and $t_2$ are considered as time parameters, with $t_1 < t_2$) for instance in a medical database in which one fixed *decision* attribute $d \in Att_1 \cap Att_2$ is selected to state a certain disease related to all the resting *condition* attributes (i.e., symptoms) $C_i = Att_i \setminus \{d\}$. In this case the increase $Att_1 \setminus \{d\} \subseteq Att_2 \setminus \{d\}$ corresponds to the fact that during the researches on the disease some symptoms which have been neglected at time $t_1$ become relevant at time $t_2$ under some new investigations.

(2) *reduct*, in which $IS_{t_1} = (X, Att_1, F_1)$ and $IS_{t_2} = (X, Att_2, F_2)$ are under the conditions that $Att_2 \subset Att_1$ and $\forall x \in X$, $\forall a_2 \in Att_2$: $F_2(x, a_2) = F_1(x, a_2)$. In this case it is of a certain interest to verify if the corresponding partitions are invariant $\pi_{Att_2}(IS_{t_2}) = \pi_{Att_1}(IS_{t_1})$, or not. In the former case one can consider $IS_{t_2}$ as the result of the reduction of the initial attributes $Att_1$ obtained by the suppression from $IS_{t_1}$ of the superfluous attributes $Att_1 \setminus Att_2$.

From a general point of view, a *reduction procedure* can be formalized by a (strictly) monotonically decreasing sequence of attribute families $RP := \{\mathcal{A}_t \subseteq Att : t \in \mathbb{N} \text{ and } \mathcal{A}_t \supset \mathcal{A}_{t+1}\}$, with $\mathcal{A}_0 = Att$. In this case it holds the following diagram, linking the family $\mathcal{A}_t$ with the generated partition $\pi(\mathcal{A}_t)$ whose co–entropy is $E(\mathcal{A}_t)$:

$$\begin{array}{ccccccccc}
\mathcal{A}_0 = Att \supset & \mathcal{A}_1 & \supset \ldots \supset & \mathcal{A}_t & \supset & \mathcal{A}_{t+1} & \ldots & \supset \mathcal{A}_T = \emptyset \\
\downarrow & \downarrow & & \downarrow & & \downarrow & & \downarrow \\
\pi(\mathcal{A}_0) & \preceq \pi(\mathcal{A}_1) & \preceq \ldots \preceq & \pi(\mathcal{A}_t) & \preceq & \pi(\mathcal{A}_{t+1}) & \ldots & \preceq \{X\} \\
\downarrow & \downarrow & & \downarrow & & \downarrow & & \downarrow \\
E(\mathcal{A}_0) & \leq E(\mathcal{A}_1) & \leq \ldots \leq & E(\mathcal{A}_t) & \leq & E(\mathcal{A}_{t+1}) & \ldots & \leq \log|X|
\end{array}$$

The first row constitutes the *attribute channel*, the second row the *partition channel* (whose upper bound is the trivial partition $\pi_t = \{X\}$), and the last row the *granularity* (or *information*) channel (whose upper bound corresponds to the maximum of roughness $\log|X|$) of the reduction procedure. After the finite number of steps $T = |Att|$, one reaches the empty set $\mathcal{A}_T = \emptyset$ with corresponding $\pi(\mathcal{A}_T) = \pi_t = \{X\}$, the trivial partition, and $E(\mathcal{A}_T) = \log|X|$. In this reduction context, the link between the situation at step $t$ and the corresponding one at $t+1$ relatively to the co–entropy is given by equation (8) which assumes now the form:

$$E(\mathcal{A}_{t+1}) = E(\mathcal{A}_t) + H(\mathcal{A}_t|\mathcal{A}_{t+1}) \tag{11}$$

From a general point if view, a practical procedure of reduction consists of starting from an initial attribute family $\mathcal{A}_0$, and according to some algorithmic criterium $\mathbb{A}$lg, step by step, one "constructs" the sequence $\mathcal{A}_t$, with this latter a subset of the previous $\mathcal{A}_{t-1}$. It is possible to fix *a priori* a suitable approximation value $\epsilon$ and then to stop the procedure at the first step $t_0$ such that $\log|X| - E(\mathcal{A}_{t_0}) \leq \epsilon$. This assures that for any other further step $t > t_0$ it is also $\log|X| - E(\mathcal{A}_t) \leq \epsilon$. The family of attributes $\mathcal{A}(t_0)$ is the *$\epsilon$–approximate reduct* with respect to the procedure $\mathbb{A}$lg. Note that in terms of approximation the following order chain holds: $\forall t > t_0$, $E(\mathcal{A}_t) - E(\mathcal{A}_{t_0}) \leq \log|X| - E(\mathcal{A}_{t_0}) \leq \epsilon$. On the other hand, for any triplet of steps $t_0 < t_1 < t_2$ it is

$$H(\mathcal{A}_{t_1}|\mathcal{A}_{t_2}) = E(\mathcal{A}_{t_2}) - E(\mathcal{A}_{t_1}) \leq \log|X| - E(\mathcal{A}_{t_0}) \leq \epsilon$$

*Example 4.* In the complete information system illustrated in table 1 let us consider the following five families of attributes $\mathcal{A}_0 = Att = \{Price,\ Rooms,\ Down-Town,\ Furniture,\ Floor,\ Lift\} \supset \mathcal{A}_1 = \{Price,\ Rooms,\ Down-Town,\ Furniture,\ Floor\} \supset \mathcal{A}_2 = \{Price,\ Rooms,\ Down-Town,\ Furniture\} \supset \mathcal{A}_3 = \{Price,\ Rooms,\ Down-Town\} \supset \mathcal{A}_4 = \{Price,\ Rooms\} \supset \mathcal{A}_5 = \{Price\}$ and the corresponding probability partitions $\pi(\mathcal{A}_1) = \pi(\mathcal{A}_2) = \{\{1,2\},\ \{3,4\},\ \{5,6\},\ \{7,9\},\ \{8\},\ \{10\}\}$, $\pi(\mathcal{A}_3) = \{\{1,2\},\ \{3,4\},\ \{5,6\},\ \{7,8,9\},\{10\}\}$, $\pi(\mathcal{A}_4) = \{\{1,2\},\{3,4,5,6\},\{7,8,9,10\}\}$, and $\pi(\mathcal{A}_5) = \{\{1,2,3,4,5,6\},\{7,8,9,10\}\}$; in this case $\pi(\mathcal{A}_0)$ corresponds to the *discrete* partition $\pi_d$.

**Table 1.** Flats complete information system

| Flat | Price | Rooms | Down-Town | Furniture | Floor | Lift |
|---|---|---|---|---|---|---|
| 1 | high | 3 | yes | yes | 3 | yes |
| 2 | high | 3 | yes | yes | 3 | no |
| 3 | high | 2 | no | no | 1 | no |
| 4 | high | 2 | no | no | 1 | yes |
| 5 | high | 2 | yes | no | 2 | no |
| 6 | high | 2 | yes | no | 2 | yes |
| 7 | low | 1 | no | no | 2 | yes |
| 8 | low | 1 | no | yes | 3 | yes |
| 9 | low | 1 | no | no | 2 | no |
| 10 | low | 1 | yes | yes | 1 | yes |

We can easily observe that $\pi(\mathcal{A}_0) \prec \pi(\mathcal{A}_1) = \pi(\mathcal{A}_2) \prec \pi(\mathcal{A}_3) \prec \pi(\mathcal{A}_4) \prec \pi(\mathcal{A}_5)$ and that $E(\mathcal{A}_0) = 0.00000 < E(\mathcal{A}_1) = 0.80000 = E(\mathcal{A}_2) < 1.07549 = E(\mathcal{A}_3) < 1.80000 = E(\mathcal{A}_4) < 2.35098 = E(\mathcal{A}_5) < \log|X| = 3.32193$. Moreover, taking for instance $E(\mathcal{A}_3)$ and $E(\mathcal{A}_4)$ and according to (11), we have $H(\mathcal{A}_3|\mathcal{A}_4) = E(\mathcal{A}_4) - E(\mathcal{A}_3) = 0.72451$.

$$\begin{array}{ccccccc}
\mathcal{A}_0 = Att \supset & \mathcal{A}_1 \supset & \mathcal{A}_2 \supset & \mathcal{A}_3 \supset & \mathcal{A}_4 \supset & \mathcal{A}_5 & \supset \mathcal{A}_T = \emptyset \\
\downarrow & \downarrow & \downarrow & \downarrow & \downarrow & \downarrow & \downarrow \\
\pi(\mathcal{A}_0) & \prec \pi(\mathcal{A}_1) = & \pi(\mathcal{A}_2) \prec & \pi(\mathcal{A}_3) \prec & \pi(\mathcal{A}_4) \prec & \pi(\mathcal{A}_5) & \prec \{X\} \\
\downarrow & \downarrow & \downarrow & \downarrow & \downarrow & \downarrow & \downarrow \\
E(\mathcal{A}_0) & < E(\mathcal{A}_1) = & E(\mathcal{A}_2) < & E(\mathcal{A}_3) < & E(\mathcal{A}_4) < & E(\mathcal{A}_5) & < \log|X|
\end{array}$$

The investigation of these (attribute–partition–granularity) triplet of channels is outside the scope of the present chapter, and shall be the argument of forthcoming researches about reduction.

## 3 Entropy and Co–entropy of Coverings: The Global Approach

In this section we analyze a possible generalization of the discussion about entropy and co–entropy of partition to the case of coverings of a (finite) universe $X$, whose collection will be denoted by $\Gamma(X)$. Let us recall that a covering $\gamma := \{C_i \in \mathcal{P}(X) : i \in I\}$ of $X$ is any family of nonempty subsets whose set theoretic union is the whole space $X$. In [11] we have introduced the notion of *genuine* covering formalized as a covering $\gamma = \{B_1, B_2, \ldots, B_N\}$ for which the following holds: $\forall B_i, B_j \in \gamma$, $B_i \neq B_i \cap B_j$ or, equivalently, $B_i \subseteq B_j$ implies $B_i = B_j$. In the sequel, we will denote by $\Gamma_g(X)$ the class of all genuine coverings of $X$. Of course, if $\gamma$ is not genuine, then the procedure which for any case of $B_i, B_j \in \gamma$, with $B_i \subseteq B_j$, eliminates $B_i$ induces in a canonical way a genuine covering, denoted by $\gamma_g$. From another point of view, we shall say that a covering is *trivial* iff it contains as element the whole universe $X$.

To any covering $\gamma = \{B_1, B_2, \ldots, B_N\}$, genuine or not, it is possible to associate the mapping $n : X \mapsto \mathbb{N}$ which counts the number of occurrences of the element $x$ in $\gamma$ according to the definition

$$\forall x \in X, \quad n(x) := \sum_{i=1}^{N} \chi_{B_i}(x) \tag{12}$$

Moreover, to any subset $B_i$ of the covering $\gamma$ one can associate the corresponding *fuzzy set* $\omega_{B_1} : X \mapsto [0,1]$ defined as

$$\forall x \in X, \quad \omega_{B_i}(x) := \frac{1}{n(x)} \chi_{B_i}(x) \tag{13}$$

The fuzzy set representation (13) of any covering $\gamma$ of the universe $X$ is always an identity resolution. Indeed, it is possible to prove the following result (see [11]): $\forall x \in X, \quad \sum_{i=1}^{N} \omega_{B_i}(x) = 1$. If one denotes by $\mathbf{1}$ the identically 1 mapping ($\forall x \in X$, $\mathbf{1}(x) = 1$), then the previous identity resolution condition can be expressed as the functional identity $\sum_{i=1}^{N} \omega_{B_i} = \mathbf{1}$.

*Remark 3.* Let us note that in the particular case of a partition $\pi$ of $X$, described by the crisp identity resolution $\mathcal{C}(\pi) = \{\chi_{A_1}, \chi_{A_2}, \ldots, \chi_{A_N}\}$, where for any subset $A$ of $X$ it is $\chi_A(x) = 1$ for $x \in A$ and 0 otherwise, the number of occurrence of any point $x$ expressed by (12) is the identically 1 constant function $\forall x \in X$, $n(x) = 1$, and so the fuzzy set (13) is nothing else than the characteristic function itself $\forall x \in X$, $\omega_{A_i}(x) = \chi_{A_i}(x)$.

The measure of the generic "event" $B_i$ of the covering $\gamma$ is then defined as follows

$$m(B_i) := \sum_{x \in X} \omega_{B_i}(x) = \sum_{x \in X} \frac{1}{n(x)} \chi_{B_i}(x) \tag{14}$$

In this way, we obtain the *measure distribution* induced by the covering $\gamma$

$$\mathbf{m}(\gamma) = (m(B_1), m(B_2), \ldots, m(B_N)) \tag{15}$$

since the following hold: (md-1) every $m(B_i) \geq 0$; (md-2) $M(\gamma) = \sum_{i=1}^{N} m(B_i) = |X|$, which is the *total length* of this measure distribution generated by the covering $\gamma$. If one introduces the quantities normalized by the total length $M(\gamma)$ of the measure distribution (15)

$$p(B_i) := \frac{1}{M(\gamma)} m(B_i) = \frac{1}{|X|} \sum_{x \in X} \frac{1}{n(x)} \chi_{B_i}(x)$$

from (md–2) it follows that the vector $\mathbf{p}(\gamma) = (p(B_1), p(B_2), \ldots, p(B_N))$ defines a probability distribution induced from the covering $\gamma$, since (1) $p(B_i) \geq 0$ for any $i = 1, 2, \ldots, N$; (2) $\sum_{i=1}^{N} p(B_i) = 1$.

As usual from any pair consisting of a measure distribution $\mathbf{m}(\gamma)$ and a probability distribution $\mathbf{p}(\gamma)$ it is possible to introduce the two following quantities.

(GA-c) The co–entropy as average *granularity measure* of the covering $\gamma$:

$$E(\gamma) = \frac{1}{|X|} \sum_{i=1}^{N} m(B_i) \log m(B_i) \tag{16}$$

(UA-c) The entropy as average *uncertainty measure* of the covering $\gamma$:

$$0 \leq H(\gamma) = -\sum_{i=1}^{N} p(B_i) \log p(B_i) \leq \log |X| \tag{17}$$

Trivially, also in this case the following identity holds: $\forall \gamma \in \Gamma(X)$, $H(\gamma) + E(\gamma) = \log |X|$, which is an extension to coverings of the identity (4) involving only possible partitions $\pi \in \Pi(X)$ of the universe $X$. So, whatever be the covering $\gamma$, the "co–entropy" $E(\gamma)$ complements the original entropy $H(\gamma)$ with respect to the constant quantity $\log |X|$, which is invariant with respect to the choice of the covering $\gamma$. As stressed in equation (17), the entropy of a covering, as sum of non–negative terms, is non–negative. But differently from the partition case, the now introduced co–entropy of a covering might have negative terms in the sum, precisely when $m(B_i) < 1$. And thus, the drawback of this co–entropy is that it might be negative.

*Example 5.* In the universe $X = \{1, 2, 3, 4\}$, let us consider the *genuine* covering $\gamma = \{\{1, 2\}, \{1, 3\}, \{1, 4\}, \{2, 3\}, \{2, 4\}\}$. Then the corresponding co-entropy is negative $E(\gamma) \cong -0.31669$ whereas the entropy is positive $H \cong 2.31669$. Note that as required by the general result $E(\gamma) + H(\gamma) \cong 2 \cong \log 4$.

The main difference between partitions and coverings with respect to the entropy and co–entropy lies in the different definition of the measure $m$ according to

$$\forall A_i \in \pi : m(A_i) = \sum_{x \in X} \chi_{A_i}(x) = |A_i|$$

$$\forall B_i \in \gamma : m(B_i) = \sum_{x\in X} \frac{1}{n(x)}\chi_{B_i}(x)$$

*Remark 4.* Of course, since any partition $\pi$ is also a covering, one can apply the (16) and (17) to $\pi$ but the obtained results coincide with the standard partition co–entropy (GA) and entropy (UA) introduced in section 2.1.

### 3.1 The "Normalized" Non–negative Co–entropy

In order to overcome the possible negativity of the above introduced co–entropy, let us define as the *normalizing value* of the covering $\gamma$ the quantity

$$\hbar(\gamma) = \min\{m(B_1), m(B_2), \ldots, m(B_N))$$

The *normalized* measure distribution generated by the covering $\gamma$ is then

$$\mathbf{m}_\hbar(\gamma) = \left(\frac{m(B_1)}{\hbar(\gamma)}, \frac{m(B_2)}{\hbar(\gamma)}, \ldots, \frac{m(B_N)}{\hbar(\gamma)}\right) = \frac{1}{\hbar(\gamma)}\mathbf{m}(\gamma)$$

This "normalized" measure distribution is such that the following hold:
(md-1) every $m_\hbar(B_i) = \frac{m(B_i)}{\hbar(\gamma)} \geq 1$;
(md-2) the corresponding total measure is $M_\hbar(\gamma) = \sum_{i=1}^N m_\hbar(B_i) = \frac{|X|}{\hbar(\gamma)}$.
Thus, the *probability distribution* generated by the covering $\gamma$ is

$$\mathbf{p}_\hbar(\gamma) = \left(\frac{m_\hbar(B_i)}{M_\hbar(\gamma)}\right)_{i=1,\ldots,N} = \left(\frac{m(B_i)}{|X|}\right)_{i=1,\ldots,N} = \mathbf{p}(\gamma)$$

i.e., the probability distribution does not change passing from the original measure distribution $\mathbf{m}(\gamma)$ to the new normalized one $\mathbf{m}_\hbar(\gamma)$. In this way the covering entropy (16) does not change: $H(\gamma) = H_\hbar(\gamma)$. It is the co–entropy which assumes now the (non–negative, owing to the above **(md-1)**) form:

$$E_\hbar(\gamma) = \frac{1}{M_\hbar(\gamma)}\sum_{i=1}^N \frac{m(B_i)}{\hbar(\gamma)} \cdot \log\frac{m(B_i)}{\hbar(\gamma)} = \frac{1}{|X|}\sum_{i=1}^N m(B_i)\cdot\log\frac{m(B_i)}{\hbar(\gamma)} \quad (18)$$
$$= E(\gamma) - \log\hbar(\gamma)$$

And so, we have now the identity $H_\hbar(\gamma) + E_\hbar(\gamma) = \log\frac{|X|}{\hbar(\gamma)}$, i.e., this sum is not invariant with respect to the choice of the covering $\gamma$ of $X$. Let us stress that in the case of the example 5, in which the co–entropy $E(\gamma)$ was negative ($\cong -0.31669$), we have that the corresponding $E_\hbar(\gamma) \cong 0.26827$.

## 4 Quasi–orderings for Coverings

In [11] we have introduced some quasi–orderings (i.e., a reflexive and transitive, but in general non anti–symmetric relation [18, p. 20]) for generic coverings, as extension to this context of the ordering **(po-1)**–**(po-3)** previously discussed in the case of partitions, with the first two of the "global" kind and the last one of the "pointwise" one.

## 4.1 The "Global" Quasi–orderings on Coverings

In the present section we take into account the generalization of the only first two *global* cases. The first quasi–ordering is the extension of **(po-1)** given by the following binary relation for $\gamma, \delta \in \Gamma(X)$:

$$\gamma \preceq \delta \quad \text{iff} \quad \forall C_i \in \gamma,\ \exists D_j \in \delta : C_i \subseteq D_j \tag{19}$$

The corresponding *strict* quasi–order relation is $\gamma \prec \delta$ iff $\gamma \preceq \delta$ and $\gamma \neq \delta$. As remarked in [11], in the class of genuine coverings $\Gamma_g(X)$ the quasi–ordering relation $\preceq$ is an ordering. Another quasi–ordering on $\Gamma(X)$ which generalizes to coverings the **(po-2)** is defined by the following binary relation:

$$\gamma \ll \delta \quad \text{iff} \quad \forall D \in \delta,\ \exists \{C_1, C_2, \ldots, C_p\} \subseteq \gamma : D = C_1 \cup C_2 \cup \ldots \cup C_p \tag{20}$$

In the covering context, there is no general relationship between (19) and (20) since it is possible to give an example of two (genuine) coverings $\gamma, \delta$ for which $\gamma \preceq \delta$ but $\gamma \not\ll \delta$, and of two other (genuine) coverings $\eta, \xi$ for which $\eta \ll \xi$ but $\eta \not\preceq \xi$.

In the following example it is illustrated the "irregularity" of the co–entropies (16) and (18) with respect to both quasi–orderings $\preceq$ and $\ll$. Let us stress that both the coverings are genuine.

*Example 6.* In the universe $X = \{1,2,3,4,5,6,7,8,9,10,11,12,13,14,15\}$, let us consider the two *genuine* coverings $\gamma = \{C_1 = \{1,4,5\}, C_2 = \{2,4,5\}, C_3 = \{3,4,5\}, C_4 = \{14,15\}, C_5 = \{4,5,\ldots,13\}\}$ and $\delta_1 = \{D_1 = \{1,4,5\} = C_1,\ D_2 = \{2,4,5\} = C_2,\ D_3 = \{3,4,\ldots,13,14\} = C_3 \cup C_5,\ D_4 = \{4,5,\ldots, 14,15\} = C_4 \cup C_5, \}$. Trivially, $\gamma \prec \delta_1$ and $\gamma \ll \delta_1$. In this case $E(\gamma) = 2.05838 < 2.18897 = E(\delta_1)$ and $E_\hbar(\gamma) = 1.47342 < 1.60401 = E_\hbar(\delta_1)$, as desired.

In the same universe, let us now take the *genuine* covering $\delta_2 = \{F_1 = \{1,4,5,\ldots,12,13\} = C_1 \cup C_5,\ F_2 = \{2,4,5,\ldots,12,13\} = C_2 \cup C_5,\ F_3 = \{3,4,\ldots,12,13\} = C_3 \cup C_5,\ F_4 = \{4,5,\ldots,14,15\} = C_4 \cup C_5\}$. Trivially, $\gamma \prec \delta_2$ and $\gamma \ll \delta_2$. Unfortunately, in this case we obtain $E(\gamma) = 2.05838 > 1.91613 = E(\delta_2)$ and $E_\hbar(\gamma) = 1.47342 > 0.10877 = E_\hbar(\delta_2)$.

## 4.2 The "Pointwise" Quasi–orderings on Coverings

In [11], given a covering $\gamma$ of $X$, we have introduced two possible kinds of similarity classes induced by an object $x$ of the universe $X$: the *lower granule* $\gamma_l(x) := \cap\{C \in \gamma : x \in C\}$ and the *upper granule* $\gamma_u(x) = \cup\{C \in \gamma : x \in C\}$ generated by $x$. Of course, in the case of a trivial covering the upper granule of any point $x$ is the whole universe $X$, and so this notion turns out to be "significant" in the only case of non trivial coverings.

Thus, given a covering $\gamma$ of a universe $X$, for any $x \in X$ we can define the *granular rough approximation of $x$ induced by $\gamma$* as the pair $r_\gamma(x) := \langle \gamma_l(x), \gamma_u(x) \rangle$, where $x \in \gamma_l(x) \subseteq \gamma_u(x)$. The collections $\gamma_u := \{\gamma_u(x) : x \in X\}$ and $\gamma_l := \{\gamma_l(x) : x \in X\}$ of all such granules are both coverings of $X$, called the *upper covering* and the *lower covering* generated by $\gamma$. In particular, we obtain that for any covering $\gamma$ of $X$ the following hold:

$$\gamma_l \preceq \gamma \preceq \gamma_u \quad \text{and} \quad \gamma_l \ll \gamma \ll \gamma_u$$

We can introduce now two more quasi–ordering relations on $\Gamma(X)$ defined by the following binary relations:

$$\gamma \trianglelefteq_u \delta \quad \text{iff} \quad \forall x \in X,\, \gamma_u(x) \subseteq \delta_u(x) \;\text{ and }\; \gamma \trianglelefteq_l \delta \quad \text{iff} \quad \forall x \in X,\, \gamma_l(x) \subseteq \delta_l(x)$$

In [11] we have shown that $\gamma \preceq \delta$ implies $\gamma \trianglelefteq_u \delta$, but it is possible to give an example of two coverings $\gamma, \delta$ such that $\gamma \preceq \delta$ and for which $\gamma \trianglelefteq_l \delta$ does not hold. So it is important to consider a further quasi–ordering on coverings defined as

$$\gamma \Subset \delta \quad \text{iff} \quad \delta \trianglelefteq_l \gamma \text{ and } \gamma \trianglelefteq_u \delta. \tag{21}$$

which can be equivalently formulated as:

$$\gamma \Subset \delta \quad \text{iff} \quad \forall x \in X,\, \delta_l(x) \subseteq \gamma_l(x) \subseteq (???) \subseteq \gamma_u(x) \subseteq \delta_u(x)$$

where the question marks represent an intermediate covering granule $\gamma(x)$, which is something of "hidden" in the involved structure. This pointwise behavior can be formally denoted by $\forall x \in X, \quad r_\gamma(x) := \langle \gamma_l(x), \gamma_u(x)\rangle \sqsubseteq \langle \delta_l(x), \delta_u(x)\rangle =: r_\delta(x)$. In other words, $\sqsubseteq$ means that for any point $x \in X$ the local approximation $r_\gamma(x)$ given by the covering $\gamma$ is better than the local approximation $r_\delta(x)$ given by the covering $\delta$. So equation (21) can be summarized by $\gamma \Subset \delta$ iff $\forall x \in X,\, r_\gamma(x) \sqsubseteq r_\delta(x)$ (this latter simply written in a more compact form as $r_\gamma \sqsubseteq r_\delta$).

## 5 Pointwise Lower and Upper Entropy and Co–entropy from Coverings

Making use of the lower granules $\gamma_l(x)$ and upper granules $\gamma_u(x)$ for $x$ ranging on the space $X$ for a given covering $\gamma$, it is possible to introduce two (*pointwise* defined) LX entropies (resp., co–entropies), named the lower and upper LX entropies (resp., co–entropies) respectively (LX since we generalize in the covering context the Liang–Xu approach to quantify information in the case of incomplete information systems – see [19]) according to the following:

$$H_{LX}(\gamma_j) := -\sum_{x\in X} \frac{|\gamma_j(x)|}{|X|} \log_2 \frac{|\gamma_j(x)|}{|X|} \quad \text{for} \quad j = l, u \tag{22a}$$

$$E_{LX}(\gamma_j) := \frac{1}{|X|} \sum_{x\in X} |\gamma_j(x)| \log_2 |\gamma_j(x)| \quad \text{for} \quad j = l, u \tag{22b}$$

with the relationships (and compare with the case of partitions (4)):

$$H_{LX}(\gamma_j) + E_{LX}(\gamma_j) = \frac{\sum_{x\in X} |\gamma_j(x)|}{|X|} \cdot \log_2 |X|$$

Since for every point $x \in X$ the following set theoretic inclusions hold: $\gamma_l(x) \subseteq \gamma_u(x)$, with $1 \le |\gamma_l(x)| \le |\gamma_u(x)| \le |X|$, it is possible to introduce the *rough co–entropy approximation* of the covering $\gamma$ as the ordered pair of non–negative numbers: $r_E(\gamma) := (E_{LX}(\gamma_l), E_{LX}(\gamma_u))$, with $0 \leq E_{LX}(\gamma_l) \leq E_{LX}(\gamma_u) \leq |X| \cdot \log|X|$. For any pair of coverings $\gamma$ and $\delta$ of $X$ such that $\gamma \Subset \delta$, one has that $E_{LX}(\delta_l) \leq E_{LX}(\gamma_l) \leq (???) \leq E_{LX}(\gamma_u) \leq E_{LX}(\delta_u)$, and so we have that $\gamma \Subset \delta$ implies $r_E(\gamma) \sqsubseteq r_E(\delta)$, which expresses a condition of monotonicity of lower–upper pairs of co–entropies relatively to the quasi–ordering $\Subset$ on coverings [11, 20].

As a final remark, recalling that in the rough approximation space of coverings, partitions are the crisp sets since $\pi_l = \pi = \pi_u$ for any $\pi \in \Pi(X)$, then the pointwise entropies (22a) and co–entropies (22b) collapse in the two following pointwise entropy and co–entropy:

$$H_{LX}(\pi) := -\sum_{x\in X} \frac{|\pi(x)|}{|X|} \log_2 \frac{|\pi(x)|}{|X|} \qquad E_{LX}(\pi) := \frac{1}{|X|} \sum_{x\in X} |\pi(x)| \log_2 |\pi(x)|$$

### 5.1 Pointwise Entropy and Co–entropy from Coverings: The Case of Incomplete Information Systems

Let us now consider the case of incomplete Information Systems $\mathcal{IS} = \langle X, Att, F\rangle$. For any family $\mathcal{A}$ of attributes it is possible to define on the objects of $X$ the similarity relation $\mathcal{S}_{\mathcal{A}}$:

$$x\mathcal{S}_{\mathcal{A}}y \quad \text{iff} \quad \forall a \in \mathcal{A}, \text{ either } \quad f_a(x) = f_a(y) \quad \text{or} \quad f_a(x) = * \quad \text{or} \quad f_a(y) = *.$$

This relation generates a covering of the universe $X$ through the *granules of information* (also *similarity classes*) $s_{\mathcal{A}}(x) = \{y \in X : (x, y) \in \mathcal{S}_{\mathcal{A}}\}$, since $X = \cup\{s_{\mathcal{A}}(x) : x \in X\}$ and $x \in s_{\mathcal{A}}(x) \neq \emptyset$. In the sequel this covering will be denoted by $\gamma(\mathcal{A}) := \{s_{\mathcal{A}}(x) : x \in X\}$ and their collection by $\Gamma(\mathcal{IS}) := \{\gamma(\mathcal{A}) \in \Gamma(X) : \mathcal{A} \subseteq Att\}$. With respect to this covering $\gamma(\mathcal{A})$, and in analogy with (22), the two pointwise LX entropy and co–entropy are (see also [19]):

$$H_{LX}(\gamma(\mathcal{A})) := -\sum_{x\in X} \frac{|s_{\mathcal{A}}(x)|}{|X|} \log_2 \frac{|s_{\mathcal{A}}(x)|}{|X|} \tag{23a}$$

$$E_{LX}(\gamma(\mathcal{A})) := \frac{1}{|X|} \sum_{x\in X} |s_{\mathcal{A}}(x)| \log_2 |s_{\mathcal{A}}(x)| \tag{23b}$$

with the relationships:

$$H_{LX}(\gamma(\mathcal{A})) + E_{LX}(\gamma(\mathcal{A})) = \frac{\sum_{x\in X} |s_{\mathcal{A}}(x)|}{|X|} \cdot \log_2 |X|$$

This co–entropy (23b) behaves monotonically with respect to the quasi orderings $\preceq$ and $\ll$ [20].

## 6 Conclusions

We have discussed the role of co–entropy, as a measure of granularity, and of entropy, as a measure of uncertainty, in the context of partitions of a finite universe, with a particular interest to the monotonic and anti–monotonic behavior with respect to the standard ordering on partitions. The local measure of rough granularity is then applied to the quantitative valuation of the rough approximation of a generic subset of the universe, always related to the monotonicity property.

The extension of this approach to coverings, also in the best conditions of genuineness, shows the drawback of a pathological behavior of the global co–entropy (and so also entropy) with respect to two natural extension of quasi–orderings. A pointwise version of co–entropy and entropy, on the contrary, has the expected monotonic behavior.

## Acknowledgements

The author's work has been supported by MIUR\PRIN project "Automata and Formal languages: mathematical and application driven studies".

## References


1. Cattaneo, G.: Abstract approximation spaces for rough theories. In: Polkowski, L., Skowron, A. (eds.) Rough Sets in Knowledge Discovery 1, pp. 59–98. Physica–Verlag, Heidelberg (1998)
2. Pawlak, Z.: Rough sets. Int. J. Inform. Comput. Sci. 11, 341–356 (1982)
3. Cattaneo, G., Ciucci, D.: Investigation about Time Monotonicity of Similarity and Preclusive Rough Approximations in Incomplete Information Systems. In: Tsumoto, S., Słowiński, R., Komorowski, J., Grzymała-Busse, J.W. (eds.) RSCTC 2004. LNCS (LNAI), vol. 3066, pp. 38–48. Springer, Heidelberg (2004)
4. Pawlak, Z.: Rough sets: A new approach to vagueness. In: Zadeh, L.A., Kacprzyc, J. (eds.) Fuzzy Logic for the Management of Uncertainty, pp. 105–118. J. Wiley and Sons, New York (1992)
5. Taylor, A.: General Theory of Functions and Integration. Dover Publications, New York (1985)
6. Khinchin, A.I.: Mathematical Foundations of Information Theory. Dover Publications, New York (1957) (translation of two papers appeared in Russian in *Uspekhi Matematicheskikh Nauk* 3, 3–20 (1953) and 1, 17–75 (1965)
7. Hartley, R.V.L.: Transmission of information. The Bell System Technical Journal 7, 535–563 (1928)
8. Ash, R.B.: Information Theory. Dover Publications, New York (1990) (originally published by John Wiley & Sons, New York, 1965)
9. Reza, F.M.: An Introduction to Information theory. Dover Publications, New York (1994) (originally published by Mc Graw-Hill, New York, 1961)
10. Shannon, C.E.: A mathematical theory of communication. The Bell System Technical Journal 27, 379–423, 623–656 (1948)
11. Bianucci, D., Cattaneo, G., Ciucci, D.: Entropies and co–entropies of coverings with application to incomplete information systems. Fundamenta Informaticae 75, 77–105 (2007)
12. Wierman, M.: Measuring uncertainty in rough set theory. International Journal of General Systems 28, 283–297 (1999)

13. Liang, J., Shi, Z.: The information entropy, rough entropy and knowledge granulation in rough set theory. International Journal of Uncertainty, Fuzziness and Knowledge-Based Systems 12, 37–46 (2004)
14. Beaubouef, T., Petry, F.E., Arora, G.: Information–theoretic measures of uncertainty for rough sets and rough relational databases. Journal of Information Sciences 109, 185–195 (1998)
15. Pawlak, Z.: Rough sets: Theoretical Aspects of Reasoning about Data. Kluwer Academic Publishers, Dordrecht (1991)
16. Pawlak, Z.: Information systems - theoretical foundations. Information Systems 6, 205–218 (1981)
17. Komorowski, J., Pawlak, Z., Polkowski, L., Skowron, A.: Rough sets: A tutorial. In: Pal, S., Skowron, A. (eds.) Rough Fuzzy Hybridization, pp. 3–98. Springer, Singapore (1999)
18. Birkhoff, G.: Lattice Theory. American Mathematical Society, Providence, Rhode Island. American Mathematical Society Colloquium Publication, 3rd edn., vol. XXV (1967)
19. Liang, J., Xu, Z.: Uncertainty measure of randomness of knowledge and rough sets in incomplete information systems. Proc. of the 3rd World Congress on Intelligent Control and Automata 4, 2526–2529 (2000)
20. Bianucci, D., Cattaneo, G.: Monotonic behavior of entropies and co-entropies for coverings with respect to different quasi-orderings. LNCS (LNAI), vol. 4585, pp. 584–593. Springer, Heidelberg (to appear, 2007)

# Patterns of Collaborations in Rough Set Research

Zbigniew Suraj[1,2] and Piotr Grochowalski[1]

[1] Chair of Computer Science, Rzeszów University, Poland
{zsuraj,piotrg}@univ.rzeszow.pl
[2] Institute of Computer Science, State School of Higher Education in Jarosław, Poland

**Summary.** In this chapter we look at some details of the structure of the collaboration graph for the rough set researchers, discuss the entire graph as it exists at the present time and study its evolution over the past 25 years. Our approach is more experimental and statistical rather than theoretical. It seems that these data are interesting in their own right as a reflection of the way in which the rough set research is done, apart from the mathematical questions raised about how to model and analyze social interactions.

**Keywords:** rough sets, pattern recognition, database systems, collaboration graph.

## 1 Introduction

Each year the rough set researchers publish approximately more than one hundred and twenty research papers. Since 2003, the editors of the Rough Set Database System (RSDS, in short) [8] which is available electronically at the web site under the address http://rsds.univ.rzeszow.pl, catalogued most of them, and the RSDS's current database contains almost three thousands and three hundred items, produced by more than one thousand five hundred and sixty authors. The data used in this article cover the period from 1981 to 2005, inclusive. By studying this wealth of data, we can discern some interesting patterns of publications, and in particular some interesting patterns of collaboration.

To get at the social phenomenon of collaboration in the rough set research, we have constructed the so-called collaboration graph. The vertices of the graph are all authors in our database, and two vertices are joined by an edge if two authors have published a joint paper.

In this chapter we look at some details of the structure of the collaboration graph for the rough set researchers and discuss some of its properties. The collaboration graph (as well as the mathematical research collaboration graph [4],[5] and other social networks studied in the literature [2],[3],[7],[10]) exhibits several interesting features. Firstly, although the number of edges is fairly small (just a little larger than the number of vertices), the average path length between the vertices in the same component is small. Furthermore, there is a "giant component" of the graph that encompasses a little more than one third of all authors, and the remaining components are very tiny. Secondly, the clustering coefficient

R. Bello et al. (Eds.): Granular Computing, STUDFUZZ 224, pp. 79–92, 2008.
springerlink.com 

is rather small. The *clustering coefficient* [7] of a graph is defined as the fraction of ordered triples of the vertices *a, b, c* in which the edges *ab* and *bc* are present that have the edge *ac* present. Intuitively speaking, how often are two neighbors of a vertex adjacent to each other? In this context, the question is [4]: "What model of a random graphical evolution will produce graphs with these (and other) properties of the collaboration graph?" These and other questions will be discussed in this chapter.

All the analysis was made by special programmes developed by us. The theoretical foundations for such computations have been described, among others, in [7]. Owing to the programmes we can make the analysis in a dynamic way which means everything in the system is taken into consideration immediately after it appears in the database, the parameters are calculated up to date.

The chapter is organized as follows. Sect. 2 provides the basic information about the RSDS data as well as the construction of the collaboration graph. In Sect. 3, some properties of the collaboration graph are discussed. The evolution of the collaboration graph over time is presented in Sect. 4. Sect. 5 is devoted to open questions and directions for future work.

## 2 The Description of the RSDS System

The RSDS system has been created in order to catalogue and facilitate the access to information about publications concerning the rough sets and similar branches. It is available at: http://rsds.univ.rzeszow.pl. The access to the system is free. In order to start the system it is necessary to have any computer with an operation system and the Internet plugged in, the web browser servicing the JavaScript, cookies and frames. It contains 3266 publications segregated according to 12 types, i.e. articles, books, etc. The functionality of the system looks as follows:

- adding new data on-line or in an automatic way,
- edition of existing data,
- registration of users in the system,
- searching for different information,
- storing the data to a file in a chosen format,
- sending the files with the data to an administrator,
- descriptions of applications using the rough set theory,
- scientific biographies of the people devoted to the development of the rough set theory,
- the module of a classification of scientific publications according to a designed classificator,
- the module of a graph-statistical analysis of the content of the system,
- an interactive map of the world, showing who and where in the world works on the rough sets,
- service of the comments of the users,
- help.

Below we are going to present a detailed description of the main (most important) functionalities of the system: adding new data to the system and searching for information. The descriptions of other possibilities of the system can be found in publications [8], [9].

The system has been equipped with a menu which allows for moving around the whole system. The menu includes the main functionalities of the system which after being chosen let a user go to options characteristic to a given functionality.

### 2.1 Adding New Data to the System

If you want to add new data into the system you have to go to the section *Append*. This section is available only for the users registered in the system. In order to register into the system you have to fill in a form available in the section *Login*. If a user logs in into the system for the first time, he has to fill in a registration form available in the section *First Login* (after pressing the key *First Login*). In this form the user gives personal data. When the form has been filled in correctly and the id and the password has been defined, the user is automatically logged in into the system. On the other hand, when the user is already registered in the system and wants to log in, then, in the section *Login* he fills in the form containing the id and the password, and when they are written correctly he is logged into the system. Next, the section *Append* becomes activated for the user. For the sake of safety the system automatically remembers which publications have been added by a given user. This information is also used when the data is edited. Adding new bibliographical descriptions has been divided into two phases (see Fig. 1):

- During the first phase the user defines information describing a given publication, which is demanded by the system *BibTeX* specification, and corresponding to a particular type of a publication.

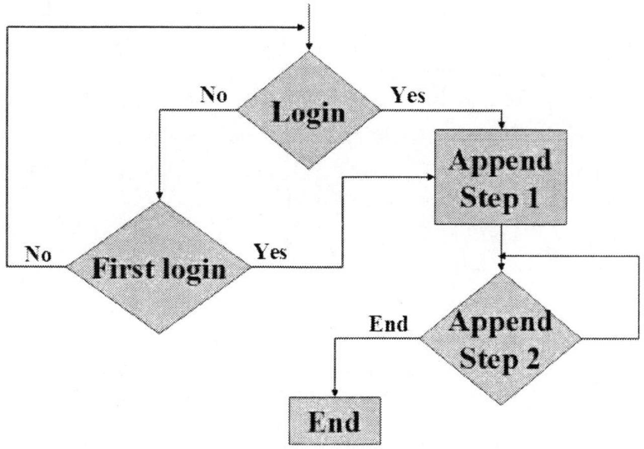


**Fig. 1.** Scheme for operation of adding data online

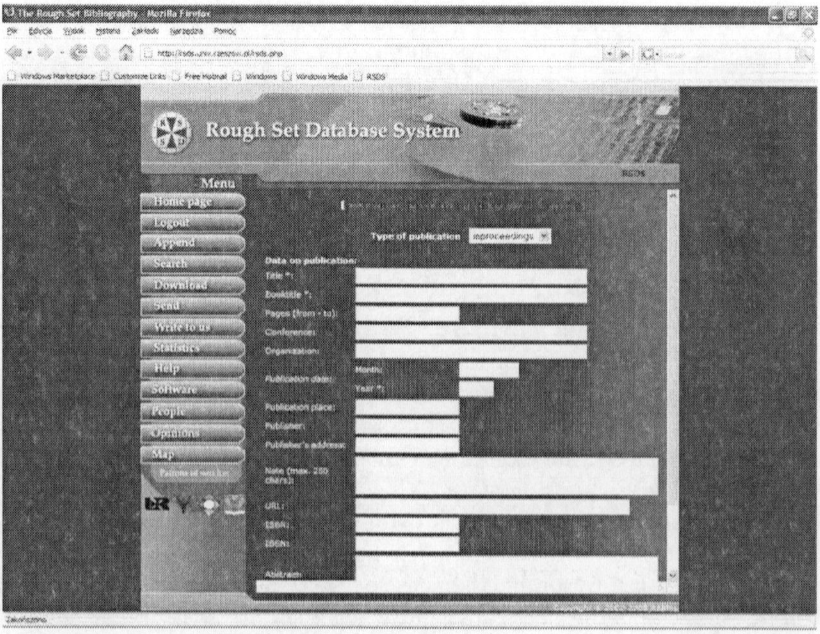

**Fig. 2.** A screenshot for appending (online) new data to the RSDS system

- During the second phase one defines information connected with authors or editors of a given publication.

At the beginning of introducing the data describing a publication the user defines the type of a publication. Depending on a chosen type, a form is generated which contains the data used for describing a given publication, i.e. a title, an editor, a year of publishing, a publishing etc. The data necessary for describing a given type are marked with an asterisk (*) (see Fig. 2). After the data has been introduced and accepted, the user is directed to the phase of introducing information about the authors/editors of a given publication. During this step, one has to introduce the authors/editors one by one, regardless of the number of them. This step will be repeated as long as the user decides he has introduced all data and accepts the whole process by pressing the key *End.* After being accepted, the data is sent to the database of the system.

### 2.2 Searching for Information

In order to search for information one has to use the section *Search* (see Fig. 3). In this section the following ways of searching has been detached: alphabetical, advanced ver.1, advanced ver.2. In the alphabetical way if searching we can distinguish searching according to: titles, authors, editors, conferences, journals, years of publications. Each of the subcategories has been adequately prepared, in order to facilitate and shorten the time of searching by the user. In searching according to:

- *Titles*, an alphabetical list of titles has been divided into successive years of publishing.
- *Authors*, the list of publications for particular authors has also been divided according to the years of publishing. In addition, the list of co-authors is being built for every author (see Fig. 4). For every author personal information (if available) has been added marked by icons (a magnifying glass, an envelope, a house).
- *Editors*, the list of publications for particular editors has also been divided into the years of publications.
- *Conferences*, in this subcategory there have been distinguished the main names of conferences and put in an alphabetical order. After choosing a particular name, the list according to the years is displayed and after choosing one year one gets the access to the publication connected with a given conference taking place in a chosen year.
- For *journals* the list of publications has been prepared in such a way, that each of the magazines has been divided according to years, which include successive numbers of magazines with adequate publications assigned.
- *Years of publishing*, in this subsection there is a division of publications according to particular years of publishing.

In the subcategories every list is being built in a dynamic way, i.e. every change in the system causes the change in the list. When the user finds an adequate publication, he has a possibility of getting its description in two formats:

- *HTML* - this is the format of displaying publications in the system (without the possibility of generating the description files),
- *BibTeX* - this is the format of displaying publications generated after clicking the link *BibTeX*. It has the possibility of generating the description files.

After generating a description of a publication in the *BibTeX* format, the possibility of adding the received description to the file created by the user or downloading the created file is activated. For the comfort of the user, two methods of downloading a created file have been implemented:

- Saving data directly on a local hard disc of the user's computer.
- Sending the file by an e-mail to a given address.

If a given user is logged into the system, he can see the link *Edit* used for editing the introduced data, next to the publications added by himself. In order to edit data one has to use a dedicated form in which one can change data, and after accepting the changes they are sent to an administrator. The user who has the privileges of the administrator of the system is able to delete repeated data.

The advanced searching ver.1 allows to search for publications according to complex conditions of searching based on the elements of a description of a publication and on the logical operators OR or AND.

The advanced searching ver.2 allows to search for publications on the basis of a defined classificator (see [9]). The primary version of the browser groups the publications into particular groups. The publications must fulfill the searching

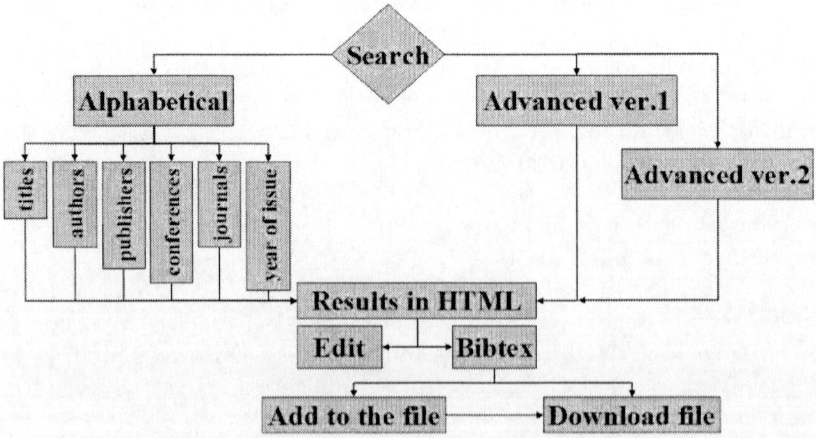


**Fig. 3.** Scheme for operation of searching for data

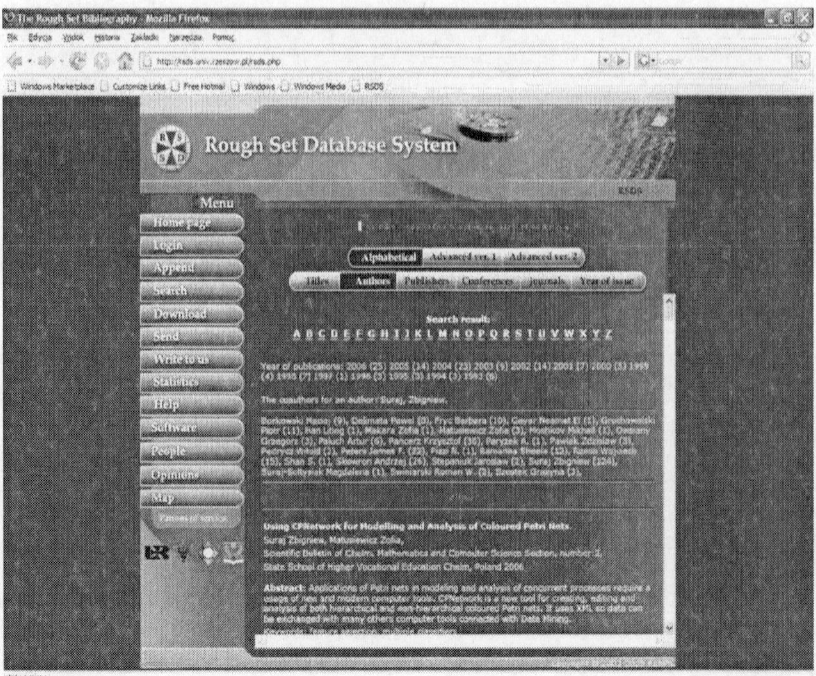


**Fig. 4.** A screenshot for alphabetical search

conditions on the basis of information added to every description. The information reflects the assignment of a given publication to a domain classificator, which has been defined by us.

## 3 The RSDS Data and the Construction of the Collaboration Graph

The current database of the RSDS contains 3079 authored items (mainly research papers), written by 1568 different authors. All publications included in the database can be classified as: article - 813 items, book - 125 items, inbook - 17 items, incollection - 282, inproceedings - 1598, manual - 2 items, masterthesis - 12, phdthesis - 14, proceedings - 67, techreport - 148 items, unpublished - 1 item. For more information about the definitions of the shorts mentioned above, we refer the reader to [6] or the web site of the RSDS at http://rsds.univ.rzeszow.pl. The authors of publications included into the database come mainly from the following countries: Canada, China, Finland, France, India, Ireland, Italy, Japan, the Netherlands, Norway, Poland, Romania, Russia, Spain, Sweden, Taiwan, the USA and the United Kingdom.

For simplicity, we call each authored item in the RSDS database a "publication", although some of them are monographs of various kinds. We ignore nonauthored items in the database such as conference proceedings - the relevant papers in the proceedings have their own entries as authored items. Moreover, we have omitted sixty items included in the database because their issue years are not known. In our analysis, we have also omitted one publication from the period 1971-1980 and sixty publications from 2006. Thus, the final number of publications which is taken into consideration in the following is equal to three thousand and twelve.

The data used in this article cover approximately the period from 1981 to 2005, inclusive, and we have broken it down approximately by a given five-year period.

The cumulative data record to the end of a given five-year period is summarized in Table 1, whose integer entries represent hundreds. The rightmost column includes all the data, and the remaining columns truncate the data after one or more five-year periods. The data are given for all authors, as well as just for authors who have collaborated.

The fourth row of Table 1 shows the average number of publications per author. The mean number of publications is about 6. The data distribution has a very long right tail, with a standard deviation of more than 76.

This database of authored items gives rise to the collaboration graph denoted by $C$, which has the set of authors as its vertices, with two authors adjacent if they are among the authors of some paper - in other words, if they have published a joint paper, with or without other co-authors. Using 25 years of data from our database, we find that this graph after verification currently has about 1456 vertices and 2466 edges.

We corrected a few anomalies in $C$ by hand before analyzing it. For example, we removed the author that in the RSDS database is identified as "et al.", who was on the author list of a number of papers, including one with no co-authors. Depending on some experience with our database over the past several years, we are confident that problems of this kind do not significantly distort the true image of the collaboration graph.

**Table 1.** The cumulative data record to the end of a given five-year period

| Year of completion | 1981-1985 | 1981-1990 | 1981-1995 | 1981-2000 | 1981-2005 |
|---|---|---|---|---|---|
| Number of publications | 35 | 121 | 610 | 1612 | 3012 |
| Number of authors | 12 | 47 | 238 | 674 | 1568 |
| Mean publications/author | 4 | 3.87 | 4.18 | 6.62 | 5.63 |
| Std. dev. publication/author | 4.93 | 5.25 | 8.64 | 62.84 | 76.75 |
| Mean authors/publication | 0.37 | 0.5 | 0.63 | 0.79 | 0.95 |
| Std. dev. authors/publication | 0.76 | 0.83 | 0.97 | 1.09 | 1.19 |
| Percentage share of publications with $n$ co-authors | | | | | |
| Luck of authors (a publication under edition only) | 2.86% | 2.48% | 5.41% | 5.4% | 5.91% |
| $n = 0$ | 68.57% | 58.68% | 47.38% | 41.44% | 34.99% |
| $n = 1$ | 20% | 28.93% | 33.28% | 32.69% | 31.81% |
| $n = 2$ | 5.71% | 6.61% | 8.52% | 12.84% | 17.23% |
| $n > 2$ | 2.86% | 3.31% | 5.41% | 7.63% | 10.06% |
| Number of authors sharing common publications | 12 | 36 | 195 | 591 | 1456 |
| Their percentage share | 100% | 76.6% | 81.93% | 87.69% | 92.86% |
| Mean co-authors/author | 2.17 | 2.13 | 2.38 | 2.82 | 3.15 |
| Mean co-authors/author sharing a common publication | 2.17 | 2.78 | 2.9 | 3.21 | 3.39 |

## 4 The Properties of the Collaboration Graph

To really get at the social phenomenon of collaboration in the rough set research, we have constructed the collaboration graph $C$, which has 1568 vertices and 2466 edges. The average degree of a vertex in $C$ (the average number of co-authors per a rough set researcher) is about 3. There are 112 isolated vertices (the number of vertices not joined with any edge to other vertices) in $C$ (7%) where is the number of authors who do not collaborate with other authors, which we will ignore for the purposes of this analysis. After all, these are not collaborating rough set researchers. That leaves 1456 vertices with a degree of at least 1. Viewed in this way, the average degree (number of co-authors for a rough set researcher who collaborates) is about 4.

Let us first look at the degrees of the vertices - the distribution of the numbers of co-authors the rough set researchers have. The data show that 23% of the collaborating rough set researchers have just one co-author, 27% have two, 20% have three, 13% have four, and 17% have five or more. More than 15 rough set researchers have written with more than 20 colleagues apiece, with Andrzej Skowron's 56 co-authors as the most extreme case. Again, the social interactions have increased over the years, no doubt due to electronic communication and the proliferation of conferences; Table 1 shows that the mean number of collaborators per rough set researcher in one half of a decade grew from about 2 in the 1980s to more than 3 in the 2005s.

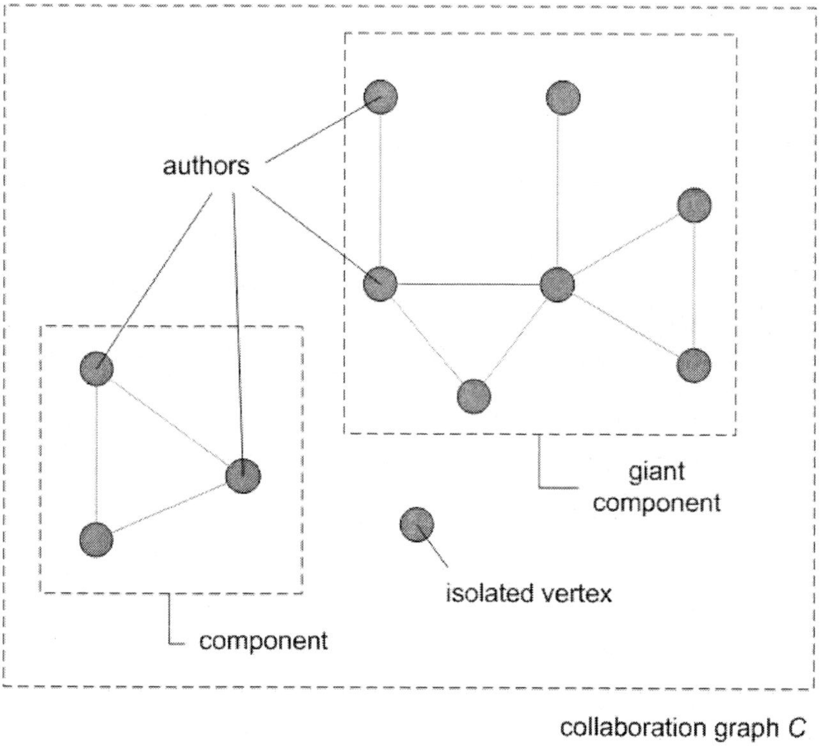


**Fig. 5.** The structure of the exemplary collaboration graph

Other graphical properties of $C$ also provide the insight into the interconnectedness of the rough set researchers [7]. For example, the collaboration graph has one giant component (the largest consistent subgraph, i.e. the graph in which for every vertex there is a path to any other vertex) with 574 vertices and 1258 edges; the remaining 882 nonisolated vertices and 1208 edges split into 227 components, having from 2 to a maximum of 27 vertices. The components of the graph present the groups of authors collaborating with each other. These groups can include people closely collaborating with each other, or authors can be included in the groups thanks to people who collaborate closely. These components can also be used to define who ought to be contacted in order to get to a particular author. The structure of the exemplary collaboration graph is presented in Fig. 5.

Next, we concentrate just on the giant component of $C$ and consider a distribution of distances between the vertices (number of edges in the shortest path joining the vertices). The average distance between two vertices is 4.54, with a standard deviation of about 1.41. A distance between two vertices (authors) in a given group means that if we make a sphere from a given vertex (author), with a radius equal to an average distance between two vertices, we would receive information about people who closely collaborate with this particular author.

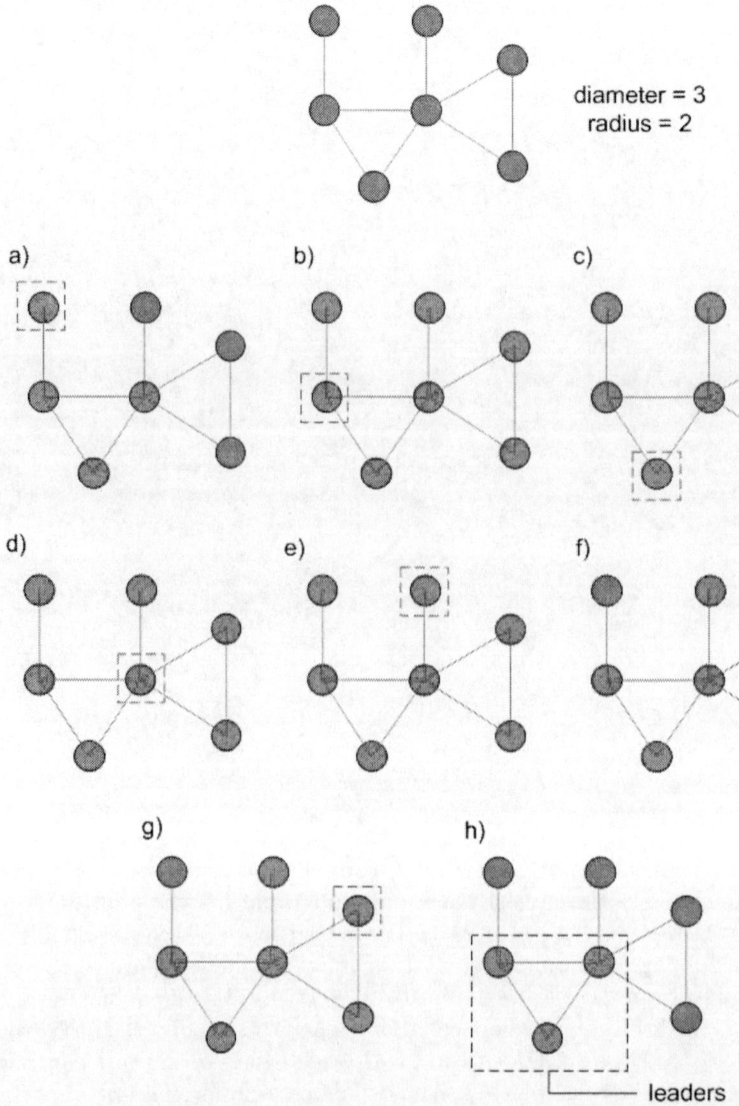


**Fig. 6.** The stages of the exemplary process of appointing the leader (leaders) of the group

The diameter of the giant component (maximum distance between two vertices) is 10, and the radius (minimum eccentricity of a vertex, with an eccentricity defined as the maximum distance from that vertex to any other) is 6. The diameter denotes how far from a particular author there is a person the least closely related to him, i.e. how far are the people who work for the group the least. The radius denotes people from the "first ranks", i.e. the people who collaborate most closely. However, if from every vertex of a component we made a sphere with a radius equal to a radius of the component, then, all these spheres would have a common part on some vertex (vertices). The vertex (vertices) from a common part denotes (denote) a leader (leaders) of a given group. Finding the authors who are a diameter far from the leader, means finding the "satellites" of a given group, while finding the authors who are a radius far from the leader means finding the very first ranks of a given group (see Fig. 6).

As a final measure, we compute the clustering coefficient of $C$ to be $3,88 \cdot 10^{-6}$. In other words, how often are two neighbours of a vertex adjacent to each other? That is 2000 times higher than one would expect for a traditional random graph with 1568 vertices and 2466 edges, another indication of the need for better models [3].

## 5 The Evolution of the Collaboration Graph over Time

Tables 2 and 3 give various statistics on the publication habits of the rough set researchers over time, organized roughly into given five-year periods (all throughout the chapter).

Table 2 shows the percentage share of authors with a given number of publications who have written various numbers of papers. It can be seen from this table that just slightly more than one third of all publishing rough set researchers have published more than one publication, and that almost two thirds of us have written only one publication, and that about one tenth of authors have written more than five publications. The other extreme is that, four people have written more than 140 publications apiece, including Andrzej Skowron with 269 publications.

Table 3 summarizes the data for a given five-year period, giving a better view of how things have changed over years. The third row of the table shows the explosion in the number of practicing rough set researchers during the period

**Table 2.** The percentage share of authors with a given number of publications

| Number of publications | Percentage of authors |
|---|---|
| 1 | 62.37% |
| 2 | 13.33% |
| 3 | 7.23% |
| 4 | 2.49% |
| 5 | 1.99% |
| 6-10 | 4.55% |
| 11-20 | 2.93% |
| 21-50 | 1.56% |
| 51-100 | 0.81% |
| 101-200 | 0.25% |
| $> 200$ | 0.06% |

**Table 3.** The data record for a given five-year period

| In years | 1981-1985 | 1986-1990 | 1991-1995 | 1996-2000 | 2001-2005 |
|---|---|---|---|---|---|
| Number of publications | 35 | 86 | 489 | 1002 | 1400 |
| Number of authors | 12 | 45 | 219 | 540 | 1089 |
| Mean publications/author | 4 | 2.98 | 3.71 | 5.29 | 4.02 |
| Std. dev. publication/author | 4.93 | 3.39 | 7.01 | 43.36 | 42.74 |
| Mean authors/publication | 0.37 | 0.56 | 0.66 | 0.89 | 1.14 |
| Std. dev. authors/publication | 0.76 | 0.84 | 1 | 1.14 | 1.27 |
| Percentage share of publications with $n$ coauthors | | | | | |
| Luck of authors (a publication under edition only) | 2.86% | 2.33% | 6.13% | 5.39% | 6.5% |
| $n = 0$ | 68.57% | 54.65% | 44.58% | 37.82% | 27.57% |
| $n = 1$ | 20% | 32.56% | 34.36% | 32.34% | 30.79% |
| $n = 2$ | 5.71% | 6.98% | 9% | 15.47% | 22.29% |
| $n > 2$ | 2.86% | 3.49% | 5.93% | 8.98% | 12.86% |
| Number of authors sharing common publications | 12 | 33 | 180 | 480 | 1021 |
| Their percentage share | 100% | 73.33% | 82.19% | 88.89% | 93.76% |
| Mean coauthors/author | 2.17 | 1.96 | 2.27 | 2.76 | 3.05 |
| Mean coauthors/author sharing a common publication | 2.17 | 2.67 | 2.77 | 3.1 | 3.26 |

we consider. We infer from the fourth row of Table 3 that in 1980s the mean number of publications per author was 3, that this figure grew to more than 5 in the period 1996-2000, and next, that it reached about 4 in 2005.

As Table 3 shows, the average number of authors per publication has gone from almost 0.4 in 1980s to more than 1 in 2005. During 1980s 26% of all publishing rough set researchers wrote joint papers, whereas 94% of those who published in 2005 collaborated at least once during a given five-year period. In the 1980s, nearly 69% of all papers were solo works, with only 3% of papers having three or more authors. If we look once again at the items in the database, we find that by the early 2000s, less than 32% of all publications had just one author, and the number of publications with three or more authors had grown to about 13%.

## 6 Open Questions and Directions for Future Work

The Rough Set Database System provides a wonderful opportunity for further study of publishing patterns of the rough set researchers, both as individuals and as a highly and intricately connected corpus. For instance, it would be interesting to look at the bipartite graph $B$, whose vertices of one type are the papers and vertices of the other type are the authors, with an edge between a paper and each of its authors, and study such things as the number of papers the rough set researchers write, and when in their careers they write them; or turn the tables and look at the "collaboration graph" of papers, rather than authors. We can also analyze the subgraphs of $C$ restricted to various branches (subfields) of the rough set theory and its application or specific subjects. Moreover, we can study the differences among the rough set researchers in different subfields, in order to see the extent to a person's publication record over the first six years gives an indication of future productivity, or to notice significant differences in publication or collaboration patterns among the rough set researchers at different types of institutions or in different countries as well as to provide some comparisons with the suitable characteristics concerning the mathematical research (see e.g. [4]).

We can ask many different questions when examining the patterns of collaboration and try to find the answers. For instance, what are the common elements of the rough set society influencing the increase of collaboration? Among different pieces of information which we obtain thanks to our research it turnes out that the development of collaboration is definitely influenced by the actual trends in research, the exchange (in the wide meaning of this word) of information about the conducted research by particular authors, which results in the number of workshops, conferences etc.

Is it possible to forsee the future productivity of a given author on the basis of the analysis of created publications? In our opinion, it is not possible to forsee the future productivity of a given author on the basis of information about the number of publications in the previous years, because the knowledge about the author that we possess is not sufficient to come to such conclusions. In order to define such predictions we would need additional information describing e.g. the author himself, information that can be parameterized such as age etc. Besides, most authors, which is shown in the analysis (see Table 2), have created a small number of publications (less than to 10), and from such number of publications it is difficult to make any conclusions for the future. In spite of these facts, we have tried to forsee the productivity of the authors on the basis of information we possess. Unfortunately, at the moment we are not able to give any information or tell if it will bring a demanded effect.

Which of the subdomains of the rough set theory influence the collaboration between the authors the most? In order to find the answer to this question we would have to modify the existing collaboration graph or define a new one on the basis of information included in the classificator defined by us (see [9]). This classificator allows to describe every publication, in a formal way to which subdomain of the rough set theory it can be classified considering the problems we presented. This compared to authors gives information about the subdomains in which particular authors create their works. Having such information it is possible to answer the question asked before, however, because in the system only a small number of publications have been classified in the way presented above, it is impossible to answer the question at the moment. On the other hand, bibliographical information such as a title, a year of publishing etc. is not sufficient to classify automatically which subdomains a given publication belongs to. Such a decision can be made by an author (authors) of a publication or by a person who knows the content of the work so as to avoid the mistakes, and the process is considerably long.

## Acknowledgment

The authors wish to thank their colleagues from the Research Group on Rough Sets and Petri Nets for their help in searching for data. Their deepest thanks go to Katarzyna Garwol from Rzeszów University and Iwona Pituch from University of Information Technology and Management in Rzeszów for their support in the creation of the RSDS system. The research has been partially supported by the grant 3 T11C 005 28 from the Ministry of Scientific Research and Information Technology of the Republic of Poland.

## References

1. Aiello, W., Chung, F., Lu, L.: A random graph model for power law graphs. Experimental Mathematics 10, 53–66 (2001)
2. Barabasi, A.L.: Linked: The New Science of Networks. Perseus, New York (2002)
3. Buchanan, M.: Nexus: Small Worlds and the Groundbreaking Science of Networks. W.W. Norton, New York (2002)
4. Grossman, J.W.: Patterns of Collaboration in Mathematical Research. SIAM New 35(9) (2002)
5. Grossman, J.W.: The Evolution of the Mathematical Research Collaboration Graph (manuscript)
6. Lamport, L.: LaTeX: A Document Preparation System. Addison-Wesley, Reading (1986)
7. Newman, M.E.J., Strogatz, S.H., Watts, D.J.: Random graphs with arbitrary degree distributions and their applications. Phys. Rev. E. 64 (2001)
8. Suraj, Z., Grochowalski, P.: The Rough Set Database System: An Overview. In: Peters, J.F., Skowron, A. (eds.) Transactions on Rough Sets III. LNCS, vol. 3400, pp. 190–201. Springer, Heidelberg (2005)
9. Suraj, Z., Grochowalski, P.: Functional Extension of the RSDS System. In: Greco, S., Hata, Y., Hirano, S., Inuiguchi, M., Miyamoto, S., Nguyen, H.S., Słowiński, R. (eds.) RSCTC 2006. LNCS (LNAI), vol. 4259, pp. 786–795. Springer, Heidelberg (2006)
10. Watts, D.J., Strogatz, S.H.: Collective dynamics of "small-world" networks. Nature 393, 440–442 (1998)

# Visualization of Local Dependencies of Possibilistic Network Structures

Matthias Steinbrecher and Rudolf Kruse

Department of Knowledge Processing and Language Engineering
Otto-von-Guericke University of Magdeburg
Universitätsplatz 2, 39106 Magdeburg, Germany
kruse@iws.cs.uni-magdeburg.de

**Summary.** In this chapter an alternative interpretation of the parameters of a Bayesian network motivates a new visualization method that allows for an intuitive insight into the network dependencies. The presented approach is evaluated with artificial as well as real-world industrial data to justify its applicability.

## 1 Introduction

The ever-increasing performance of database systems enables today's business organizations to collect and store huge amounts of data. However, the larger the data volumes grow the need to have sophisticated analyzation methods to extract hidden patterns does alike. The research area of Data Mining addresses these tasks and comprises intelligent data analysis techniques such as classification, prediction or concept description, just to name a few.

The latter technique of concept description tries to find common properties of conspicuous subsets of given samples in the database. For example, an automobile manufacturer may plan to investigate car failures by identifying common properties that are exposed by specific subsets of cars.

Good concept descriptions should have a reasonable length, i. e., they must not be too short in order not to be too general. Then again, long descriptions are too restrictive since they constrict the database samples heavily, resulting in only a few covered sample cases. Since we have to assume that the database entries expose hundreds of attributes, it is essential to employ a feature selection approach that reduces this number to a handy subset of significant attributes.

In this chapter, we assume the database entries having nominal attributes[1] with one distinguished attribute designating the class of each data sample. We will use probabilistic and possibilistic network induction methods to learn a dependence network from the database samples. Further, we only draw our attention to the class attribute and its conditioning attributes, which are its direct parents in the network, i. e., the subset of attributes that have a direct arc connecting it with the class attribute. Since most network induction algorithms allow

[1] For the treatment of metric attributes, a discretization phase has to precede the analysis task.

R. Bello et al. (Eds.): Granular Computing, STUDFUZZ 224, pp. 93–104, 2008.
springerlink.com 

for the restriction of the number of parent attributes to some upper bound, we are in a favorable position to control the length of the concept descriptions to be generated.

We then show that the network structure alone does not necessarily provide us with a detailed insight into the dependencies between the conditioning attributes and the class attribute. Emphasis is then put on the investigation of the network's local structure, that is, the entries of its potential tables. Finally, a new visualization method for these potential tables is presented and evaluated.

The remainder of this chapter is structured as follows: Section 2 presents a brief review of the methods of probabilistic and possibilistic networks, mostly for introducing the nomenclature used in the following sections. In section 3 arguments for the importance of visualizing the network parameters are produced. This will lead to a concrete application and analysis in section 4. The chapter concludes with section 5, giving an outlook of intended further investigations.

## 2 Background

For the formal treatment of sample cases or objects of interest, we identify each sample case with a tuple $t$ that exposes a fixed number of attributes $\{A_1, \ldots, A_n\}$, each of which can assume a value with the finite respective domain $\mathrm{dom}(A_i) = \{a_{i1}, \ldots, a_{ir_i}\}, i = 1, \ldots, n$. Let $\Omega$ denote the set of all possible tuples, then we can model a database $D$, which constitutes the starting point of analysis, as a weight function $w_D : \Omega \to \mathbb{N}$ that assigns to each tuple $t \in \Omega$ the number of occurences in the database $D$. The total number of tuples or sample cases in $D$ is $N = \sum_{t\in\Omega} w_D(t)$. The fact $w_D(t) = 0$ states, that the tuple $t$ is not contained in $D$. With this definition, the weight function can be considered an extended indicator function: The respective indicator function $\mathbb{1}_D$ would be defined as

$$\forall t \in \Omega : \mathbb{1}_D(t) = \min\{w_D(t), 1\}.$$

From $w_D$ we can derive the following probability space $\mathbb{P}_D = (\Omega, \mathcal{E}, P)$ with the components defined as follows:

$$\mathcal{E} = 2^{\Omega},$$
$$\forall t \in \Omega : p(t) = \tfrac{w_D(t)}{N} \quad \text{and} \quad \forall E \in \mathcal{E} : P(E) = \textstyle\sum_{t\in E} p(t)$$

In the following, we only have one database at the time, so we drop the index $D$ and refer simply to $w$ as the source of all information and assume the space $\mathbb{P}_D$ to be the implicit probability space underlying all consequent probabilistic propositions. Therefore, a given database of sample cases represents a joint probability distribution. Even though the number of tuples in the database is small compared to $|\Omega|$, we have to look for means of further reducing the size of the joint distribution.

One prominent way are *Graphical Models*, which can be destinguished further between Markov Networks [10] and Bayesian Networks [12], the latter of which is introduced in the next section.

### 2.1 Bayesian Networks

From the database oriented point of view, reducing one large, high dimensional database table can be accomplished by decomposing it into several lower dimensional subtables. Under certain conditions one can reconstruct the initial table using the natural join operation. These certain conditions comprise the *conditional relational independence* between the attributes in the initial table. Attributes $A$ and $B$ are relationally independent given a third attribute $C$, if once any value of $C$ is held fixed, the values of $A$ and $B$ are freely combinable.

The probabilistic analog consists of decomposing the high dimensional joint probability distribution into multiple distributions over (overlapping) subsets of attributes. If these sets of attributes are *conditionally probabilistically independent* given the instantiations of the attributes contained in the overlap, a lossless reconstruction of the original joint distribution is possible via the chain rule of probability:

$$\forall \tau \in \mathcal{S}_n : P(A_1, \ldots, A_n) = \prod_{i=1}^{n} P(A_{\tau(i)} \mid A_{\tau(i-1)}, \ldots, A_{\tau(1)})$$

$\mathcal{S}_n$ denotes the symmetric group of permutations of $n$ objects. The description which attributes are involved in a conditional independence relation is encoded in a directed acyclic graph (DAG) in the following way: The nodes of the graph correspond to the attributes. Let parents($A$) denote the set of all those nodes that have a directed link to node $A$. Then, given an instantiation of the attributes in parents($A$), attribute $A$ is conditionally independent of the remaining attributes. Formal: Let $X = \{A_1, \ldots, A_k\}$, $Y = \{B_1, \ldots, B_l\}$ and $Z = \{C_1, \ldots, C_m\}$ denote three disjoint subsets of attributes, then $X$ and $Y$ are conditionally probabilistically independent given $Z$, if the following equation holds:

$$\begin{aligned}
&\forall a_1 \in \mathrm{dom}(A_1) : \cdots \forall a_k \in \mathrm{dom}(A_k) : \\
&\quad \forall b_1 \in \mathrm{dom}(B_1) : \cdots \forall b_l \in \mathrm{dom}(B_l) : \\
&\quad\quad \forall c_1 \in \mathrm{dom}(C_1) : \cdots \forall c_m \in \mathrm{dom}(C_m) : \\
&\quad\quad\quad P(A_1 = a_1, \ldots, A_k = a_k, B_1 = b_1, \ldots, B_l = b_l \mid C_1 = c_1, \ldots, C_m = c_m) \\
&\quad\quad\quad\quad = P(A_1 = a_1, \ldots, A_k = a_k \mid C_1 = c_1, \ldots, C_m = c_m) \\
&\quad\quad\quad\quad\ \cdot P(B_1 = b_1, \ldots, B_l = b_l \mid C_1 = c_1, \ldots, C_m = c_m)
\end{aligned} \tag{1}$$

If a network structure is given, each attribute $A_i$ is assigned a *potential table*, i. e., the set of all conditional distributions, one for each distinct instantiation of the attributes in parents($A_i$). The general layout of such a table is shown in figure 1. Each column (like the one shaded in gray) corresponds to one specific parent attributes' instantiation $Q_{ij}$. Each entry $\theta_{ijk}$ is read as

$$P(A_i = a_{ik} \mid \mathrm{parents}(A_i) = Q_{ij}) \;=\; \theta_{ijk}$$

The learning of Bayesian Networks consists of identifying a good candidate graph that encodes the independencies in the database. The goodness of fit is

| $A_i$ | $Q_{i1}$ | $\cdots$ | $Q_{ij}$ | $\cdots$ | $Q_{iq_i}$ |
|---|---|---|---|---|---|
| $a_{i1}$ | $\theta_{i11}$ | $\cdots$ | $\theta_{ij1}$ | $\cdots$ | $\theta_{iq_i1}$ |
| $\vdots$ | $\vdots$ | $\ddots$ | $\vdots$ | $\ddots$ | $\vdots$ |
| $a_{ik}$ | $\theta_{i1k}$ | $\cdots$ | $\theta_{ijk}$ | $\cdots$ | $\theta_{iq_ik}$ |
| $\vdots$ | $\vdots$ | $\ddots$ | $\vdots$ | $\ddots$ | $\vdots$ |
| $a_{ir_i}$ | $\theta_{i1r_i}$ | $\cdots$ | $\theta_{ijr_i}$ | $\cdots$ | $\theta_{iq_ir_i}$ |

**Fig. 1.** A general potential table

estimated by an evaluation measure. Therefore, usual learning algorithms consist of two parts: a search method and the mentioned evaluation measure which may guide the search. Examples for both parts are studied in [4, 9, 3].

### 2.2 Possibilistic Networks

While probabilistic networks like Bayesian Networks are well-suited to handle uncertain information, they lack the ability to cope with imprecision. Imprecision in the application discussed arises when tuples in the database have missing values.

The interpretation of possibility, especially the notion of *degrees of possibility* is based on the context model [8] where possibility distributions are induced by *random sets* [11]. A random set needs a sample space that it is referencing to. In the studied case this will be $\Omega$. Further, a random set defines a family of (neither necessarily disjoint nor nested) subsets $C = \{c_1, \dots, c_m\}$ of $\Omega$, called *contexts*. These contexts are the sample space of a probability space $(C, 2^C, P_\Gamma)$ and are understood as the physical frame conditions under which the contained elements, namely the $\omega \in \Omega$, are considered possible. This family is defined via $\gamma : C \rightarrow 2^\Omega$. With these ingredients, the tuple $\Gamma = (\gamma, P)$ constitutes an imperfect description of an unknown state $\omega_0 \in \Omega$. The degree of possibility is then defined as the one-point coverage [11] of $\Gamma$, namely:

$$\pi_\Gamma : \Omega \rightarrow [0,1] \quad \text{with} \quad \pi_\Gamma(\omega) = P_\Gamma(\{c \in C \mid \omega \in \gamma(c)\})$$

The imperfection named above now incorporates imprecision as well as uncertainty: imprecision enters via the set-valued context definitions while uncertainty is modeled by the probability space over the contexts. Relations and probability distributions can be seen as the two extremes of a possibility distribution: if there is no imprecision, i. e., all contexts contain only one element, a possibility distribution becomes a probability distribution. In contrast to this, when there is only one context $c^\star$ with $\gamma(c^\star) = R \subseteq \Omega$ then for each $\omega \in \Omega$ we have

$$\pi_\Gamma(\omega) = \begin{cases} 1 & \text{if } \omega \in R \\ 0 & \text{otherwise} \end{cases}$$

and thus the uncertainty disappears.

In the interpretation from [2] we can derive a possibility measure $\Pi$ from the distribution $\pi_\Gamma$ in the following way:

$$\Pi : 2^\Omega \to [0,1] \quad \text{with} \quad \Pi(E) = \max_{\omega \in E} P_\Gamma(\{c \in C \mid \omega \in \gamma(c)\})$$

A possibilistic analog for the conditional probabilistic independence constitutes the *possibilistic non-interactivity*[5], which is defined as follows: Let $X = \{A_1, \ldots, A_k\}$, $Y = \{B_1, \ldots, B_l\}$ and $Z = \{C_1, \ldots, C_m\}$ denote three disjoint subsets of attributes, then $X$ and $Y$ are conditionally possibilistically independent given $Z$, if the following equation holds:

$$\begin{aligned}
&\forall a_1 \in \mathrm{dom}(A_1) : \cdots \forall a_k \in \mathrm{dom}(A_k) : \\
&\quad \forall b_1 \in \mathrm{dom}(B_1) : \cdots \forall b_l \in \mathrm{dom}(B_l) : \\
&\quad\quad \forall c_1 \in \mathrm{dom}(C_1) : \cdots \forall c_m \in \mathrm{dom}(C_m) : \\
&\quad\quad\quad \Pi(A_1 = a_1, \ldots, A_k = a_k, B_1 = b_1, \ldots, B_l = b_l \mid C_1 = c_1, \ldots, C_m = c_m) \\
&\quad\quad\quad\quad = \min\{\Pi(A_1 = a_1, \ldots, A_k = a_k \mid C_1 = c_1, \ldots, C_m = c_m), \\
&\quad\quad\quad\quad\quad\quad \Pi(B_1 = b_1, \ldots, B_l = b_l \mid C_1 = c_1, \ldots, C_m = c_m)\}
\end{aligned} \tag{2}$$

where $\Pi(\cdot \mid \cdot)$ denotes the conditional possibility measure defined as follows:

$$\begin{aligned}
&\Pi(A_1 = a_1, \ldots, A_k = a_k \mid B_1 = b_1, \ldots, B_l = b_l) \\
&\quad = \max\{\pi_\Gamma(\omega) \mid \omega \in \Omega \ \wedge \ \bigwedge_{i=1}^{k} A_i(\omega) = a_i \ \wedge \ \bigwedge_{i=1}^{l} B_i(\omega) = b_i\}
\end{aligned} \tag{3}$$

With these prerequisites a possibilistic network is a decomposition of a multivariate possibility distribution:

$$\forall \tau \in \mathcal{S}_n : \Pi(A_1, \ldots, A_n) = \min_{i=1}^{n} \Pi(X_{\tau(i)} \mid X_{\tau(i-1)}, \ldots, X_{\tau(1)})$$

Learning possibilistic networks follows the same guidelines as the induction of probabilistic networks. Again, a usual learning task consists of two components: a search heuristic and an evaluation measure. Examples for the former are the same as for Bayesian Networks, examples for the latter can be found in [6].

## 3 The Quantitative Component: Visualization

The result of the network learning task consists of a directed acyclic graph (DAG) representing the observed probabilistic or possibilistic (in)dependencies between the attributes exposed by the database samples. An example is depicted in figure 2.

This graph can be interpreted as the *structural* or *qualitative* or *global* component of such a network. This view is justified since the graph structure describes

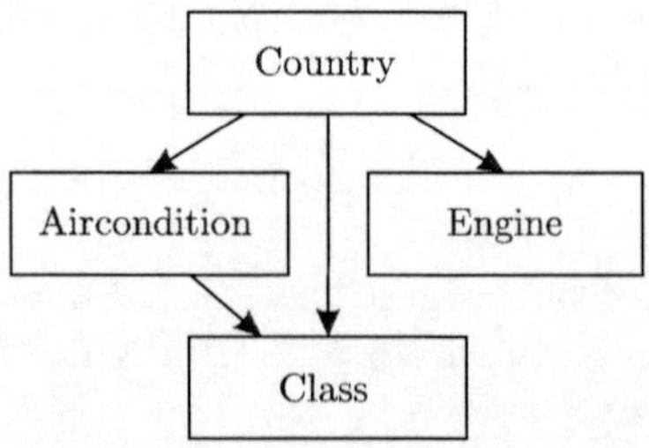


**Fig. 2.** An example of a probabilistic network

the identified (in)dependencies between the entirety of attributes. The graph allows us to deduce statements like the following:

- Attributes Country and Aircondition have some (statistical) influence on the Class attribute.
- Engine does not seem to have a reasonable impact on the Class attribute. It is merely governed by attribute Country.[2]

Although these statements certainly convey valuable information about the domain under consideration, some questions remain unanswered. Combined into one question, it is desirable to know *which combinations* of the conditioning attributes' *values* have *what kind of impact* on *which class values*? The emphasized words denote the entities that carry much more information about the data volume under analysis. Fortunately, this information is already present in form of the *quantitative* or *local* component of the induced networks, namely the potential tables of the nodes.

Since the goal stated in section 1 was to find concept descriptions based on concepts designated by the class attribute, we only need to consider the class attribute's potential table. Therefore, the actual problem to solve is: How can a potential table (containing either probabilistic or possibilistic values) be represented graphically, incorporating the entities mentioned above?

The remainder of this section will deal with the didactical introduction of a visualization method for probabilistic potential tables. Then, this method will be transferred to the possibilistic case.

Figure 1 shows a general potential table. In the case studied here, the attribute $A_i$ corresponds to the class attribute $C$. However, we will continue to refer to it as $A_i$, since we can use the visualization for presenting any attribute's potential table. Each of the $q_i$ columns of the table corresponds to a distinct instantiation of the conditioning attributes. Therefore, the database can be partitioned into $q_i$ disjoint subsets according to these conditioning attributes instantiations. Every fragment, again, is then split according the $r_i$ values of

[2] Since these networks are computationally induced, we refrain from using the notion *causality* here. It is for an expert to decide whether the extracted dependencies carry any causal relationships.

attribute $A_i$. The relative frequencies of the cardinalities of these resulting sets comprise the entries of the potential table, namely the $\theta_{ijk}$.

We can assign to each table entry $\theta_{ijk}$ a set of database samples $\sigma_{ijk} \subseteq \Omega$ which corresponds to all samples having attribute $A_i$ set to $a_k$ and the parent attributes set to the $j$-th instantiation (out of $q_i$ many). Since we know the entire potential table, we can compute probabilities such as $P(A_i = a_{ik})$ and $P(\text{parents}(A_i) = Q_{ij})$. With these ingredients each table entry $\theta_{ijk}$ can be considered an association rule [1]:

$$\texttt{If } \text{parents}(A_i) = Q_{ij} \texttt{ then } A_i = a_{ik} \texttt{ with confidence } \theta_{ijk}.$$

Therefore, all association rule measures like recall, confidence, lift,[3] etc. can be evaluated on each potential table's entry.

With these prerequisites, we are able to depict each table entry as a circle, the color of which depends on the class variable. As an example we consider the class attribute $C$ to have two parent attributes $A$ and $B$. All three attributes are binary. The domain of the class attribute will be assigned the following colors: $\{c_1, c_2\} = \{\circ, \bullet\}$. The (intermediate) result is shown in figure 3(a). In the next step (figure 3(b)) we enlarge the datapoints to occupy an area that corresponds to the absolute number of database samples represented, i. e., $|\sigma_{ijk}|$.

Finally, each datapoint has to be located at some coordinate $(x, y)$. For this example we choose

$$x = \text{recall}(\sigma_{ijk}) \quad \text{and} \quad y = \text{lift}(\sigma_{ijk})$$

The result is shown in figure 3(c). A data analysis expert can now examine the chart and extract valuable information easily in the following ways: At first, since he is likely to be interested only in sample descriptions belonging to one specific class (e.g. class=failure), his focus is put on the black (filled) circles in the diagram. If he is interested in highly conspicuous subsets of sample cases, the circles at the very top are auspicious candidates since they possess a high lift. Put briefly, the rule of thumb for an expert may read:

> "Large circles in the upper right corner are promising candidate subsets of samples that could most likely yield a good concept description."

An example with meaningful attributes is postponed to section 4. For the remainder of this section, we will discuss the applicability of the presented visualization that was based on probabilistic values and measures to the possibilistic domain.

---

[3] These measures are defined as follows: $\forall \theta_{ijk}$ :

$$\begin{aligned}
\text{recall}(\sigma_{ijk}) &= P(\text{parents}(A_i) = Q_{ij} \mid A_i = a_{ik}) \\
\text{conf}(\sigma_{ijk}) &= P(A_i = a_{ik} \mid \text{parents}(A_i) = Q_{ij}) = \theta_{ijk} \\
\text{lift}(\sigma_{ijk}) &= \frac{\text{conf}(\sigma_{ijk})}{P(A_i = a_{ik})}
\end{aligned}$$

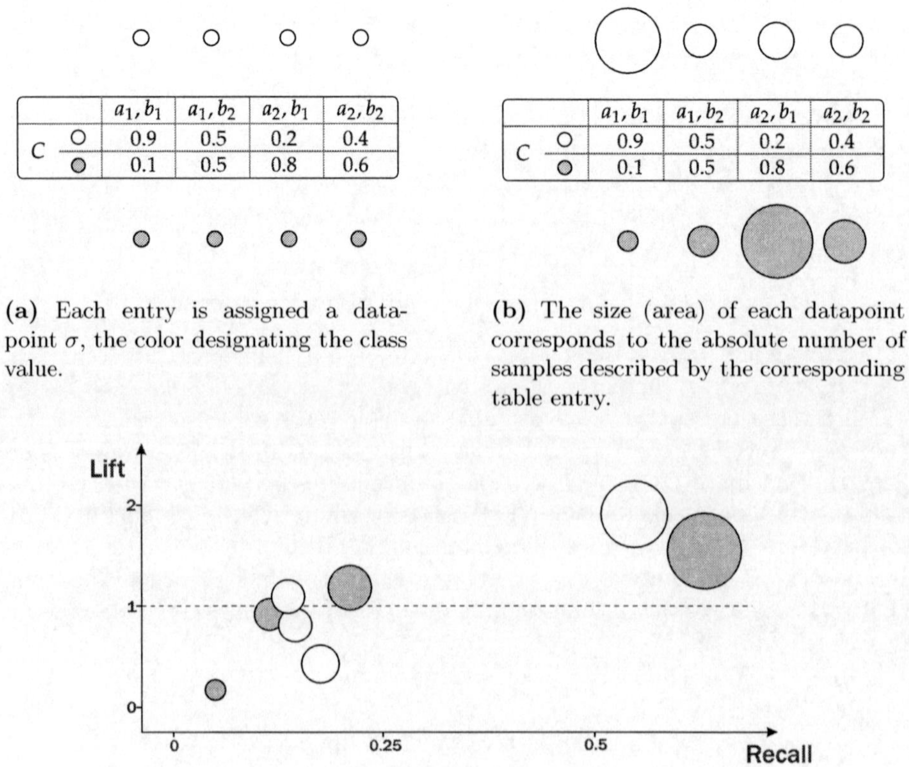


| | | $a_1, b_1$ | $a_1, b_2$ | $a_2, b_1$ | $a_2, b_2$ |
|---|---|---|---|---|---|
| C | ○ | 0.9 | 0.5 | 0.2 | 0.4 |
| | ● | 0.1 | 0.5 | 0.8 | 0.6 |

**(a)** Each entry is assigned a datapoint $\sigma$, the color designating the class value.

| | | $a_1, b_1$ | $a_1, b_2$ | $a_2, b_1$ | $a_2, b_2$ |
|---|---|---|---|---|---|
| C | ○ | 0.9 | 0.5 | 0.2 | 0.4 |
| | ● | 0.1 | 0.5 | 0.8 | 0.6 |

**(b)** The size (area) of each datapoint corresponds to the absolute number of samples described by the corresponding table entry.

**(c)** The location of the center of each datapoint $\sigma$ is set to the coordinates $(x, y) = (\text{recall}(\sigma), \text{lift}(\sigma))$.

**Fig. 3.** We assume the class attribute $C$ to have the two parent (conditioning) attributes $A$ and $B$. All three attributes are binary with the respective domains $\{a_1, a_2\}$, $\{b_1, b_2\}$ and $\{c_1 = \circ, c_2 = \bullet\}$.

### 3.1 The Possibilistic Case

The above-mentioned circles are serving as visual clues for subsets of samples and were located at coordinates which are computed by probabilistic (association rule) measures. Of course, these measures can be mathematically carried over to the possibilistic setting. However, we have to check whether the semantics behind these measures remain the same. For the following considerations, we assume the following abbreviations for the sake of brevity: A subset of sample cases $\sigma$ is defined by the class value $a_{ik}$ and the instantiation of the parent attributes $Q_{ij}$:

$$\sigma = (Q_{ij}, a_{ik}) \overset{\text{Abbrev}}{=} (A, c)$$

Since the definition of the conditional possibility is symmetric, i. e., $\forall A, B : \Pi(A \mid B) = \Pi(B \mid A) = \Pi(A, B)$, the definitions for recall, confidence and support would coincide. Therefore, we define them as follows:

$$\text{supp}^{\text{poss}}(\sigma) = \Pi(A, c) \qquad \text{recall}^{\text{poss}}(\sigma) = \frac{\Pi(A, c)}{\Pi(c)}$$

$$\text{conf}^{\text{poss}}(\sigma) = \frac{\Pi(A, c)}{\Pi(A)} \qquad \text{lift}^{\text{poss}}(\sigma) = \frac{\Pi(A, c)}{\Pi(A)\Pi(c)}$$

The justification for this type of definition is as follows: As the degree of possibility for any tuple $t$, we assign the total probability mass of all contexts that contain $t$ [7]. With this interpretation, the term $\Pi(A = a)$ refers to the maximum degree of possibility of all sets of tuples, for which $A(t) = a$ holds, i. e., $\Pi(A = a) = \max\{p(t) = \frac{w(t)}{N} \mid t \in \Omega \wedge A(t) = a\}$. This probabilistic origin allows us to look at the possibility of an event $E$ (i. e., a set of tuples) as an upper bound of elementary events' probablitities contained in $E$ [2].

## 4 Application and Results

For testing purposes, we firstly created an artificial dataset where some conspicuity was manually put into the data in order to verify whether these dependencies were found and, most importantly, whether these peculiarties become obvious in the visualization. Then, of course, the presented technique was evaluated on real-life data the (anonymized) results of which we will present as well.

### 4.1 Manually-Crafted Dataset

The artificial dataset was generated by a fictitious probabilistic model the qualitative structure of which is shown in figure 2. The conspicuity to be found was that a single aircondition type had a higher failure rate in two specific counties, whereas this type of aircondition accounted for the smallest proportion of all airconditions.

As learning algorithm we used the well-known K2 algorithm [4] with the K2 metric as evaluation measure. Note that this example visualizes the potential tables of a Bayesian Network (the one shown in figure 2), i. e., it represents probabilistic values.

Figure 4 shows all sets of sample cases that are marked defective by the class attribute. Since in this artificial model both attributes Aircondition and Country have a domain of five values each, there are 25 different parent instantiations and thus 25 circles in the chart. As one can cleary see, there are two circles standing out significantly. Because we chose the lift to be plotted against the y-axis, these two sets of sample cases expose a high lift value, stating that the respective parent instantiations (here: combination of Country and Aircondition) make the failure much more probable. Since both circles account for only a small portion of all tuples in the database, they have small recall, indicated by being located at the left side of the chart.

### 4.2 Real-Life Dataset

The real-life application which produced empirical evidence that the presented visualization method greatly enhances the data analysis process took place during a cooperative project at the DaimlerChrysler Research Center. As a

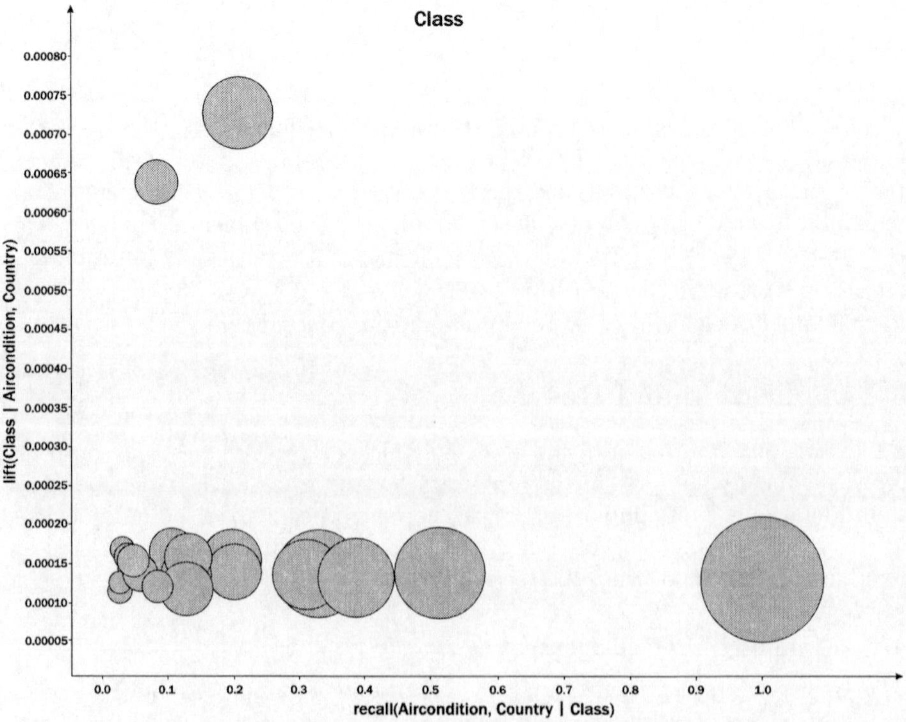


**Fig. 4.** The two outstanding circles at the top of the chart indicate two distinct sets of samples having a much higher failing rate than the others. They reveal the two intentionally incorporated dependencies, i. e., one specific type of arcondition is failing more often in two specific countries.

leading manufacturer of high-quality automobiles, one of the company's crucial concerns is to maintain the high level of quality and reliability of their products. This is accomplished by collecting extensively detailed information about every car sold and to analyze complaints in order to track down the fault promptly. Since these data volumes are highly confidential, we are not allowed to present specific attribute names and background information. Nonetheless, the charts generated by visualizing the induced possibilistic networks will provide a fairly good insight into the everyday usage of the presented visualization method.

Figure 5 shows a possibilistic chart of the binary class variable. In this case, the non-faulty datasets are depicted as well (unfilled circles). As one can easily see, we find a relatively large circle in the upper right corner. The size of this circle tells that it represents a reasonable number of affected cars, while the high lift states, that the selected parent instantiation should be subject of a precise investigation. In fact, the consultation of a production process expert indeed revealed a causal relationship.

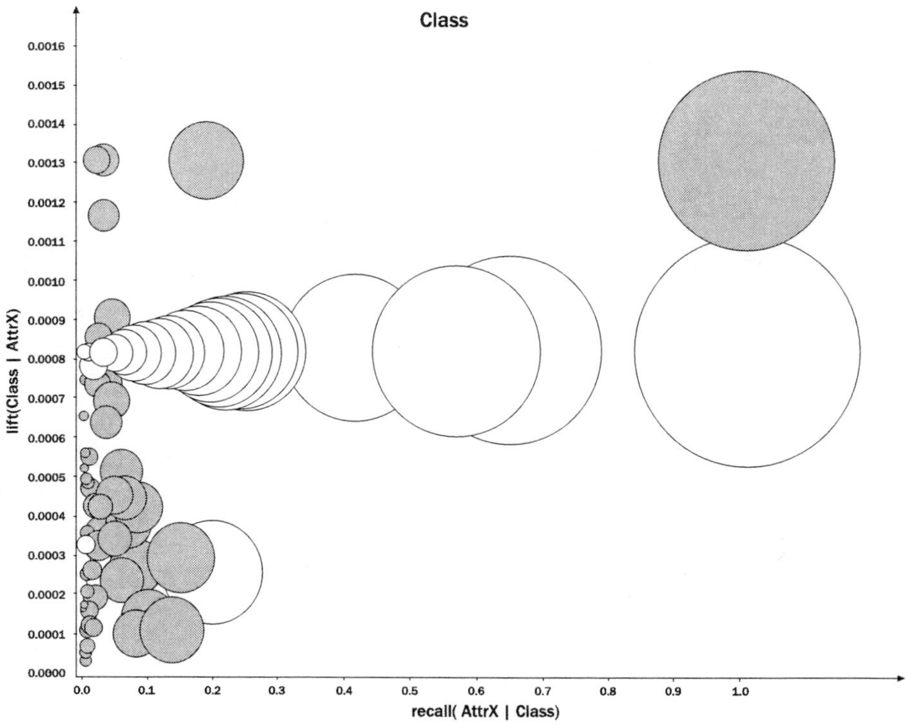


**Fig. 5.** The large circle in the top right corner indicates a set of vehicles whose specific parents attributes' values lead to a higher failure rate. An investigation by experts revealed a real causal relationship.

### 4.3 Practical Issues on the Visualization

As it can be seen from figure 5 and 4, the circles show a fairly large overlap which may lead to large circles covering and thus hiding smaller ones. In the real-world application — from which the figures are taken — there are several means of increasing the readability of the charts. On the one hand, all circles can be scaled to occupy less space while the user can zoom into a smaller range of the plot. Further, the circles can be made transparent which reveals accidentally hidden circles.

## 5 Conclusion and Future Work

In this chapter, we presented a brief introduction to both probabilistic and possibilistic networks, the latter due to its natural ability of handling imprecise data becoming increasingly interesting for industrial applications since real-world data often contains missing data. We argued further that the learning of such a network only reveals the qualitative part of the contained dependencies, yet the more meaningful information being contained inside the potential tables, i. e., the quantitative part of the network. Then, a new visualization technique was presented that is capable of displaying high-dimensional, nominal potential tables containing probabilistic as well as possibilistic parameters. This plotting method was evaluated in an industrial setting enabling production experts to easier identify extreme data samples.

Since the presented technique only dealt with datasets that represented the current state of the database at a specific (but fixed) moment in time, it would be interesting to extend the visualization to temporal aspects, that is, time series. Then, it would be possible not only to use the mentioned association rule measures but also their derivatives in time to make trends visible.

## References

1. Agrawal, R., Imielinski, T., Swami, A.: Mining association rules between sets of items in large databases. In: Proc. of the ACM SIGMOD Conference on Management of Data, pp. 207–216 (1993)
2. Borgelt, C.: Data Mining with Graphical Models. PhD Thesis, Otto-v.-Guericke-Universität Magdeburg, Germany (2000)
3. Borgelt, C., Kruse, R.: Some experimental results on learning probabilistic and possibilistic networks with different evaluation measures. In: ECSQARU/FAPR 1997. Proc. of the 1st International Joint Conference on Qualitative and Quantitative Practical Reasoning, pp. 71–85 (1997)
4. Cooper, G., Herskovits, E.: A Bayesian method for the induction of probabilistic networks from data. Journal of Machine Learning (1992)
5. Dubois, D., Prade, H.: Possibility theory. Plenum Press, New York (1988)
6. Gebhardt, J., Kruse, R.: Learning possibilistic networks from data. In: Proc. 5th Int. Workshop on Artificial Intelligence and Statistics, pp. 233–244 (1995)
7. Gebhardt, J., Kruse, R.: A possibilistic interpretation of fuzzy sets by the context model. In: IEEE International Conference on Fuzzy Systems, pp. 1089–1096 (1992)
8. Gebhardt, J., Kruse, R.: Int. Journal of Approximate Reasoning 9, 283–314 (1993)
9. Heckerman, D., Geiger, D., Maxwell, D.: Learning Bayesian networks: The combination of knowledge and statistical data. Technical Report MSR-TR-94-09 85–96, Microsoft Research, Advanced Technology Division, Redmond, WA (1994)
10. Lauritzen, S., Spiegelhalter, D.: Journal of the Royal Statistical Society. Series B 2(50), 157–224 (1988)
11. Nguyen, H.: Information Science 34, 265–274 (1984)
12. Pearl, J.: Probabilistic reasoning in intelligent systems: Networks of plausible inference. Morgan Kaufmann, San Mateo, California (1988)

# Two Fuzzy-Set Models for the Semantics of Linguistic Negations

Silvia Calegari, Paolo Radaelli, and Davide Ciucci

Dipartimento di Informatica, Sistemistica e Comunicazione
Università degli Studi di Milano Bicocca
Via Bicocca degli Arcimboldi 8, 20126 Milano (Italy)
{calegari,radaelli,ciucci}@disco.unimib.it

**Summary.** Two methods based on fuzzy sets are proposed in order to handle the understanding of linguistic negations. Both solutions assign an interpretation of the negated nuances of the natural language (i.e. the humans use adverbs and adjectives to make their requests) depending on the context. The first approach is a modification of Pacholczyk's model able to handle a non-predetermined chain of hedges. The second one is a new framework based on the idea to give two different semantics for the "not" particle, depending on whether it is used to change the meaning of a linguistic modifier or to alter a fuzzy set.

## 1 Introduction

Nowadays a main open issue in Computational Intelligence is to deal with the statements expressed in natural languages by humans. For example, one of the key topics in the development of the Semantic Web [1] is to enable machines to understand users requests and exchange meaningful information across heterogeneous applications. The aim is thus to allow both the user and the system to communicate concisely by supporting information exchange based on semantics. In this area of research a crucial topic is to understand the right interpretation of the linguistic negation.

It is, indeed, very hard to find a unique formal interpretation of negations and consequently, to enable a system to understand the right sense of this type of information. Let us notice that different meanings may be associated with a sentence like "Sophie is not very tall", such as "Sophie is extremely small" or "Sophie is quite tall". Therefore, it is needed to deal with all these possible meanings which depend on the context and on the interpretation given to the nuances that the negation brings. More formally, we take into account the delicate case where the meaning of a negated statement "$x$ is not $m_\alpha A$" has the form "x is $m_\beta B$" where $A$ and $B$ are modelled by a suitable fuzzy set and $m_\alpha$, $m_\beta$ define a conceptual modifier [2, 3, 4]. In the previous example, $x$ = "Sophie", $m_\alpha$ = "very" and $A$= "tall", whereas we have two possible interpretations for $B$ and $m_\beta$: $B$ = "small", $m_\beta$= "extremely" and $B$ = "tall", $m_\beta$= "quite".

In this chapter, we propose two solutions to the problem of interpreting a negated statement. The first one is a modification of the model developed by

R. Bello et al. (Eds.): Granular Computing, STUDFUZZ 224, pp. 105–120, 2008.
springerlink.com 

Pacholczyk in [5, 6, 7, 8, 9], whose main drawback is the static representation of the possible interpretations of a negation. Indeed, it allows to establish the meaning of a negation according to the context, but only on a fixed set of modifiers. Our purpose is to be able to use, and this means correctly interpreting, a non-predetermined combination of modifiers. For example, in our proposal, we can handle a dynamic chain of modifiers, such as "little, very very, very very little" and so on.

The second model we have studied contains a new logical framework based on linguistic considerations. The idea behind this model is to distinguish the way a negation is used inside a sentence: to alter the meaning of a property or the meaning of a linguistic modifier, and to handle the two uses differently.

The rest of the chapter is organized as follows: Section 2 introduces the two models. In Section 3 the differences of the two models are reported. Section 4 presents an example in order to compare the methodologies that we have proposed. Finally, in Section 5 some conclusions are reported.

## 2 Proposals to Handle the Linguistic Negation

In order to give a meaning, suited to a given context, to a linguistic negation, two different models are introduced and analyzed. The first part of the section is devoted to modifiers of fuzzy linguistic variables which play an important role in both approaches.

### 2.1 Concept Modifiers

First of all, we define an hedge as a function which alters the fuzzy value of a given property. Let us denote the collection of all hedges by $\mathcal{H}$. A chain of hedges $M = h_q h_{q-1} \ldots h_1 \; \forall i \; h_i \in \mathcal{H}$, is called a *concept modifier* [10], and $\mathcal{M}$ denotes the collection of all concept modifiers. For example, "very very" is a concept modifier composed by the hedge "very" repeated twice.

Hedges (and concept modifiers) are divided into two groups: precision modifiers, which alter the shape of a given fuzzy set, and translation modifiers, which translate a fuzzy set.

To any precision modifier is associated a value $\beta > 0$ which is used as an exponent in order to modify the fuzzy value of the assigned property [3, 11].

**Definition 1.** *A* precision modifier *is represented by an exponential function* $p : [0,1] \mapsto [0,1]$ *applied to a given fuzzy set* $f : X \mapsto [0,1]$ *in order to alter its membership value as* $p(f(x)) := f(x)^\beta$, *where* $\beta > 0$.

According to the value of $\beta$, precision modifiers can be classified in two groups: *concentration* and *dilation*. The effect of a concentration modifier is to reduce the grade of a membership value where in this case $\beta > 1$; whereas a dilation hedge has the effect of raising a membership value, that is $\beta \in (0, 1)$. For instance, let us suppose to consider the first group where for the hedge "very" has assigned $\beta = 2$. So, if "Cabernet has a dry taste with value 0.8", then "Cabernet has a very dry taste" will have value ($0.8^2 = 0.64$).

A translation modifier does not modify the shape of a fuzzy set (as precision modifiers do), but only translate it through a value $\gamma > 0$.

**Definition 2.** *A* translation modifier *is represented by a function* $t : [0,1] \mapsto [0,1]$ *applied to a given fuzzy set* $f : X \mapsto [0,1]$ *in order to alter its membership value as* $t(f(x)) := f(x \pm \gamma)$, *where* $\gamma > 0$ *is such that* $(x \pm \gamma) \in X$.

For instance, let us suppose to have for the translation modifier "extremely" applied to the fuzzy set tall a value of $-3$. This means that this hedge translates towards right the fuzzy set function. So, if "that man is tall" with value 0.75, then "that man is extremely tall" will have, for instance (it depends on the definition of the fuzzy set "tall"), value $f(182-3) = 0.39$.

So, as can be easily seen, precision modifiers have the same effect whatever the fuzzy value they are applied to. On the contrary, translation modifiers, eavily depend on the domain of application.

One of the problems to solve [12] is which semantic interpretation corresponds to a chain of hedges and how to handle requests based on this type of statements. Our solution is to establish a unique set of precision and translation modifiers, (the set $\mathcal{M}$) for all properties and use the algorithm proposed by Khang et al. in [11] (see below) in order to give the semantic interpretation of a chain of hedges. This algorithm allows to define a concept modifier whose length is not known a priori. For example, given a finite set of fuzzy modifiers like $\{little, very\}$ a possible set of combinations will be $\{very\ very\ little, little, very\ very, \ldots\}$. In this way a dynamic set of modifiers, not predictable by the expert, can be obtained.

Given the total number of hegdes, it is possible to calculate the maximum number of the combinations of the modifiers. Let $n = |\mathcal{M}|$, then, the cardinality of a chain of at most $k$ modifiers is $m := \sum_{i=1}^{k} n^i$.

In the present work, we limit the length of the combinations to two elements, in order to have a semantic meaning close to the expressions used by humans and to limit the computational load.

*Example 1.* Let us suppose to have a translation set defined as $\{vaguely, extremely\}$ and the precision set defined as $\{little, very\}$ applied on the property "MEDIUM height" (see Fig. 1). In conformity to the previous formula, we have $m := (n + n^2) = 4 + 4^2 = 20$ different concept modifiers.

### Khang et al.'s Algorithm

This algorithm allows to give a semantic interpretation of a chain of hedges of unknown length, taking into account both precision and translation modifiers.

It is supposed that the sets of precision and translation modifiers are totally ordered by the relations $p_\alpha < p_\beta \Leftrightarrow \alpha < \beta$ and $t_\gamma < t_\delta \Leftrightarrow \gamma < \delta$ respectively. Furthermore, for using this algorithm, every hedge need to be classified as positive or negative w.r.t. the others: in the case of precision modifier, it means an increase/decrease in the fuzzy value of the property; whereas in the case of translation modifiers, a shift to left or right of the fuzzy set.

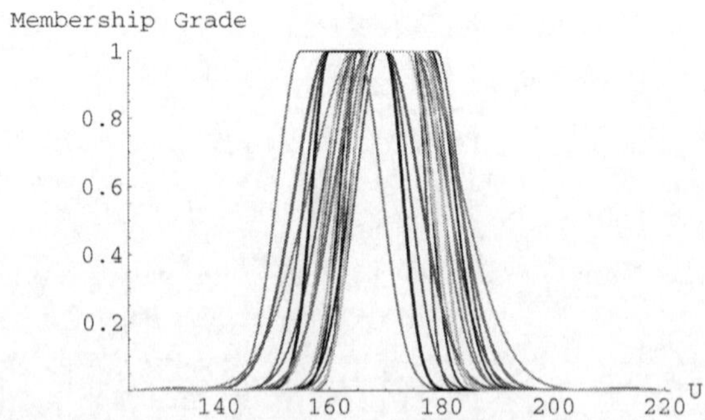


**Fig. 1.** Number of modifiers $m := (n^2 + n) = 20$ where $n := |\{little, very\}| + |\{nearly, exactly\}|$

Formally, if $\mathcal{H}$ identifies the set of all the hedges and $\mathcal{MP}_c$ the set of nuanced properties (see Section 2.2), then the "*sign*" function is defined as $sign : \mathcal{H} \times (\mathcal{H} \cup \mathcal{MP}_c) \mapsto \{-1, 1\}$,

$$sign(h_i, h_{P_c}) = \begin{cases} -1, & \text{if } h_i \text{ is negative w.r.t. } h_{P_c} \\ 1, & \text{if } h_i \text{ is positive w.r.t. } h_{P_c} \end{cases} \tag{1}$$

where $h_i \in \mathcal{H}$ and $h_{\mathcal{P}_c} \in \mathcal{H} \cup \mathcal{MP}_c$.

Now, we explain how to understand if an hedge $h_i$ is positive or negative w.r.t. a nuanced property. The procedure is the same for both translation and precision modifiers.

First of all, when the expert of domain defines the set $\mathcal{H}$ he/she has to state for each hedge $h$ also its sign $S(h)$, i.e., for the hedge "vaguely" he assign $S(vaguely) = -1$. Then, the sign of an hedge $h_i$ w.r.t. each other hedge $h_j$ is computed as $sign(h_i, h_j) = S(h_i) \cdot S(h_j)$. That is, $sign(h_i, h_j)$ is positive if $h_i$ and $h_j$ have the same effect (are both positive or both negative), and negative on the contrary. Finally, given a chain of hedges $H = h_1, \ldots, h_n$ and a property $P$ the sign of $H$ relative to $P$ is recursively computed as

$$sign(h_n, P) = S(h_n) \tag{2a}$$

$$sign(h_1 \ldots h_{(n-1)}, h_n P) = sign(h_1, h_2) \cdot sign(h_2; h_3, \ldots, h_n P) \tag{2b}$$

Let us note that each hedge (chain of hedges) behaves in the same way for all properties.

*Example 2.* Consider $\mathcal{H} = \{very, little\}$ the matrix which defines the sign of an hedge w.r.t. other hedges is defined as in Table 1.

The "sign" $S$ column is defined by the expert whereas the other values are computed as explained above.

Using this matrix it is possible to state the *sign* function for every chain of hedge composed by this specific set $\mathcal{H}$. In this example the length of the concept modifier $\mathcal{M}$ has been limited to two elements. The result is shown in Table 2 where the property $P$ is omitted for simplicity.

**Table 1.** *Sign* matrix for the set $\mathcal{H}$

| | S | very | little |
|---|---|---|---|
| very | 1 | 1 | -1 |
| little | -1 | -1 | 1 |

**Table 2.** Calculus of the *sign* of concept modifiers

| $\mathcal{M}$ | sign |
|---|---|
| very | 1 |
| little | -1 |
| very very | 1 |
| little very | -1 |
| very little | 1 |
| little little | -1 |

This method can be used also to determine the sign of a concept modifier made of an hybrid chain of hedges. An hybrid chain is composed by precision and translation hedges jointly. For instance, the chains of modifiers like "very very vaguely" or "extremely little" have an hybrid behaviour on the semantic of the statement to analyse. Let us suppose of having to calculate the value of a chain of modifiers like $h_{p_1}, \ldots h_{p_n}, h_{t_1}, \ldots, h_{t_n}$. In this case the algorithm is applied twice: the first time in order to obtain the value $\gamma$ for the sub-chain of translation hedges $h_{t_1}, \ldots h_{t_n}$ and the second time for calculating the exponent $\beta$ for the chain $h_{p_1}, \ldots h_{p_n}$. Then, the final value is obtained applying consecutively the two functions of modifiers: $p(t(f(\mathcal{P}_c)))) := f(\mathcal{P}_c + \gamma)^\beta$.

*Example 3.* Given the set $\mathcal{H} = \{very, vaguely\}$ and a sign for each of these elements, let us consider the chain of hedges "very very vaguely". The sign of this concept modifier is obtained splitting the problem in two sub-chains in order to individualize precision and translation hedges. So, it is found the sign for the hedge "vaguely" and the sign for the chain "very very".

The pseudo-code of the algorithm to calculate the membership modifier $\phi$ of a concept modifier in $\mathcal{M}$ is now reported.

**Table 3.** Calculus of the *sign* of an hybrid chain of hedges

| $\mathcal{M}$ | sign |
|---|---|
| very | 1 |
| vaguely | 1 |
| very very | 1 |

As previously stated, the *sign* function indicates how the candidate $\beta$ value has to alter the fuzzy value, i.e., if $k_i$ increases ($sign = 1$) or decreases ($sign = -1$) the value of $\phi_{i-1}$. In this way, in the i-th step, $\phi_i$ will be correlated w.r.t. all the hedges $k_i k_{i-1} ... k_1$ examined until now. The values $(lo_i, up_i)$ define the interval in which the $\phi$ value can be obtained.

The algorithm takes also in account the cases of most changing positive modifiers like "very very... very" and of most changing negative modifier like "little little... little" if precision modifiers are considered. So, in the positive situation $f(x_i) = upper^i$ and $f(x_{i+1}) = upper^{i+1}$ in order to extend the interval to $(lo_i, \infty)$. In an inverse manner, in the negative case, it has been assigned $f(x_i) = lower^i$ and $lo_i = lower^{i+1}$ for extending the interval to $(0, up_i)$. In the case of translation modifiers, we set $f(x_i) = upper{*}i$ and $f(x_{i+1}) = upper{*}(i+1)$ for the positive case, and $f(x_i) = lower{*}i$ and $lo_i = lower{*}(i+1)$ for the negative one.

## 2.2 The *Reference Frame* Model

The main idea of this model is to define the negative expression "x is not A" as an affirmative assertion "x is P", where "P" is a property defined in the same domain of "A". Of course there can be different possibilities about "P". All these alternatives are called the *reference frame* of the negation relative to a given concept $x$. In particular, we consider the case in which the meaning of the negated statement is obtained through a modified property $mP$, where $P$ is different from $A$. For example, the interpretation of the statement "Sophie is not tall" could be "Sophie is enough medium".

Let us introduce the following notations: $\mathcal{C}$ is the set of distinct concepts $c$, $\mathcal{D}_c$ is the domain associated with a concept $c$, $\mathcal{M}$ is the set of modifiers, $\mathcal{P}_c$ is the set of basic properties, represented by fuzzy sets, associated to the concept $c$, and $\mathcal{MP}_c$ is the set of all nuanced (modified) properties associated with $c$.

Given a concept $c$, the *reference frame* of a linguistic negation is defined as the function $Neg : \mathcal{MP}_c \mapsto P(\mathcal{MP}_c)$, $Neg(Q) = \mathcal{MP}_c \backslash \{Q\}$. That is, given a (negated) nuanced property it returns, as a possible interpretation of the negation "not Q", all the nuanced properties except $Q$.

The advantage of having different interpretations of the linguistic negation is the possibility to cope with the semantic richness provided by natural languages. Indeed, humans use the linguistic nuances (i.e., linguistic adverbs like "very" or "more or less") in order to better specify their requests.

Furthermore, this approach can deal with both types of modifiers (and chain of modifiers): precision and translation.

*Remark 1.* Pacholczyk's model [5, 6, 13] defined a static representation of the modifiers sets: to any property is associated a pre-defined and fixed set of modifiers given by the expert during the domain definition. Thus, this approach requires that users can make their requests using only these specific modifiers associated to the properties. But in this way the users should know all the modifiers sets of all properties before writing a query.

---

*MEMBERSHIP MODIFIER ($\mathcal{M}$:input, $\phi$:output)*

```
{
  up = upper; lo = lower; φ = 1; mpos = 1; mneg = 1; sign = sign(h_i, h_P);
  for i = 1 to q do {
      compute j such that k_i = h_j^- or k_i = h_j ;
      if i > 1 then {
        sign = sign × sign(k_i, k_{i-1});
        if sign == 1 then {
          φ_i = φ_{i-1} + (up_{i-1} - φ_{i-1})/(2p) × (2j - 1);
          up_i = φ_{i-1} + (up_{i-1} - φ_{i-1})/(2p) × (2j);
          lo_i = φ_{i-1} + (up_{i-1} - φ_{i-1})/(2p) × (2j - 2);
          if (mpos == 1 ∧ j == p) then {
            φ_i = f(x_i); up_i = f(x_{i-1});
          }
          else mpos = 0;
          mneg = 0;
        }
        else {
            φ_i = φ_{i-1} - (φ_{i-1} - lo_{i-1})/(2p) × (2j - 1);
            up_i = φ_{i-1} - (φ_{i-1} - lo_{i-1})/(2p) × (2j - 2);
            lo_i = φ_{i-1} - (φ_{i-1} - lo_{i-1})/(2p) × (2j);
            if (mneg == 1 ∧ j == 1) then {
              φ_i = f(x_i); lo_i = f(x_{i-1});
            }
            else mneg = 0;
            mpos = 0;
        }
      }
  }
}
```

---

Figure 5 shows the difference between the static Pacholczyk's model and its new dynamic behaviour.

Thus, we obtain a family of solutions, which, in general, can contain several possibilities. However, if desired, it is possible to give some methods in order to reduce the choice among all the plausible meanings of a negation. We are going to outline one of them.

As a first step, it is possible to reduce the number of elements of the reference frame through the combination of neighbourhood and similarity relations [5, 6, 13]. These relations allow to determine only a subset of the reference frame as the family of possible solutions. In particular, only the properties which are *far* from the property to negate are kept. The new reference frame is denoted as $Neg_{\rho,\varepsilon}(A)$ and consists of the properties which are *$\rho$-compatible with $A$ with a tolerance threshold* $\varepsilon$ as formally explained in the following definition.

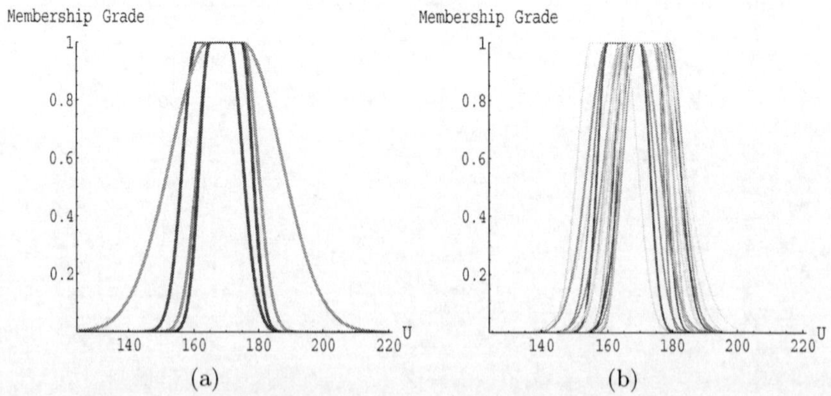


**Fig. 2.** (a) Number of modifiers applying the Pacholczyk's model ($n := 4$). (b) Number of modifiers applying the dynamic modification of Pacholczyk's model ($m := (4^2 + 4)$).

**Definition 3.** *Let $\rho$ and $\epsilon$ such that $0 \leq \epsilon \leq \rho \leq 1$. Let us define the function $Neg_{\rho,\epsilon} : \mathcal{MP}_c \mapsto P(\mathcal{MP}_c)$ as the collection of nuanced properties such that:*

*(N1) $\forall A \in \mathcal{MP}_c$ if $N \in Neg_{\rho,\epsilon}(A)$ then $N \in Neg(A)$, i.e., $Neg_{\rho,\epsilon}(A) \subseteq Neg(A)$;*
*(N2) $\forall N \in Neg_{\rho,\epsilon}(A)$ it holds $\forall y \in \mathcal{D}_c$, $\mu_A(y) \geq \rho$ implies $\mu_N(y) \leq \epsilon$;*
*(N3) $\forall N \in Neg_{\rho,\epsilon}(A)$ it holds $\forall y \in \mathcal{D}_c$, $\mu_N(y) \geq \rho$ implies $\mu_A(y) \leq \epsilon$.*

*where $\mu_J : \mathcal{D}_c \mapsto [0, 1]$, with $J = A$ or $J = P$, is the fuzzy set which defines property $J$.*

By the definition 3 we obtain a family of solutions. To give only one interpretation of the negation another step is required. A choice of a nuanced property $mP \in Neg_{\rho,\varepsilon}(A)$ defining "$x$ is $mP$" as the meaning of "$x$ is not $A$" can be made according to the following algorithm:

1. The property $P$ is chosen as the one with maximum membership value on $x$: $\mu_P(x) = \max_Q\{\mu_Q(x)|nQ \in Neg_{\rho,\varepsilon}(A)\}$;
2. The modifier $m$ of $P$ is selected in two steps:
   a) $MP_{(x,P)} = \{n \in \mathcal{M}|\max_n\{\mu_{nP}(x)|nP \in Neg_{\rho,\varepsilon}(A)\}\}$
   b) Once defined as $x_1, ..., x_N$ the points of symmetry of the functions $\mu_{nP}$, $n \in MP_{(x,P)}$, it is chosen the one with minimum distance from $x$: $\min_{x_i}\{(||x, x_1||, ..., ||x, x_i||, ..., ||x, x_N||)\}$.

*Remark 2.* The methodology used by Pacholczyk in order to select a single solution cannot be used. His idea was to choose the property with the highest membership function on $x$ and having the lowest complexity, where the complexity of a property A is equal to the number of nuances (or modifiers) which can assume. Obviously, this definition is not useful here since we have assumed a unique set of modifiers for all properties.

## 2.3 A Two-Level Approach

The second method to manage the meaning of linguistic negations that we are going to present is based on the idea that the linguistic particle "not" can be used to alter the meaning of the linguistic modifier, rather than those of a fuzzy set. In other words, we are going to consider the term "not exceptionally", in a phrase like "not exceptionally high" as a single linguistic modifier, rather than a sequence of an intensifier followed by a negation.

This approach is motivated by the consideration that "not exceptionally" in the example is used to denote entities that are anyway high, even if modestly. Small entities are usually considered to not be part of the "not exceptionally high" set. Thus, the traditional fuzzy way to handle such an expression (where the fuzzy set representing "not exceptionally high" is the negation of the fuzzy set representing "exceptionally high") fails to give the expected meaning, since it cannot differentiate (for example) between moderately high elements and very low ones. If we consider the phrase "not exceptionally" as a single linguistic modifier, we are allowed to define a new linguistic hedge to represent the meaning of the phrase, and to assign to it a meaning like "quite high but not extremely".

Our proposal assumes that the negative linguistic particle can carry out two different linguistic roles: it can be used to alter the meaning associated to a fuzzy set, altering its membership function (working as the usual fuzzy negation); or it can work as a sort of second-level modifier, which alters the semantic of a linguistic modifier. In order to differentiate these two uses, we introduce a new operator called *mmodifier* to represent this second role of "modifier of modifier". mmodifiers are functions that can be applied to a linguistic modifier in order to alter its meaning, just like linguistic modifiers can alter the semantic of a fuzzy-set. within this approach, the term "not" can have two different representations, on the basis of the contest it is expressed: it can be used as the standard fuzzy set negation (to alter the semantic of a fuzzy set membership function $f(x)$ into a function $f'(x) = 1 - f(x)$), or it can be a mmodifier that affects the semantic of the linguistic hedge it is applied to.

In order to formalize the definition of mmodifier, we need to give a formalization of the linguistic modifiers that can be handled by a mmodifier. This formalization is an extension of those proposed by She et al. in [2]. Within this representation, each linguistic modifier is univocally defined by two parameters: a *type* and an *intensity*. The *type* of a modifier identifies at which "family" the modifier belongs and the general effects it gives when it is applied to a fuzzy set. Concentrators and dilators, for example, represent two different types of modifiers. While the type of modifier defines the general effects of each modifier, each individual linguistic hedge has an *intensity* value that is used to represent the strength of the modifier itself. For example, two modifiers like "greatly" and "very" can belong to the same type (being two concentrators), but the first one will have a higher intensity than the second one, giving more extreme results when it is applied.

Each modifier type is related to a *modifier function*, which defines how the results obtained to a fuzzy set's membership function must be changed when the correspondent modifier is applied to the set. Modifier functions are parametric with respect to the intensity values of the modifiers, thus a couple $\langle \text{type}, \text{intensity} \rangle$ univocally defines the effects of any modifier. The formal definition of a membership function is the following:

**Definition 4.** *The* modifier function *for a modifier type t is a function $M_t(i, x)$ : $\Re^+ \times [0, 1] \mapsto [0, 1]$, where $i \in \Re^+$ is an intensity value and $x \in [0, 1]$ is a fuzzy membership value.*

We limited our study to four linguistic hedges categories: concentrators, dilators, contrast intensifiers [14] and the *negatively* hedge described in [15] (a modifier that has the opposed effect than a contrast intensifier modifier). Each category corresponds to one type of modifier.
The functions used as modifier functions for the four types of modifiers are the ones suggested in [2]:

- Modifier function for dilators is
$$f_{\text{dil}}(i, x) = x^{\frac{1}{i}} \tag{3}$$
- Modifier function for concentrators is
$$f_{\text{con}}(i, x) = x^{i} \tag{4}$$
- Modifier function for contrast intensifier is
$$f_{\text{pos}}(i, x) = \begin{cases} \left(\frac{1}{2}\right)^{1-i} x^i & \text{if } x < \frac{1}{2} \\ 1 - \left(\frac{1}{2}^{1-i} (1-x)^i\right) & \text{if } x \geq \frac{1}{2} \end{cases} \tag{5}$$
- Modifier function for negatively hedges is
$$f_{\text{neg}}(x, i) = \begin{cases} \left(\frac{1}{2}\right)^{1-\frac{1}{i}} x^{\frac{1}{i}} & \text{if } x < \frac{1}{2} \\ 1 - \left(\frac{1}{2}^{1-\frac{1}{i}} (1-x)^{\frac{1}{i}}\right) & \text{if } x \geq \frac{1}{2} \end{cases} \tag{6}$$

For example, assuming that the modifier "very" is a concentrator with an intensity value of 2, its effects are described by the function

$$f_{\text{very}}(x) = f_{\text{con}}(x, 2) = x^2 \tag{7}$$

Another concentrator, like "exceptionally", could have an intensity value of 3 and will end up with the formula $f_{\text{except.}}(x) = x^3$.

Given this formalization of a linguistic modifier, a mmodifier is defined as follows:

**Definition 5.** *A* mmodifier *is a function $K : \Re^+ \mapsto \Re^+$, that is used to modify the intensity value of a linguistic modifier in order to derive a new linguistic modifier from it.*

The modifier obtained by applying a mmodifier to a linguistic modifier is a new modifier that maintains the same type of the old one. More formally, we can say that

**Definition 6.** *The application of a mmodifier $K$ to a modifier $M$, with type $t$ and intensity $i$, is a new linguistic modifier $K(M)$, whose modifier function is $M'(x) = f_t(K(i), x)$*

Despite the fact that the application of a mmodifier cannot change the type of a linguistic modifier, modifier functions have been selected in a way that is possible to transform any modifier function into the one of the opposite type (i.e. concentrators and dilators, or contrast identifiers and negatively hedges) simply replacing the $i$ parameter with its inverse $\frac{1}{i}$. In fact, any modifier of any type with an intensity $i$, $0 < i < 1$ works like a modifier of the opposite type. For example, a hypothetical dilator with an intensity of $\frac{1}{2}$ will effectively work like a concentrator, lowering the membership value of the elements of the fuzzy-set it is applied to. This property is used to define the mmodifier "not", whose semantic requires to invert the meaning of the original modifier. Not mmodifier's function is the following:

$$K_{\text{not}}(i) = 1 - \frac{1}{i} \tag{8}$$

When applied to, for example, an intensifier with intensity 2 (like the modifier "very" in the previous example) it gives a new "intensifier" with the modifier function

$$f'(x) = f_{\text{con}}(x, 1 - \frac{1}{2}) = \sqrt{x} \tag{9}$$

whose formula is identical to those of a concentrator of intensity 2. The same mmodifier can be applied to other modifiers, for example the "exceptionally" modifier described above. In this case, the resulting modifier function will be

$$f''(x) = x^{1-\frac{1}{3}} = \sqrt[3]{x^2} \tag{10}$$

In addition to the mmodifier "not", it is possible to define other mmodifiers that work in an analogous way. For example, it it possible to define a "very" mmodifier (which is different from the concentrator with the same name), to formulate expressions like "very few tall" or similar ones. The formula proposed for the "very" mmodifier is the following:

$$K_{\text{very}}(i) = 2i \tag{11}$$

With this formula mmodifiers, for example, "very very" (in an expression like "He is very, very tall") will be a concentrator with an intensity of 4, instead of 2.

## 3 Comparison Between the Two Models

In this section we examine and discuss the differences between the two methods proposed in Sections 2.2 and 2.3.

The mmodifier method works by computing a new intensity value for an altered linguistic modifier, given the intensity value of the original modifier. The function which represents each family of modifiers and mmodifiers is well defined and different modifiers of the same family are distinguished from each other only by their intensity value. These assumptions make the second method very efficient both in terms of computational costs and in time needed to define the fuzzy model to use.

Our reference frame model, however, requires to provide the initial group of modifiers applicable to a base fuzzy set, in order to select the best suited modifier. This implies a longer time required in order to define the behaviour of the base modifiers, and a higher computational cost for the execution of the algorithm.

On the other hand, the linguistic negation approach allows a better control over the modifiers definition, allowing, for example, to model different modifiers of the same family. Linguistic negation model is also applicable when using translation modifiers rather than precision modifiers, while mmodifier approach only works when using precision ones. Although it is possible,in theory, to apply the same technique even on translation modifiers, a deeper study is needed to define what formulas to assign to modifiers and mmodifiers in order to obtain effects comparable to the formulas used with precision modifiers.

Another important difference between the two proposals is that the linguistic negation method provides different results on the basis of the element to be evaluated, while mmodifier approach give always the same semantic to an altered modifier independently of the entities in exam.

This different behaviour makes the mmodifier approach better suited to situations which require comparing the membership of two or more elements, for example in the case of information filtering. Linguistic modifier method, on the other hand, can be useful for discriminating the effective meaning of a modifier when referred to a particular element (for example, to understand if a sentence like "Matt is not very tall" means that Matt has a normal height or is quite small). This can be useful, for instance, when dealing with semantic interpretation or document annotation problems.

## 4 Case of Study

In this section we are going to illustrate the procedure to evaluate the meaning of a negated expression within the two frameworks. We'll show the process involved in finding the most suitable interpretation for the sentence "that man is not very high", when this expression is referred to an individual with a height of 179 centimeters.

We start illustrating the method described in Section 2.2. According to it, we have that $c$ is the *height* concept, $\mathcal{D}_c$ is the domain of the people relative to the concept $c$, the set of the properties is $\mathcal{P}_c = \{low, medium, high\}$ and the set of modifiers is $\mathcal{H} = \{little, very, vaguely, extremely\}$ obtained considering the precision modifier set $\{little, very\}$ and the translation modifiers set $\{vaguely, extremely\}$ jointly. So, the set of modifiers is $\mathcal{M}$ = {*little little, little*

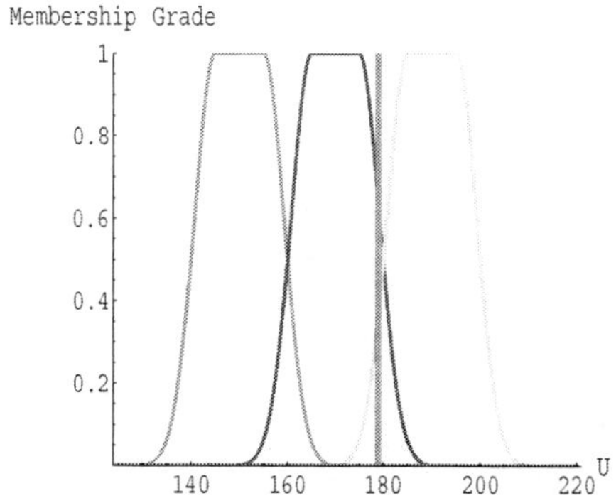


**Fig. 3.** The concept frame used in the example, where it is marked the case of $x = 179$

*very, little, little vaguely, little extremely, very, very very, very little,*...} where the length of the chain of the modifiers has been limited to two elements as defined in Section 2.1. The concept frame we used is shown in Figure 3.

Figure 4(a) shows all modifiers applied to the three candidate properties ("low", "medium" and "high") in order to assign the meaning of the sentence negated. According to the definition of reference frame, they are all the nuanced properties except "very high".

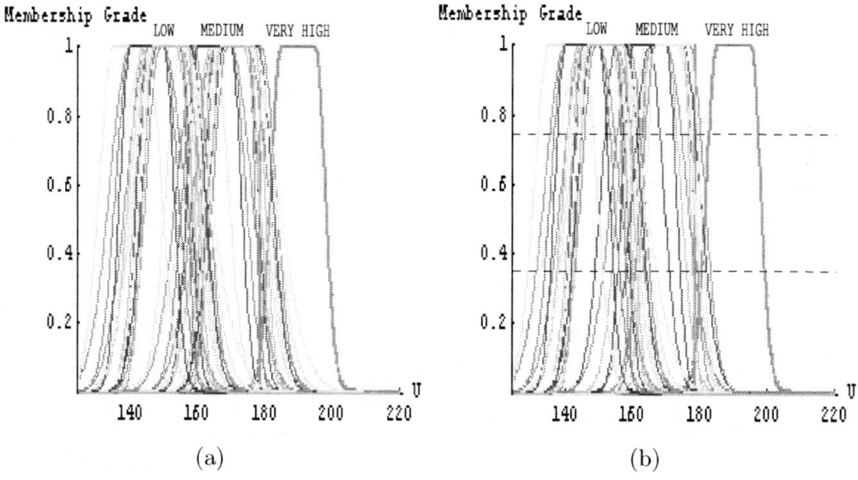


**Fig. 4.** (a) All the modifiers applying the *reference frame* model. (b) The family of solutions obtained.

In order to define the restricted reference frame (see Definition 3 of Section 2.2), the first problem is to understand which values have to be assigned to the $\rho$ and $\varepsilon$ variables. Following the considerations given by Pacholczyk [5, 6, 13, 7, 8, 9], we have set $\rho = 0.75$ and $\epsilon = 0.35$. In detail, we have that

$Neg_{0.75,0.35}$(very high man) contains all the nuances of "low", all the nuances of "medium" except {*little little, little, little extremely, extremely extremely, extremely little, extremely vaguely*} and the following nuances of "high": {*very vaguely, vaguely, vaguely vaguely, vaguely very*}. In Figure 4(b) is reported the family of all these solutions.

In order to have a unique interpretation of the sentence negated we apply the algorithm of Section 2.2. By the first step of the algorithm we have that the chosen property is "high" (see Figure 3). Then, according to the second step, the chosen modifier is *very vaguely.* In this case we have to calculate the value of an hybrid chain of hedges as reported in Section 2. In detail, this chain is composed by a precision and a translation modifier. Applying the Khang et al.'s algorithm, we obtain $\gamma$ and $\beta$ values to use consecutively as $f(x \pm \gamma)^{\beta}$. In Table 4 are reported the signs and $\phi$ values of the concept modifier *very extremely.*

**Table 4.** Calculus of the *sign* and $\phi$

| $\mathcal{M}$ | sign | $\phi$ |
|---|---|---|
| very | +1 | $\beta = 2.0$ |
| vaguely | +1 | $\gamma = 5.0$ |

So, the intended meaning of the statement "that man is not very high" is "that man is very vaguely high" with value 0.94, i.e., $0.94 = f(179 + 5.0)^{2.0} = 0.97^{2.0}$.

Now, we can compare the approach given by the dynamic modification of the linguistic negation model given in this chapter with the original method proposed by Pacholczyk [5, 6, 13, 7, 8, 9]. The first difference is given by the static number of modifiers used. Indeed, this approach allows to use the sets of

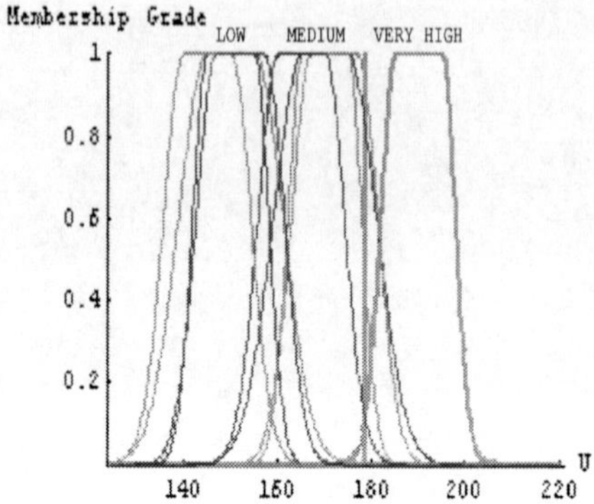


(a)

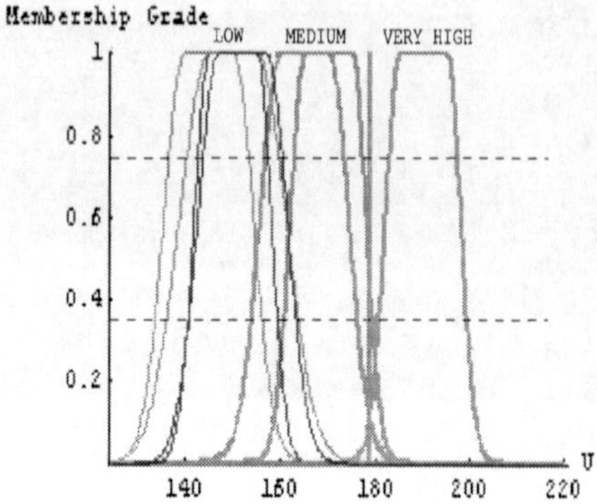


(b)

**Fig. 5.** (a) All the modifiers using Pacholczyk's model. (b) The family of solutions obtained.

precision and translation modifiers $\mathcal{M}$ ={*very,little,vaguely, extremely*} given by the expert of domain on the "low" and "medium" properties, and the same set $\mathcal{M}$ except the modifier "very" on the property "high" (see Figure 5(a)). The choice of the family of solution is given in the same way previously proposed by Definition 3. Figure 5(b) shows the family of modifiers obtained deleting the set {*little, extremely*} for the property "medium" and deleting all the modifiers except *vaguely* for the property "high".

The other difference is given by the strategy used in order to obtain one interpretation of the negation. Indeed, Pacholczyk strategy choices the solution leading to the most significant membership degree and having the weakest complexity [5, 6, 13, 7, 8, 9]. Thus, "high" property and "vaguely" modifier are the elements obtained using the original method. In this case the meaning of the sentence "that man is not very high" is "that man is vaguely high" with value 0.97, i.e., $f(179+5) = 0.97$.

The possibility of having a major number of modifiers in the new model than in the original one proposed by Pacholczyk, let us to enrich the semantic of the model, allowing a better interpretation of the negation nearer to the natural language used by humans.

Using the second approach, the membership function related to the term "not very high" is obtained by applying the mmodifier function to the linguistic modifier "very". Thus, we apply the obtained modifier function to the fuzzy membership function which represents the semantic of the adjective "high". What we obtain is the function used to represent in fuzzy terms the meaning of the property "not very high". Differently than the previous example, this function is the only one that represents the property's semantic for all the peoples' height which constitute our example domain. If we call $f_{\text{high}}$ the membership function related to the fuzzy-set named "high", the membership function related to "not very high" will be $\sqrt{f_{\text{high}}(x)}$, according to what we said in Section 2.3. Since an element of height 179 has a membership value of 0.38 of the membership degree of high, its membership function for "not very high" will be $\sqrt{0.38} = 0.62$. This value sound reasonable with respect to the usual meaning of the expression "not very high", which indicates an element *slightly lower* than an ordinary "high" element.

## 5 Conclusions

In this chapter we discussed the importance of the linguistic negation's interpretation. In order to solve the problem two methods have been proposed, both based on fuzzy sets theory. With the first method, we can have a family of possible interpretations, whereas in the other a unique solution is obtained. A case of study has also been presented showing the differences between these approaches. As a future development we plan to integrate these methods in the fuzzy ontology framework [3]. This will also enable us to perform a deeper analysis of the methods, comparing their results in real situations.

## References

1. Berners-Lee, T., Hendler, T., Lassila, J.: The semantic web. Scientific American 284(5), 34–43 (2001)
2. Shi, H., Ward, R., Kharma, N.: Expanding the Definitions of Linguistic Hedges. In: Joint 9th IFSA World Congress and 20th NAFIPS International Conference (2001)
3. Calegari, S., Ciucci, D.: Integrating Fuzzy Logic in Ontologies. In: Proceedings of ICEIS 2006, pp. 66–73 (2006)
4. Calegari, S., Ciucci, D.: Fuzzy ontology, fuzzy description logics and fuzzy-owl. In: Proceedings of WILF 2007. LNCS, Springer, Heidelberg (accepted, 2007)
5. Pacholczyk, D.: A new approach to linguistic negation of nuanced information in knowledge-based systems. In: Giunchiglia, F. (ed.) AIMSA 1998. LNCS (LNAI), vol. 1480, pp. 363–376. Springer, Heidelberg (1998)
6. Pacholczyk, D.: A new approach to linguistic negation based upon compatibility level and tolerance threshold. In: Polkowski, L., Skowron, A. (eds.) RSCTC 1998. LNCS (LNAI), vol. 1424, Springer, Heidelberg (1998)
7. Pacholczyk, D., Quafafou, M., Garcia, L.: Optimistic vs. pessimistic interpretation of linguistic negation. In: Scott, D. (ed.) AIMSA 2002. LNCS (LNAI), vol. 2443, pp. 132–141. Springer, Heidelberg (2002)
8. Pacholczyk, D., Hunter, A.: An extension of a linguistic negation model allowing us to deny nuanced property combinations. In: Hunter, A., Parsons, S. (eds.) ECSQARU 1999. LNCS (LNAI), vol. 1638, pp. 316–327. Springer, Heidelberg (1999)
9. Pacholczyk, D., Levrat, B.: Coping with linguistically denied nuanced properties: A matter of fuzziness and scope. In: Proceeding of ISIC - IEEE, pp. 753–758 (1998)
10. Zadeh, L.A.: A fuzzy-set-theoretic interpretation of linguistic hedges. Journal of Cybernetics 2, 4–34 (1972)
11. Khang, T.D., Störr, H., Hölldobler, S.: A fuzzy description logic with hedges as concept modifiers. In: Third International Conference on Intelligent Technologies and Third Vietnam-Japan Symposium on Fuzzy Systems and Applications, pp. 25–34 (2002)
12. Abulaish, M., Dey, L.: Ontology Based Fuzzy Deductive System to Handle Imprecise Knowledge. In: Proceedings of InTech 2003, pp. 271–278 (2003)
13. Pacholczyk, D., Quafafou, M.: Towards a linguistic negation approximation based on rough set theory. In: Proceedings of ICAI, pp. 542–548 (2002)
14. Zadeh, L.A.: The concept of a linguistic variable and its application to approximate reasining — part I, II and III. Information Sciences 8(9), 199–251, 301–357, 43–80 (1975)
15. Zadeh, L.A.: A fuzzy-set-theoretic interpretation of linguistic hedges. Journal of Cybernetics 2(3), 4–34 (1972)

# A Coevolutionary Approach to Solve Fuzzy Games

Wanessa Amaral and Fernando Gomide

Department of Computer Engineering and Automation
Faculty of Electrical and Computer Engineering
State University of Campinas, 13083-970 Campinas, SP, Brazil

**Summary.** This chapter addresses fuzzy games within the framework of coevolutionary computation. In fuzzy games, payoff matrices elements are fuzzy numbers and the players must find strategies to optimize their payoffs. The co-evolutionary approach suggested herein is a heuristic procedure that maintains a population of players, each of which having a particular strategy. Both, zero and nonzero sum games are solved by the coevolutionary procedure. Contrary to mathematical programming-based and other heuristic solution procedures, the coevolutionary approach produces a set of solutions whose strategies achieve comparable payoffs.

## 1 Introduction

In this chapter we introduce a coevolutionary approach to solve non cooperative two-player games with fuzzy payoffs. Non-cooperative games assume that there is no communication between the players or, if there is, players do not agree on bidding strategies and act rationally. Game theory research plays an important role in decision making theory and in many practical situations especially in economics [1], mechanism design and market analysis [2], multi-agent systems [3], deregulated energy markets [4] and biology [5], to mention a few.

In actual situations, however, it is difficult to know the values of payoffs exactly. Recently, considerable attention has been given to game problems with fuzzy payoff matrices to approach both, zero and non-zero sum games.

In parallel to theoretical developments, methods for solving fuzzy game problems have also been developed. For instance, Campos [6] proposes five different ranking functions to turn a fuzzy payoff matrix into a linear matrix for solving zero sum fuzzy game problems. Lee-Kwang and Lee [7] present a method to rank fuzzy numbers and use the ranking method to solve decision-making problems. Maeda [8] shows that equilibrium strategies of fuzzy games can be characterized by Nash equilibrium strategies of a family of parametric bi-matrix games with crisp payoffs.

Genetic algorithms have been used to study learning in games, Axelrod [9] being among the pioneers. More recently links between genetic algorithms, learning and evolutionary games have been reported. Borges [10] proposed an approach where the iterative prisoners dilemma's possible cooperation and defection moves

R. Bello et al. (Eds.): Granular Computing, STUDFUZZ 224, pp. 121–130, 2008.
springerlink.com 

are modeled as fuzzy sets. The use of co-evolution to address oligopolistic markets is considered by Chen et al. [11] to analyze different equilibrium models during strategic interaction between agents in markets, including Cournot, Bertrand, Stackelberg and others schemes. The use of co-evolution allows the study of the behavior of many populations of individuals. In coevolutionary approaches, fitness of a single individual of a population relates to individuals of different populations. Therefore populations interact and evaluation of an individual of a population depends of the state of the evolutionary process of the remaining populations. Coevolutionary frameworks also give more information to decision makers because they do not produce just the equilibrium solution, but a family of near equilibrium solutions for the game because co-evolution considers populations of candidate solutions instead of a single solution.

This chapter presents a coevolutionary algorithm to solve non-cooperative fuzzy game problems with mixed strategies. The algorithm differs from previous works addressed in the literature [9] [11] because it allows pure and mixed strategies, using float representations and co-evolution to obtain a family of solutions, a more realistic approach when dealing with real world applications.

After this introduction, the next section briefly reviews the basic concepts of fuzzy sets and games, and section 3 details the co-evolutionary approach suggested herein. Section 4 discusses experimental results. Finally, conclusions and future work are summarized in the last section.

## 2 Fuzzy Sets and Games

In this section, basic notions on fuzzy sets and fuzzy games are briefly reviewed. An introduction to fuzzy sets is given in Pedrycz and Gomide [12]. For a comprehensive treatment of fuzzy and multiobjective games see, for example, Nishizaki and Sakawa [13].

A fuzzy set $\tilde{a}$ is characterized by a membership function mapping the elements of a universe of discourse $X$ to the unit interval [0, 1]:

$$\tilde{a} : X \rightarrow [0, 1] \tag{1}$$

A fuzzy number is a fuzzy set $\tilde{a}$ that has the set of real numbers $\mathbb{R}$ as its universe of discourse and possess the following properties:

1. There exists a unique real number x such that $\tilde{a}(x) = 1$
2. $\tilde{a}_\alpha$ must be a closed interval for every $\alpha \in [0, 1]$
3. The support of $\tilde{a}$ must be bounded.

where $\tilde{a}_\alpha = \{x \mid \tilde{a}(x) \geq \alpha\}$, is the $\alpha$-cut of $\tilde{a}$. The $\alpha$-cut $\tilde{a}_\alpha$ of $\tilde{a}$ is a set consisting of real numbers whose membership values are equal to or exceed a certain threshold level $\alpha$.

This chapter considers games with fuzzy payoff matrices, namely, games whose elements of the payoff matrices are fuzzy numbers, that is:

$$\tilde{A} = \begin{bmatrix} \tilde{a}_{11} & \cdots & \tilde{a}_{1n} \\ \vdots & \ddots & \vdots \\ \tilde{a}_{m1} & \cdots & \tilde{a}_{mn} \end{bmatrix} \tag{2}$$

where $\tilde{a}_{ij}$ is a fuzzy number. Here we assume the reader to be familiar with the notion of payoff matrices for two-person zero-sum games. See [14] for an introduction.

### 2.1 Zero-Sum Games

Let I and II be the players of the game, and let $\tilde{A}$ be the corresponding payoff matrix. A mixed strategy $x = (x_1, \ldots, x_m)$ for player I is a probability distribution on the set of his pure strategies, represented by:

$$X = \left\{ x = (x_1, \ldots, x_m) \in \mathbb{R}^m \mid \sum_{i=1}^{m} x_i = 1,\ x_i \geq 0,\ i = 1, 2, \ldots, m \right\} \tag{3}$$

where $\mathbb{R}^m$ is the set of m-dimensional real numbers [13]. A mixed strategy $y = (y_1, \ldots, y_n)$ for player II is defined similarly. The expected payoff $P$ of the game is given by the function:

$$P(x, y) = \sum_{i=1}^{m} \sum_{j=1}^{n} x_i\, a_{ij}\, y_j = \mathbf{xAy} \tag{4}$$

A game is zero-sum if and only if the total amount that a player gains is the same as the other player loses. Thus, the game has only one payoff matrix.

For a two-person zero-sum game, the worst possible expected payoff for player I is $\nu_I = \min_{y \in Y} \mathbf{xAy}$ . Thus, player I aims to maximize $\nu_I(x)$. Similarly, the worst possible expected payoff for player II is $\nu_{II} = \max_{x \in X} \mathbf{xAy}$ and player II aims to minimize $\nu_{II}(x)$.

It is well known that, for zero-sum games, the von Neumann max-min theorem holds [13]. For a two-person zero-sum game $A$ it follows that:

$$\max_{x \in X} \min_{y \in Y} \mathbf{xAy} = \min_{y \in Y} \max_{x \in X} \mathbf{xAy} \tag{5}$$

A pair of strategies $(x^*, y^*)$ satisfying expression (5) is an equilibrium solution.

### 2.2 Nonzero-Sum Games

Two payoff matrices $A$ and $B$, one for each player, represent two-person nonzero-sum games. Because of this fact, these games are often called bi-matrix games.

$$A = \begin{bmatrix} a_{11} & \cdots & a_{1n} \\ \vdots & \ddots & \vdots \\ a_{m1} & \cdots & a_{mn} \end{bmatrix} \text{ and } B = \begin{bmatrix} b_{11} & \cdots & b_{1n} \\ \vdots & \ddots & \vdots \\ b_{m1} & \cdots & b_{mn} \end{bmatrix} \tag{6}$$

Equilibrium solutions for nonzero-sum game are pairs of $m$-dimensional vectors $x^*$ and $n$-dimensional vectors $y^*$, such that:

$$\begin{aligned} \mathbf{x}^* \mathbf{A} \mathbf{y}^* &\geq \mathbf{x} \mathbf{A} \mathbf{y}^* \\ \mathbf{x}^* \mathbf{B} \mathbf{y}^* &\geq \mathbf{x}^* \mathbf{B} \mathbf{y} \end{aligned} \tag{7}$$

## 3 Coevolutionary Algorithm to Solve Fuzzy Games

Evolutionary computation encompasses different population-based heuristic search methods that use reproduction, mutation, recombination and adaptation as operators to find solutions to complex optimization and related problems [15].

Evolutionary algorithms are powerful but their implementation can become difficult for certain types of problems. This is the case in game theory problems because it is not obvious how to choose a key component of evolutionary procedures, the fitness function. In game problems there are at least two populations evolving at the same time, and each population affects the evolution of the other. In games scenario, the fitness values for any population depend not only on the performance of the population individuals, but also on the interaction between the populations involved. In such circumstances the use of coevolutionary algorithms is more appropriate. Generally speaking, coevolutionary computational algorithms and models can be regarded as a special form of agent-based procedures in the sense of systems of interacting agents. The ability to capture the independent decision-making behavior and interactions of individual agents provide a powerful platform to model fuzzy games. In coevolutionary algorithms, several evolutionary processes take place simultaneously and fitness of each single element in a population is related to the fitness of the remaining individuals. Different populations interact and evaluation of individuals depends on the state of the evolution process as a whole, not individually [15]. Interaction of a single individual with different individuals of the different populations is part of the evaluation.

The coevolutionary algorithm addressed in this chapter considers a population of players, each of which with a particular strategy. Initially, the strategy of each player is chosen randomly. At each further generation, players play games and their scores are memorized. Similarly as in genetic algorithms, mutation and crossover operators are used to evolve the populations, and some players are selected to be part of the next generation. The coevolutionary algorithm searches for mixed strategies and do not record past moves. Players strategies emerge as a result of the evolutionary process.

The algorithm uses real-valued chromosomes, each of which representing a single mixed strategy. The values are such that they satisfy (3).

The performance of an individual is evaluated using a fitness function. Let $\tilde{A}$ be the fuzzy payoff matrix of a zero-sum game and let $x$ and $y$ be the strategies for players I and II, respectively. The fitness function is as follows:

$$F_{x_i} = \frac{\sum_{j=1}^{n} x_i \; a(\lambda)_{ij} \; y_j}{n} \quad \text{and} \quad F_{y_j} = \frac{\sum_{i=1}^{m} x_i \; a(\mu)_{ij} \; y_j}{m} \tag{8}$$

where $m$ and $n$ are the number of moves for players I and II, and $F_{x_i}$ and $F_{y_j}$ the fitness values for elements $x_i$ and $y_j$ respectively. The parameters $\lambda$ and $\mu \in [0,1]$ are set according to membership degrees as in (9):

$$\begin{aligned} A(\lambda) &= M + (1-2\lambda)H \\ A(\mu) &= M + (1-2\mu)H \end{aligned} \tag{9}$$

where $M$ is the center of the fuzzy number and $H$ the deviation parameter [8] [12]. Thus, the parameters $\lambda$ and $\mu$ are set according to the desired membership degree, as illustrated in Fig. 1.

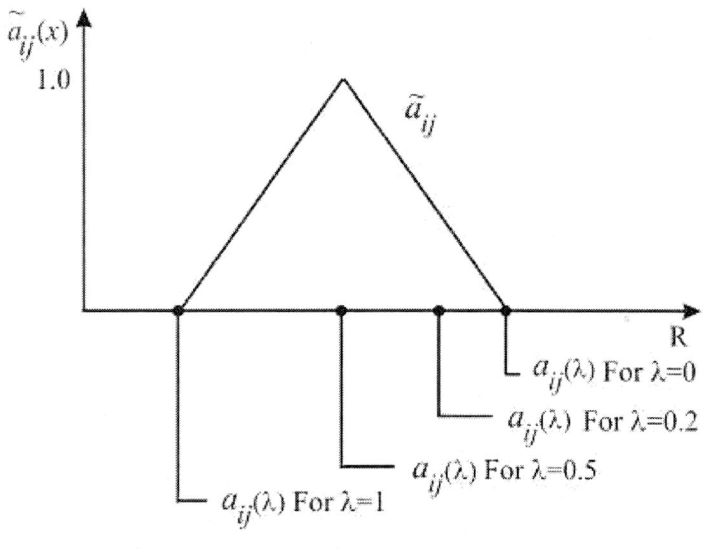


**Fig. 1.** Relation between the elements of $A(\lambda)$ and $\tilde{A}$

In two-player games each population has individuals to represent players I and II strategies. They evolve simultaneously and, as (8) suggests, the fitness of individuals of a population depends on the strategies of the other population individuals.

Local search is used to signalize the algorithm that an individual $I$ is placed in an efficient location of the search space. Local search finds individuals that are close to $I$, the neighbor individuals of $I$. Neighbor individuals differ by a small value, which can be different for different problems. In our case, the neighbor individuals are close to $I$ within 0.5 units or less. The algorithm selects neighbor individuals randomly and computes their fitness. If the average fitness of neighbor individuals is better than the fitness of $I$, then we add some units to the fitness of $I$ to signalize to the evolutionary algorithm that $I$ is probably placed in a good location in the search space.

Selection procedures are extremely important in evolutionary processes. In this work we adopt the tournament selection mechanism. A small subset of individuals is chosen at random and the best individual of this set is selected. This process is repeated $m$ times if the population size is $m$.

The coevolutionary algorithm works combining selection with diversity mechanisms to increase effectiveness of crossover and mutation. For instance, assume that a game has $n$ possible decisions $x_1, x_2, \ldots, x_z, \ldots, x_n$, and let $k$ be the original individual and $k+1$ the mutated individual. The mutated individual should satisfy the following:

$$x_z^{k+1} = \begin{cases} x_{z+1}^k, & \text{if } z \neq n \\ x_1^k, & \text{if } z = n \end{cases} \tag{10}$$

Crossover is the interchange of genetic material between two good solutions, intended to produce offspring with some similarity to their parents. One commonly used operator for crossover in real-valued representations is the arithmetic crossover [15], a linear combination of two individuals $x_a$ and $x_b$ defined as follows:

$$x = x_a \cdot \alpha + x_b \cdot (1 - \alpha) \tag{11}$$

where the real value $\alpha \in [0, 1]$ is chosen randomly. The coevolutionary algorithm presented here uses the arithmetic crossover.

After evaluation of the population individuals, the best fitting individual is selected and crossover and mutation operators are applied to create the new generation. The new generation keeps solutions close to the best solutions found in the previous generation.

The same procedure can be easily adapted to handle non-zero sum games. The only difference is the definition of the fitness function, which should be as follows:

$$F_{x_i} = \frac{\sum_{j=1}^{n} x_i\, a(\lambda)_{ij}\, y_j}{n} \quad \text{and} \quad F_{y_j} = \frac{\sum_{i=1}^{m} x_i\, b(\mu)_{ij}\, y_j}{m} \tag{12}$$

where $n$, $m$, $\lambda$ and $\mu$ are similar to (8), $F_{x_i}$ and $F_{y_j}$ are the fitness values for elements $x_i$ and $y_j$ which are individuals of population for players I and II, respectively.

The following procedure summarizes the coevolutionary algorithm to obtain mixed strategies for fuzzy games:

1. Start creating $m$ ($m$ is the population size) vectors with random values satisfying (3) for each population. This is the first generation. Each population represents strategies of a specific player.

2. A tournament is performed. Elements of each population play against the elements of the other population and their fitness is computed using (8) and (12), for zero and non-zero sum games, respectively. Next, the local search is ran, and individuals located in a good region of the search space receive larger fitness.
3. Select, for each population, the $n$ players with the highest fitness. Apply crossover and mutation operators to the individuals selected.
4. Create next generation with the individuals selected in step 3) and their children, namely, solutions close to the individuals selected in step 3).
5. If the stop condition holds, then end, else go to step 2).

The stop condition is problem-dependent. In this work we use the simplest, namely, the maximum number of generations.

## 4 Numerical Experiments

First, we consider a zero-sum game problem addressed in [6]. The fuzzy payoff matrix is:

$$\tilde{A} = \begin{bmatrix} 180,5 & 156,5 \\ 90,5 & 180,5 \end{bmatrix} \tag{13}$$

We consider the entries of $\tilde{A}$ as symmetric triangular fuzzy numbers.

**Table 1.** Solutions for the zero-sum game

| Experiment | $x$ | $y$ | Payoff |
|---|---|---|---|
| I | (0.7941, 0.2059) | (0.1550, 0.8450) | 161.0233 |
| II | (0.5975, 0.4025) | (0.1289, 0.8711) | 162.8394 |
| III | (0.8968, 0.1032) | (0.1279, 0.8721) | 160.0422 |

Table 1 shows a sample of mixed strategies evolved for players I and II. These values are the best results of three experiments, that is, the highest fitness value for the three times the algorithm was ran. Table 1 shows several solutions, one optimum solution for each experiment. Note that the payoff values are very close for all the strategies. This gives the decision-maker more information to choose his play to consider, e.g., information not explicitly accounted by the formal game model.

Figure 2 shows how the average fitness of each population evolves along 300 generations. Steep variations in the figure are due to mutation. The experiments were performed using populations of 60 individuals each, mutation and crossover rate set at 0.01 and 0.5, respectively. Different experiments were carried out to verify how co-evolution behaves when mutation and crossover rates change. Large mutation and crossover rates introduce larger *noise* in population

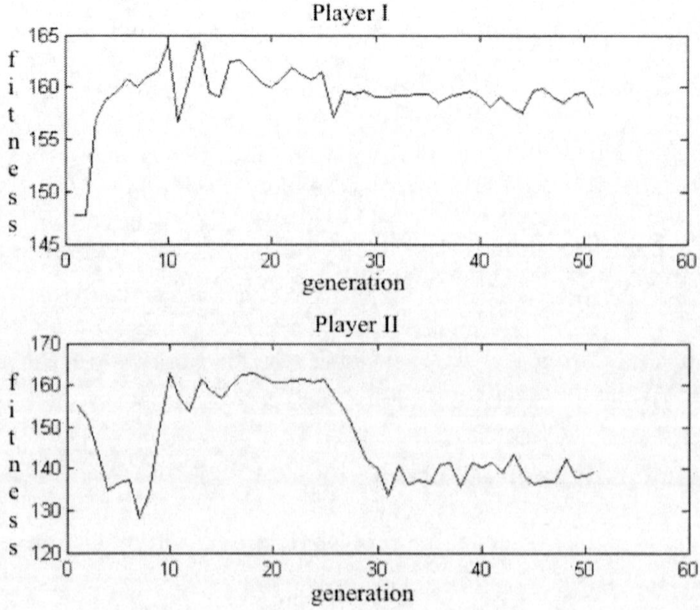


**Fig. 2.** Evolution of average fitness of populations for the zero-sum game

and make convergence very difficult. On the other hand, with too small rates the algorithm may converge to wrong solutions. Experimentally, small mutation rates and moderate crossover rates have shown to perform successfully.

A non zero-sum game example was also solved. We transformed the original matrices into fuzzy matrices, assuming symmetric triangular fuzzy numbers with modal values set equal to the original matrices values. The modal values of the fuzzy numbers of $\tilde{A}$ and $\tilde{B}$ are the entries of the original matrices. The deviation parameter value was set at 0.2.

$$\tilde{A} = \begin{bmatrix} 1, 0.5 & 0, 1 \\ 2, 1.5 & -1, 0.5 \end{bmatrix} \quad \text{and} \quad \tilde{B} = \begin{bmatrix} 3, 1.5 & 2, 1 \\ 0, 0.5 & 1, 1 \end{bmatrix} \tag{14}$$

Table 2 shows the solutions obtained. Again we note different solutions with similar payoffs. Therefore, similarly as with the zero-sum game example, decision-makers obtain more information because they get several strategies with similar payoffs. Fig. 3 shows the average fitness of each population along 80 generations.

**Table 2.** Solutions for the non-zero sum game

| $x$ | $y$ | Payoff for player I | Payoff for player II |
|---|---|---|---|
| (0.5496, 0.4504) | (0.5552, 0.4448) | 0.6050 | 1.6046 |
| (0.5616, 0.4384) | (0.5438, 0.4562) | 0.5823 | 1.6286 |

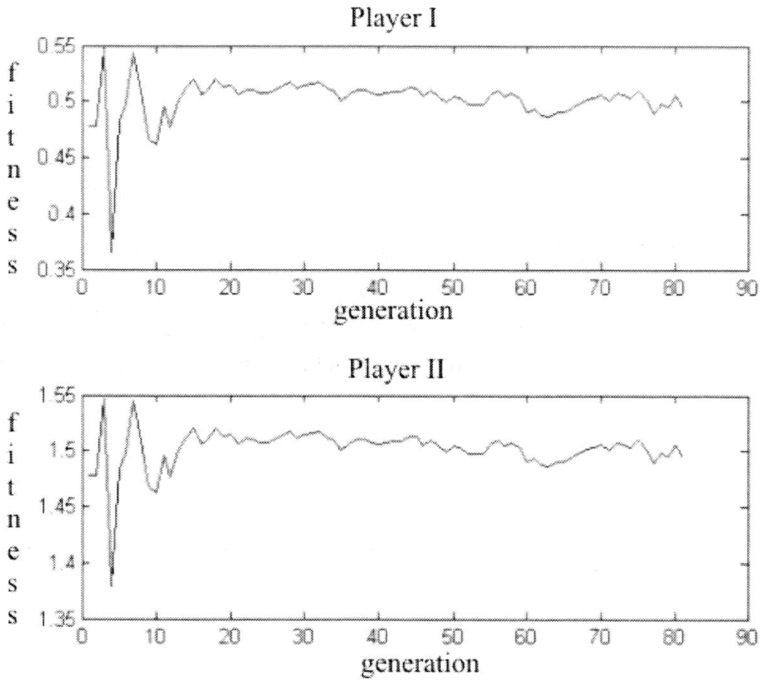


**Fig. 3.** Average fitness of population for the non-zero sum game

Further experiments with the coevolutionary approach show that the exploration of the search space is more effective because the algorithm avoids local optima exploring diversity. This fact indicates that the algorithm can be useful to solve complex decision problems involving non-linear payoffs and discrete search spaces.

Since the coevolutionary algorithm evolves decisions with similar payoffs, it provides the decision maker not only with the optimal solution, but also a set of alternative solutions. This is useful in real-world decision-making scenarios where information other than payoffs and rationality usually play a significant role. We note that the coevolutionary approach solves both, zero and non-zero sum fuzzy games.

## 5 Conclusions

This chapter presented a coevolutionary algorithm to find equilibrium solutions of two-person non-cooperative games with fuzzy payoff matrices and mixed strategies.

The coevolutionary approach uses different populations of candidate solutions, and individuals are evaluated by fitness functions that depend on different individuals of different populations. Each population is associated with a player.

Conventional genetic operators, namely, selection, mutation and recombination, can be effectively used, but the choice of the corresponding rates influence the algorithm behavior. With proper choice of mutation and crossover rates, experiments with game problems addressed in the literature were performed. The coevolutionary algorithm evolved both, the theoretically optimal solution and alternative solutions with payoffs values closer to the optimal ones. This is an interesting characteristic in practice because the algorithm develops a set of nearly optimal solutions instead of just a single one.

Despite the promising results achieved so far, further work still needs to be done. For instance, iterative games are often closer to what actually happens in practical applications, such as in energy markets and multi-agent computer systems for example. Solution of fuzzy games considering finite memory of past moves is an issue that deserves further investigation. Another consideration concerns extension of the algorithm to handle $n$-person fuzzy games.

## Acknowledgement

The authors would like to thank the anonymous referees for their invaluable comments that helped to improve the chapter. The second author is grateful to CNPq, the Brazilian National Research Council, for its support via grant 304857/2006-8.

## References

1. Mas-Collel, A., Whiston, W., Green, J.: Microeconomic Theory. Oxford University Press, Oxford (1995)
2. Nisan, N., Ronen, A.: Games and Economic Behavior 35, 166–196 (2001)
3. Weiss, G.: Multiagent systems: A modern approach to distributed artificial intelligence. MIT Press, Cambridge (1999)
4. Green, R.: Competition in generation: The economic foundations. Proceedings of the IEEE 88(2), 128–139 (2000)
5. Smith, J.: Evolution and the Theory of Games. Cambridge University Press, Cambridge (2000)
6. Campos, L.: Fuzzy Sets and Systems 32, 275–289 (1989)
7. Lee-Kwang, H., Lee, J.: IEEE Trans. on Fuzzy Systems 7, 677–685 (1999)
8. Maeda, T.: Fuzzy Sets and Systems 139, 283–296 (2003)
9. Axelrod, R.: The evolution of cooperation. Basic Books, New York (1984)
10. Borges, P., Pacheco, R., Khator, Barcia, R.: A fuzzy approach to the prisoner's dilemma. BioSystems (1995)
11. Chen, H., Wong, K., Nguyen, D., Chung, C.: IEEE Trans. Power Systems 21, 143–152 (2006)
12. Pedrycz, W., Gomide, F.: Fuzzy systems engineering: Toward human-centric computing. Wiley Interscience, Hoboken, New Jersey (2007)
13. Nishizaki, I., Sakawa, M.: Fuzzy and multiobjective games for conflict resolution. Physica-Verlag, New York (2001)
14. Osborne, M., Rubistein, A.: A Course in Game Theory. MIT Press, Cambridge (1994)
15. Michalewicz, Z.: Genetic Algorithms + Data Structures = Evolution Programs. Springer, Heidelberg (1996)

# Rough Set Approach to Video Deinterlacing Systems

Gwanggil Jeon[1], Rafael Falcón[2], and Jechang Jeong[1]

[1] Department of Electronics and Computer Engineering, Hanyang University
17 Haengdang-dong, Seongdong-gu, Seoul, Korea
{windcap315,jjeong}@ece.hanyang.ac.kr

[2] Computer Science Department, Central University of Las Villas
Carretera Camajuaní km 5 1/2, Santa Clara, Cuba
rfalcon@uclv.edu.cu

**Summary.** A deinterlacing algorithm that is based on rough set theory is researched and applied in this chapter. The fundamental concepts of rough sets, with upper and lower approximations, offer a powerful means of representing uncertain boundary regions in image processing. However, there are a few studies that discuss the effectiveness of the rough set concept in the field of video deinterlacing. Thus, this chapter proposes a deinterlacing algorithm that will choose the most suitable method for being applied to a sequence, with almost perfect reliability. This proposed deinterlacing approach employs a size reduction of the database system, keeping only the essential information for the process. Decision making and interpolation results are presented. The results of computer simulations show that the proposed method outperforms a number of methods presented in the literature.

**Keywords:** rough set theory, deinterlacing, information system, reduct, core.

## 1 Introduction

Interpolation is a method of constructing new data points from a discrete set of known data points [1]. Sometimes, interpolation is called as resampling, which is an imaging method to increase (or decrease) the number of pixels in a digital image. Interpolation is used in many image-processing applications such as image enhancement, deinterlacing, scan-rate conversion, etc. Among these applications, deinterlacing is a very active research area. The current analog television standards, such as NTSC, PAL, and SECAM, are still widely used in the television industry and they will be included in future DTV standards. However, the sampling process of interlaced TV signals in the vertical direction does not satisfy the Nyquist sampling theorem [2], and the linear sampling-rate conversion theory cannot be utilized for effective interpolation [3]. This causes several visual artifacts which decrease the picture quality of the interlaced video sequence.

Deinterlacing methods can be roughly classified into three categories: spatial domain methods, which use only one field; temporal domain methods, which use

R. Bello et al. (Eds.): Granular Computing, STUDFUZZ 224, pp. 131–147, 2008.
springerlink.com 

multiple fields; and spatio-temporal domain methods [4] [5]. The most common method in the spatial domain is Bob, which is used on small LCD panels [6]. Weave is the most common method in the temporal domain [7]. There exist many edge-direction-based interpolation methods. The edge line average (ELA) algorithm was proposed to interpolate pixels along the edges in the image [8]. ELA utilizes only the spatial domain information. However, the amount of data limits the interpolation by causing missing pixels at complex and motion regions. Thus, spatio-temporal line average (STELA) was proposed in order to expand the window to include the temporal domain [5].

Making essential rules is not an easy task, since various features offer several attributes for the nature of a sequence. Thus, the conventional deinterlacing method cannot be applied to build an expert system. In order to create an expert system, rough set theory [9] is applied to classify the deinterlacing method. In this theory, prior knowledge of the rules is not required, but the rules are rather automatically discovered from a database. Rough set theory provides a robust and formal way of manipulating the uncertainty in information systems. Sugihara and Tanaka proposed a new rough set approach which deals with ambiguous and imprecise decision system [11]. Rough set theory has been applied to several engineering fields such as knowledge discovery [11], feature selection [12], clustering [13], image recognition and segmentation [14], quality evaluation [15] and medical image segmentation [16]. It has proved to be a profitable tool in real-world applications as well, such as semiconductor manufacturing [17], landmine classification [18] and power system controllers [19]. The rough set methodology has been used in image processing; however its application to video deinterlacing has not been investigated. The deinterlacing technique causes a mode decision problem, because the mode decision method may affect interpolation efficiency, complexity as well as objective and subjective results. We propose the study involving deinterlacing systems that are based on Sugihara's extended approach to rough set theory.

In this chapter, a decision making algorithm that utilizes rough sets to video deinterlacing problems is introduced. The way decision making is carried out for deinterlacing is intrinsically complex due to the high degree of uncertainty and the large number of variables involved. Our proposed deinterlacing algorithm employs four deinterlacing methods: Bob, Weave, ELA and STELA.

The rest of the chapter is structured as follows. In Section 2, basic notions of rough set theory will be discussed. In Section 3, we will briefly review some of the conventional deinterlacing methods. The details of the proposed rough-set-based deinterlacing algorithm (RSD) are given in Section 4. Experimental results and conclusions are finally outlined in Sections 5 and 6.

## 2 Rough Set Theory: Fundamental Ideas

Rough sets, introduced by Pawlak et al., are a powerful tool for data analysis and characterization of imprecise and ambiguous data. They have successfully been used in many application domains, such as machine learning and expert systems

[9]. Let $U \neq \emptyset$ be a universe of discourse and $X$ be a subset of $U$. An equivalence relation $R$ partitions $U$ into several subsets $U/R = \{X_1, X_2, \ldots, X_n\}$ in which the following conditions are satisfied:

$$X_i \subseteq U, \quad X_i \neq \emptyset\, \forall i, \quad X_i \bigcap X_j \neq \emptyset\, \forall i, j \;\; and \;\; \bigcup_{i=1,2\ldots n} X_i = U$$

Any subset $X_i$, which is called a category or class, represents an equivalence class of $R$. A category in $R$ containing an object $x \in U$ is denoted by $[x]_R$. For a family of equivalence relations $P \subseteq R$, an indiscernibility relation over $P$ is denoted as IND(P) and defined as follows:

$$IND(P) = \bigcap_{R \in P} IND(R) \tag{1}$$

The set $X$ can be approximated according to the basic sets of $R$, namely a lower approximation and an upper approximation. Such sets are used to represent the uncertainty of the knowledge that the set $X$ describes. Suppose a set $X \subseteq U$ represents a vague concept, then the R-lower and R-upper approximations of $X$ are defined.

$$\underline{R}X = \{x \in U : [x]_R \subseteq X\} \tag{2}$$

The above expression is the set of all elements $x$ belonging to $X$ whose related objects according to $R$ also belong to $X$. This is called the "lower approximation".

$$\overline{R}X = \{x \in U : [x]_R \cap X \neq \emptyset\} \tag{3}$$

On the other hand, expression (3) defines the set of all objects that relate in any degree to any element of $X$. In rough set theory (RST), a decision table is utilized for describing the objects of a universe. The decision table can be seen as a two-dimensional table. Each row is an object and each column is an attribute. Attributes can be divided into condition attributes and decision attributes. Generally, it cannot be said that all of the condition attributes are essential to the purpose of describing the objects. It is a fact that the classification accuracy rises when the surplus of attributes is removed from the decision system.

RST classifies attributes in the decision system into three types according to their role: *core* attributes, *reduct* attributes and *superfluous* attributes. Here, the minimal set of condition attributes which fully describe all objects in the universe is called a *reduct*. One decision system might have several different reducts at the same time. The intersection of those reducts is the *core* of the decision system and the attributes within the core are the ones that actually exercise an influence over the overall classification.

In the conventional rough set theory, it is assumed that the given values with respect to a decision attribute are certain. That is, each object $x$ has only one decision value in the set of decision values. However, there exist some cases in which this assumption is not appropriate to real decision making problems. Sugihara and Tanaka considered the situations that decision values $d(x)$ are given to each object $x$ as interval values [11].

Let $Cl_n(n = 1, \ldots, N)$, be the n-th class with respect to a decision attribute. It is supposed that for all $s$, $t$, such that $t > s$, each element of $Cl_t$ is preferred to each element of $Cl_s$. The interval decision classes (values) $Cl_{[s,t]}$ are defined as:

$$Cl_{[s,t]} = \bigcup_{s \leq r \leq t} Cl_{[r]} \tag{4}$$

It is assumed that the decision of each $x \in U$ belongs to one or more classes, that is, $d(x) = Cl_{[s,t]}$. By $Cl_{[s,t]}$, a decision maker expresses ambiguous judgments to each object $x$. Based on the above equations, the decisions $d(x)$ with respect to the attribute set $P$ can be obtained by the lower and upper approximations as shown below:

$$\underline{P}\{d(x)\} = \bigcap_{y \in R_p(x)} d(y) \tag{5}$$

$$\overline{P}\{d(x)\} = \bigcup_{\{d(z) \supseteq d(y) | y \in R_p(x)\}} d(z) \tag{6}$$

$\underline{P}\{d(x)\}$ means that $x$ certainly belongs to common classes which are assigned to all the elements of the equivalence classes $R_P(x)$. $\overline{P}\{d(x)\}$ means that $x$ may belong to the classes which are assigned to each element of the equivalence classes $R_P(x)$, respectively. It is obvious that the following inclusion relation $\underline{P}\{d(x)\} \subseteq d(x) \subseteq \overline{P}\{d(x)\}$ holds. Equations (5) and (6) are based on the concept of greatest lower and least upper, respectively.

## 3 Conventional Deinterlacing Methods

In this section, we briefly describe three of the previously mentioned algorithms for deinterlacing purposes.

**Bob** is an intra-field interpolation method which uses the current field to interpolate the missing field and to reconstruct one progressive frame at a time. Let $x(i, j-1, k)$ and $x(i, j+1, k)$ denote the lower and upper reference lines, respectively. The current pixel $x_{Bob}(i, j, k)$ is then determined by:

$$x_{Bob}(i, j, k) = \frac{x(i, j-1, k) + x(i, j+1, k)}{2} \tag{7}$$

**Inter-field** deinterlacing is a simple deinterlacing method. The output frame $x_{Weave}(i, j, k)$ is defined as (8):

$$x_{Weave}(i, j, k) = \begin{cases} x(i, j, k) & \text{j mod 2 = n mod 2} \\ x(i, j, k-1) & \text{otherwise} \end{cases} \tag{8}$$

where $(i, j, k)$ designates the position, $x(i, j, k)$ is the input field defined for $j \ mod \ 2{=}n \ mod \ 2$ only and $k$ is the field number. It is well-known that the video quality of the inter-field interpolation is better than that of the intra-field interpolation in a static area. However, the line-crawling effect occurs in motion areas.

The **ELA** algorithm utilizes directional correlations among pixels to linearly interpolate a missing line. A 3-by-2 localized window is used to calculate directional correlations and interpolate the current pixel. The measurement $I_C(m)$ is the intensity change in the direction represented by $m$. $I_C(m)$ is then used to calculate the direction of the highest spatial correlation. The edge direction $\theta$ is computed. The current pixel $x(i,j,k)$ is then reckoned.

$$I_C(m) = |x(i+m, j-1) - x(i-m, j+1)|, \quad -1 \leq m \leq 1 \tag{9}$$

$$\theta = argmin\ I_C(k) \quad -1 \leq k \leq 1 \tag{10}$$

$$x_{ELA}(i,j,k) = \{x(i+\theta, j-1, k) + x(i-\theta, j+1, k)\} >> 1 \tag{11}$$

The **STELA** algorithm performs the edge-based line averaging on the spatio-temporal window [5]. Fig. 1 shows the block diagram of the **STELA** algorithm. First, a 2-D input signal is decomposed into the low-pass and high-pass filtered signals. The high-pass filtered signal is obtained by subtracting the low-pass filtered signal from the input signal. Then, each signal is processed separately to estimate the missing scan lines of the interlaced sequence.

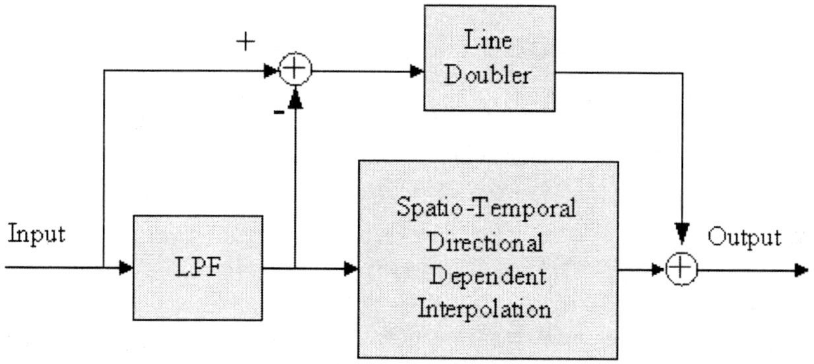


**Fig. 1.** The block diagram of the **STELA** algorithm

The interpolation method uses a spatio-temporal window with four scan lines and determines the minimum directional change, then chooses the median from the average value of the minimum directional change, pixel values of previous and next frames and pixel values of top and bottom fields in current frame.

The line doubling method to fill the missing scan lines processes the residual high frequency components of the signal. In the final stage of the **STELA** algorithm, the results of the line double and the directional dependent interpolation are added to fill the missing lines.

The edge direction (ED) detector utilized directional correlations among pixels in order to linearly interpolate a missing line. A 3D localized window was used to calculate directional correlations and to interpolate the current pixel, as

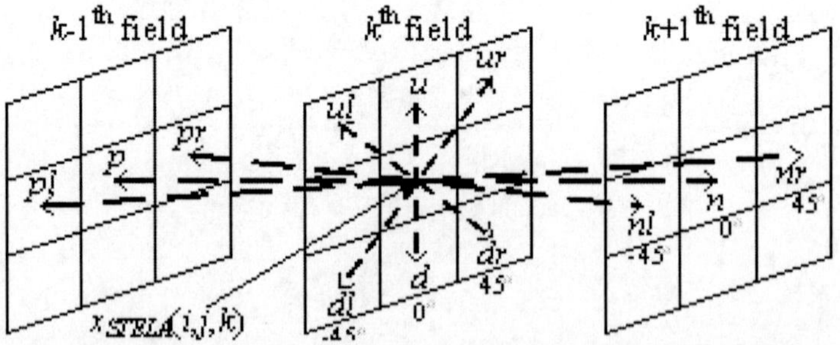


**Fig. 2.** Spatio-temporal window for the direction-based deinterlacing

shown in Fig. 2. Here, $\{u, d, r, l, p$ and $n\}$ represent $\{up, down, right, left,$ *previous* and *next*$\}$, respectively. For the measurement of the spatio-temporal correlation of the samples in the window, six directional changes are provided. They are reckoned as:

$$I_{C_1} = |ul - dr| \quad I_{C_2} = |u - d| \quad I_{C_3} = |ur - dl|$$

$$I_{C_4} = |pl - nr| \quad I_{C_5} = |p - n| \quad I_{C_6} = |pr - nl|$$

Then, the output of the directional-based algorithm is obtained by:

$$x_{STELA}(i, j, k) = Med(A, u, d, p, n) \tag{12}$$

Here, $A$ is the average value of two samples with the minimum directional change. This scheme can increase the edge-detection consistency by checking the past and future edge orientation at the neighboring pixel.

## 4 Rough Set-Based Deinterlacing: Attributes Definition

In this chapter, it is assumed that an image can be classified according to four main parameters: TD, SD, TMDW and SMDW (see expressions (13) – (16)). The characteristics of TMDW and SMDW are described in [20], where $\beta$ is an amplification factor that affects the size of membership functions resulting in TMDW and SMDW varying between 0 and 255. The number of pixels of temporal and spatial window are $N_{W_T}$ and $N_{W_S}$, respectively. Each provide six and $x(i, j, k)$ denotes the intensity of the pixel which will be interpolated in our work. $i$ refers to the column number, $j$ refers to the line number, and $k$ stands for the field number as graphically portrayed in Fig. 3.

$$TD = |x(i, j, k - 1) - x(i, j, k + 1)| \tag{13}$$

$$SD = |x(i, j - 1, k) - x(i, j + 1, k)| \tag{14}$$

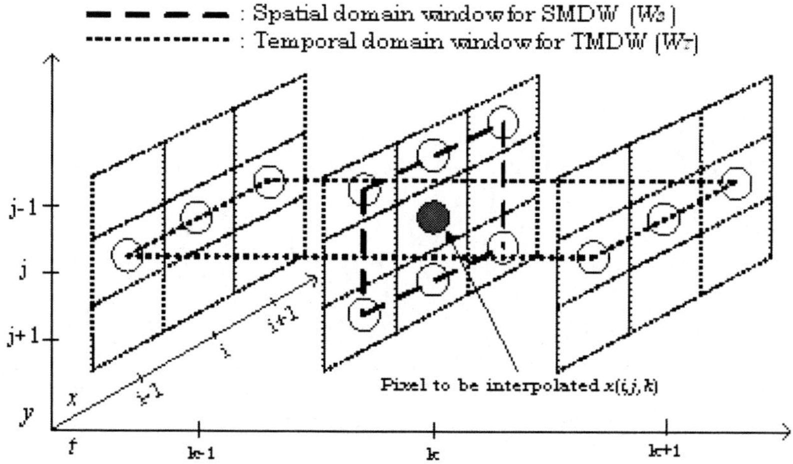


**Fig. 3.** Illustration of the spatial domain ($W_S$) and temporal domain ($W_T$) windows

$$TMDW = \frac{\max\limits_{(i,j,k)\in W_T} x(i,j,k) - \min\limits_{(i,j,k)\in W_T} x(i,j,k) \times N_{W_T}}{\sum\limits_{(i,j,k)\in W_T} x(i,j,k)} \times \beta \tag{15}$$

$$SMDW = \frac{\max\limits_{(i,j,k)\in W_S} x(i,j,k) - \min\limits_{(i,j,k)\in W_S} x(i,j,k) \times N_{W_S}}{\sum\limits_{(i,j,k)\in W_S} x(i,j,k)} \times \beta \tag{16}$$

The temporal domain maximum difference over the window (TMDW) parameter and the spatial domain maximum difference over the window (SMDW) parameter represent the spatial and temporal entropy. Temporal difference (TD) or spatial difference (SD) is the pixel difference between two values across the missing pixel in each domain. The continuous values of the features have been discretized by following expert criterion into a symbol table. We assume that the pixels with low SD or low SMDW values are classified into the plain area and the others are classified into the complex area. Moreover, the pixels with low TD or TMDW are classified within the static area while the remaining pixels are classified into the motion area. Based on this classification system, a different deinterlacing algorithm is activated, in order to obtain the best performance. In all, twelve pixels around the missing pixel $x(i, j, k)$ must be read before attributes are extracted. The extracted attributes are normalized at the position of each missing pixel.

The categorization step of the attribute involves converting the attributes from numerical to categorical. At this point, some data may be lost during the conversion from analog to digital information. None of the frequency-based nor the boundary-based methods are optimal. Instead, the numerical range is determined according to the frequencies of each category boundary. According to the experts, each state can be classified into the most suitable class among four possible regions. These regions can be selected for the decision making for the video deinterlacing system: they are plain-stationary region, complex-stationary region, plain-motion region or complex-motion region. The first step of the algorithm is to redefine the value of each attribute according to a certain metric. Using 100 frames ($2^{nd}$ to $101^{st}$ frames) of each of the six CIF sequences (Akiyo, Table Tennis, Foreman, News, Mobile, and Stefan) as the training data, the decision making map can be obtained through the training process. Table 1 shows a comparison of the normalized average CPU time among the four methods. In case of TD and SD, the numerical range is linearly divided into two categories: S (small) and L (Large).

In the case of TMDW and SDMW, the numerical range is linearly divided into three categories: `S` (small), `M` (medium) and `L` (large). Since each sequence has different degrees of spatial and temporal details, it is a tough process to design consistent decision making tables. The detail required to determine $abcd/U$ is described in Table 2.

The set of all possible decisions are listed in Table 3, which were collected through several training sequences. The proposed information system is composed of $R = [a, b, c, d, m \mid \{a, b, c, d\} \rightarrow \{m\}]$ as shown in Table 3. This table is a decision table in which $a$, $b$, $c$ and $d$ are condition attributes whereas $m$ is the decision attribute. Using these values, a set of examples can be generated. The attribute $m$ represents the selected method which is the decision maker's

**Table 1.** Comparison of the normalized average CPU time among four deinterlacing methods with six above CIF test sequences

| Method | Akiyo | Table Tennis | Foreman | News | Mobile | Stefan | Average |
|---|---|---|---|---|---|---|---|
| Bob | 0.012707 | 0.008654 | 0.013235 | 0.012922 | 0.013735 | 0.015443 | 0.2917 |
| Weave | 0.011301 | 0.007593 | 0.012963 | 0.011678 | 0.013565 | 0.014434 | 0.2721 |
| ELA | 0.028740 | 0.019820 | 0.028979 | 0.029467 | 0.031508 | 0.032593 | 0.6508 |
| STELA | 0.042955 | 0.030274 | 0.045288 | 0.044065 | 0.048394 | 0.051943 | 1.0000 |

**Table 2.** Fuzzy rules for the determination of attributes $a$, $b$, $c$ and $d$

| | | |
|---|---|---|
| 1 | IF $TD$ is smaller than $2^3$ | THEN $a$ is $S$ |
| | IF $TD$ is larger than $2^3$ | THEN $a$ is $L$ |
| 2 | IF $SD$ is smaller than $2^3$ | THEN $b$ is $S$ |
| | IF $SD$ is larger than $2^3$ | THEN $b$ is $L$ |
| 3 | IF $TMDW$ is smaller than $2^2$ | THEN $c$ is $S$ |
| | IF $TMDW$ is larger than $2^2$ and smaller than $2^4$ | THEN $c$ is $M$ |
| | IF $TMDW$ is larger than $2^4$ | THEN $c$ is $L$ |
| 4 | IF $SMDW$ is smaller than $2^2$ | THEN $d$ is $S$ |
| | IF $SMDW$ is larger than $2^2$ and smaller than $2^4$ | THEN $d$ is $M$ |
| | IF $SMDW$ is larger than $2^4$ | THEN $d$ is $L$ |

**Table 3.** Selecting the method corresponding to each pattern

| $abcd(U)$ | $P$ | $AD_B$ | $AD_W$ | $AD_E$ | $AD_T$ | $C_B$ | $C_W$ | $C_E$ | $C_T$ | $m$ |
|---|---|---|---|---|---|---|---|---|---|---|
| SSSS | 35.14% | 1.33 | 1.64 | 1.38 | 1.15 | 1.84 | 2.12 | 2.52 | 2.90 | B, W |
| SSSM | 11.17% | 2.50 | 3.24 | 2.70 | 2.12 | 3.01 | 3.72 | 3.84 | 3.87 | B |
| SSSL | 0.97% | 6.12 | 6.19 | 5.40 | 4.25 | 6.63 | 6.67 | 6.54 | 6.00 | T |
| SSMS | 2.76% | 2.85 | 2.92 | 2.90 | 2.31 | 3.36 | 3.40 | 4.04 | 4.06 | B, W |
| SSMM | 5.37% | 3.09 | 3.83 | 3.54 | 2.74 | 3.60 | 4.31 | 4.68 | 4.49 | B |
| SSML | 1.59% | 5.66 | 6.41 | 5.29 | 4.11 | 6.17 | 6.89 | 6.43 | 5.86 | T, B |
| SSLS | 0.07% | 6.40 | 4.05 | 4.72 | 3.90 | 6.91 | 4.53 | 5.86 | 5.65 | W |
| SSLM | 0.35% | 7.06 | 5.27 | 5.81 | 4.69 | 7.57 | 5.75 | 6.95 | 6.44 | W |
| SSLL | 1.67% | 5.04 | 7.42 | 5.84 | 4.57 | 5.55 | 7.90 | 6.98 | 6.32 | B |
| SLSS | 3.60% | 7.64 | 4.39 | 4.65 | 3.72 | 8.15 | 4.87 | 5.79 | 5.47 | W |
| SLSM | 3.51% | 10.17 | 5.70 | 7.01 | 5.45 | 10.68 | 6.18 | 8.15 | 7.20 | W |
| SLSL | 1.45% | 14.46 | 9.46 | 10.19 | 7.97 | 14.97 | 9.94 | 11.33 | 9.72 | T, W |
| SLMS | 0.88% | 10.05 | 6.00 | 7.53 | 5.90 | 10.56 | 6.48 | 8.67 | 7.65 | W |
| SLMM | 2.44% | 9.38 | 6.80 | 8.41 | 6.86 | 9.89 | 7.28 | 9.55 | 8.61 | W |
| SLML | 1.88% | 12.57 | 10.04 | 8.88 | 7.23 | 13.08 | 10.52 | 10.02 | 8.98 | T |
| SLLS | 0.15% | 16.86 | 8.98 | 10.39 | 8.05 | 17.37 | 9.46 | 11.53 | 9.80 | W, T |
| SLLM | 0.60% | 15.39 | 8.58 | 10.28 | 7.91 | 15.90 | 9.06 | 11.42 | 9.66 | W |
| SLLL | 2.15% | 11.27 | 11.27 | 8.35 | 7.17 | 11.78 | 11.75 | 9.49 | 8.92 | T |
| LSSS | 1.50% | 4.88 | 9.76 | 4.39 | 4.14 | 5.39 | 10.24 | 5.53 | 5.89 | B, E |
| LSSM | 2.03% | 6.29 | 12.08 | 6.66 | 5.55 | 6.80 | 12.56 | 7.80 | 7.30 | B, T |
| LSSL | 0.58% | 7.25 | 25.26 | 8.61 | 7.02 | 7.76 | 25.74 | 9.75 | 8.77 | B |
| LSMS | 0.69% | 6.25 | 9.73 | 5.59 | 5.03 | 6.76 | 10.21 | 6.73 | 6.78 | E |
| LSMM | 1.90% | 6.48 | 10.82 | 7.31 | 5.74 | 6.99 | 11.30 | 8.45 | 7.49 | B |
| LSML | 0.92% | 7.31 | 17.40 | 7.89 | 6.63 | 7.82 | 17.88 | 9.03 | 8.38 | B |
| LSLS | 0.09% | 9.12 | 7.72 | 6.34 | 5.43 | 9.63 | 8.20 | 7.48 | 7.18 | T, E |
| LSLM | 0.36% | 9.11 | 10.87 | 8.06 | 6.80 | 9.62 | 11.35 | 9.20 | 8.55 | T |
| LSLL | 1.10% | 8.10 | 15.30 | 9.18 | 7.55 | 8.61 | 15.78 | 10.32 | 9.30 | B |
| LLSS | 1.15% | 14.44 | 19.41 | 14.84 | 9.89 | 14.95 | 19.89 | 15.98 | 11.64 | T |
| LLSM | 2.06% | 11.66 | 19.93 | 14.09 | 9.39 | 12.17 | 20.41 | 15.23 | 11.14 | T |
| LLSL | 1.34% | 12.48 | 28.10 | 17.06 | 11.37 | 12.99 | 28.58 | 18.20 | 13.12 | B, T |
| LLMS | 0.79% | 14.74 | 16.61 | 15.36 | 10.24 | 15.25 | 17.09 | 16.50 | 11.99 | T |
| LLMM | 2.48% | 13.38 | 18.67 | 15.51 | 10.34 | 13.89 | 19.15 | 16.65 | 12.09 | T |
| LLML | 2.49% | 13.10 | 23.06 | 16.41 | 10.94 | 13.61 | 23.54 | 17.55 | 12.69 | T |
| LLLS | 0.22% | 19.35 | 17.72 | 17.39 | 11.59 | 19.86 | 18.20 | 18.53 | 13.34 | T |
| LLLM | 1.01% | 18.91 | 16.38 | 17.90 | 11.93 | 19.42 | 16.86 | 19.04 | 13.68 | T |
| LLLL | 3.54% | 12.84 | 19.64 | 15.17 | 10.11 | 13.35 | 20.12 | 16.31 | 11.86 | T |

choice: **Bob** assigned to B, **Weave** assigned to W, **ELA** assigned to E, and **STELA** assigned to T. It is assumed that the average absolute difference between the real value and the **Bob** method utilized the interpolated value as $AD_B$ as portrayed in Table 3. In the same manner, $AD_W$, $AD_E$ and $AD_T$ were obtained.

Since each method has its own advantages and drawbacks, the **RSD** method is based on variable deinterlacing mode technique. And this procedure causes a mode decision problem, because the mode decision method may affect interpolation efficiency, complexity as well as objective and subjective results. As rate-distortion optimization (RDO) of reference software in H.264, we proposed a rule to select the suitable methods in each condition. This rule has been applied to various video sequences and supplies good performance in terms of PSNR and complexity. The goal of the rule is to select the mode having minimum average cost in a given computational CPU time.

$$C_i = AD_i + \mathcal{K} \cdot RT_i \tag{17}$$

where $i \in \{B, W, E, T\}$, $C_i$ is the cost associated to method $i$, $AD_i$ is the average absolute difference, $RT_i$ is the expected required computational CPU time, and the parameter $\mathcal{K}$ is determined before runnig the experiment (simulation results yielded $\mathcal{K} = 1.75$). It is assumed that the method having the least cost is picked up as the selected method in each condition. However, it is difficult to determine the suitable method in some conditions, such as rules SSSS, SSMS, SSML, SLSL, SLLS, LSSS, LSSM, LSLS and LLSL, because the cost difference between the two best methods is too small (less then 0.5).

**Table 4.** The information system (evaluation rules)

| Staff | $a$ | $b$ | $c$ | $d$ | $m$ | Staff | $a$ | $b$ | $c$ | $d$ | $m$ |
|---|---|---|---|---|---|---|---|---|---|---|---|
| 1 | S | S | S | S | B | 24 | L | S | S | S | B |
| 2 | S | S | S | S | W | 25 | L | S | S | S | E |
| 3 | S | S | S | M | B | 26 | L | S | S | M | B |
| 4 | S | S | S | L | T | 27 | L | S | S | M | T |
| 5 | S | S | M | S | B | 28 | L | S | S | L | B |
| 6 | S | S | M | S | W | 29 | L | S | M | S | E |
| 7 | S | S | M | M | B | 30 | L | S | M | M | B |
| 8 | S | S | M | L | T | 31 | L | S | M | L | B |
| 9 | S | S | M | L | B | 32 | L | S | L | S | T |
| 10 | S | S | L | S | W | 33 | L | S | L | S | E |
| 11 | S | S | L | M | W | 34 | L | S | L | M | T |
| 12 | S | S | L | L | B | 35 | L | S | L | L | B |
| 13 | S | L | S | S | W | 36 | L | L | S | S | T |
| 14 | S | L | S | M | W | 37 | L | L | S | M | T |
| 15 | S | L | S | L | T | 38 | L | L | S | L | B |
| 16 | S | L | S | L | W | 39 | L | L | S | L | T |
| 17 | S | L | M | S | W | 40 | L | L | M | S | T |
| 18 | S | L | M | M | W | 41 | L | L | M | M | T |
| 19 | S | L | M | L | T | 42 | L | L | M | L | T |
| 20 | S | L | L | S | W | 43 | L | L | L | S | T |
| 21 | S | L | L | S | T | 44 | L | L | L | M | T |
| 22 | S | L | L | M | W | 45 | L | L | L | L | T |
| 23 | S | L | L | L | T | | | | | | |

Table 4 shows the drawn 45 evaluation rules from the deinterlacing system. The system designer assigns the suitable methods: Bob, Weave, ELA, and STELA.

Let us make clear some notation in Table 4:

$$
\begin{aligned}
&U = \{1, 2, 3, \ldots, 42, 44, 45\} \\
&C = \{a\,(TD),\, b\,(SD),\, c\,(TMDW),\, d\,(SMDW)\} \\
&V_{TD} = V_{SD} = \{S, L\} \\
&V_{TMDW} = V_{SMDW} = \{S, M, L\} \\
&\{d\} = \{B,\, W,\, E,\, T\}
\end{aligned}
$$

From the indiscernibility relations, the lower and the upper approximations of the decision $d(x)$ for each object $x$ are obtained as follows:

$$
\begin{array}{ll}
\underline{P}\{d(1)\} = \emptyset & \overline{P}\{d(1)\} = [B, W] \\
\underline{P}\{d(21)\} = \emptyset & \overline{P}\{d(21)\} = [W, T] \\
\underline{P}\{d(2)\} = \emptyset & \overline{P}\{d(2)\} = [B, W] \\
\underline{P}\{d(24)\} = \emptyset & \overline{P}\{d(24)\} = [B, E] \\
\underline{P}\{d(5)\} = \emptyset & \overline{P}\{d(5)\} = [B, W] \\
\underline{P}\{d(25)\} = \emptyset & \overline{P}\{d(25)\} = [B, E] \\
\underline{P}\{d(6)\} = \emptyset & \overline{P}\{d(6)\} = [B, W] \\
\underline{P}\{d(26)\} = \emptyset & \overline{P}\{d(26)\} = [B, T] \\
\underline{P}\{d(8)\} = \emptyset & \overline{P}\{d(2)\} = [T, B] \\
\underline{P}\{d(27)\} = \emptyset & \overline{P}\{d(27)\} = [B, T] \\
\underline{P}\{d(9)\} = \emptyset & \overline{P}\{d(9)\} = [T, B] \\
\underline{P}\{d(32)\} = \emptyset & \overline{P}\{d(32)\} = [T, E] \\
\underline{P}\{d(15)\} = \emptyset & \overline{P}\{d(15)\} = [T, W] \\
\underline{P}\{d(33)\} = \emptyset & \overline{P}\{d(33)\} = [T, E] \\
\underline{P}\{d(16)\} = \emptyset & \overline{P}\{d(16)\} = [T, W] \\
\underline{P}\{d(38)\} = \emptyset & \overline{P}\{d(38)\} = [B, T] \\
\underline{P}\{d(20)\} = \emptyset & \overline{P}\{d(20)\} = [W, T] \\
\underline{P}\{d(39)\} = \emptyset & \overline{P}\{d(39)\} = [B, T]
\end{array}
\tag{18}
$$

If $f(x, TD) =' L'$, $f(x, SD) =' L'$, $f(x, TMDW) =' S'$ and $f(x, SMDW) =' L'$ then exactly $\emptyset$ (supported by $38, 39$)

If $f(x, TD) =' L'$, $f(x, SD) =' L'$, $f(x, TMDW) =' S'$ and $f(x, SMDW) =' L'$ then possibly $d\{x\} = [B, T]$ (supported by $38, 39$)

(19)

The other rules have a crisp decision value, e.g.:

$$\begin{array}{ll}
\underline{P}\{d(3)\} = \overline{P}\{d(3)\} = [B] & \underline{P}\{d(29)\} = \overline{P}\{d(29)\} = [E] \\
\underline{P}\{d(4)\} = \overline{P}\{d(4)\} = [T] & \underline{P}\{d(30)\} = \overline{P}\{d(30)\} = [B] \\
\underline{P}\{d(7)\} = \overline{P}\{d(7)\} = [B] & \underline{P}\{d(31)\} = \overline{P}\{d(31)\} = [B] \\
\underline{P}\{d(10)\} = \overline{P}\{d(10)\} = [W] & \underline{P}\{d(34)\} = \overline{P}\{d(34)\} = [T] \\
\underline{P}\{d(11)\} = \overline{P}\{d(11)\} = [W] & \underline{P}\{d(35)\} = \overline{P}\{d(35)\} = [B] \\
\underline{P}\{d(12)\} = \overline{P}\{d(12\} = [B] & \underline{P}\{d(36)\} = \overline{P}\{d(36)\} = [T] \\
\underline{P}\{d(13)\} = \overline{P}\{d(13)\} = [W] & \underline{P}\{d(37)\} = \overline{P}\{d(37)\} = [T] \\
\underline{P}\{d(14)\} = \overline{P}\{d(14)\} = [W] & \underline{P}\{d(40)\} = \overline{P}\{d(40)\} = [T] \\
\underline{P}\{d(17)\} = \overline{P}\{d(17)\} = [W] & \underline{P}\{d(41)\} = \overline{P}\{d(41)\} = [T] \\
\underline{P}\{d(18)\} = \overline{P}\{d(18)\} = [W] & \underline{P}\{d(42)\} = \overline{P}\{d(42)\} = [T] \\
\underline{P}\{d(19)\} = \overline{P}\{d(19)\} = [T] & \underline{P}\{d(43)\} = \overline{P}\{d(43)\} = [T] \\
\underline{P}\{d(22)\} = \overline{P}\{d(22)\} = [W] & \underline{P}\{d(44)\} = \overline{P}\{d(44)\} = [T] \\
\underline{P}\{d(23)\} = \overline{P}\{d(23)\} = [T] & \underline{P}\{d(45)\} = \overline{P}\{d(45)\} = [T] \\
\underline{P}\{d(28)\} = \overline{P}\{d(28)\} = [B] &
\end{array} \tag{20}$$

The priority of rules 38 and 39 are the same, thus we can use any method in that case (in our simulation, priority order is given to $T$, $E$, $W$ and $B$)

Rough set theory offers a mathematical way for strict treatment of data classification problems. The idea behind the knowledge base reduction is a simplification of the Table 4. The algorithm that provides the reduction of conditions is described by the following steps:

1. Remove dispensable attributes
2. Find the core of the decision table
3. Associate a table with a reduct value
4. Extract possible rules.

To simplify the decision system, the reduction of the set of condition attributes is necessary in order to define the decision categories. If we remove attribute $a$ from Table 5 then we obtain an inconsistent decision table. Hence, the attribute $a$ cannot be removed from the decision system. In the same manner, it has been observed that the remaining attributes ($b$, $c$ or $d$) are indispensable. This means that none of the condition attributes can be removed from Table 5. Hence, the set of condition attributes is m-independent.

The next step is finding whether some elementary condition categories can be eliminated, i.e., some superfluous values of condition attributes in Table 5. The core values of each decision rule in such table are presented in Table 6 whereas Table 7 depicts the final essential decision rules which can be rewritten as a minimal decision algorithm in normal form. Combining the decision rules leading to the same decision class, the following decision algorithm is achieved.

The final results, presented in Table 7, can be rewritten as a minimal decision algorithm in normal form which is based on the original rough set theory [9]. Combining the decision rules into a single decision class leads to the ensuing decision algorithm:

**Table 5.** The whole deinterlacing system ($x$ = don't care)

| $U$ | $a$ | $b$ | $c$ | $d$ | $m$ | $U$ | $a$ | $b$ | $c$ | $d$ | $m$ | $U$ | $a$ | $b$ | $c$ | $d$ | $m$ |
|---|---|---|---|---|---|---|---|---|---|---|---|---|---|---|---|---|---|
| 1 | S | S | S | M | B | 13 | S | L | x | M | W | 25 | L | x | S | M | T |
| 2 | x | S | M | M | B | 14 | S | x | M | S | W | 26 | L | x | L | S | T |
| 3 | x | S | L | L | B | 15 | S | L | x | M | W | 27 | L | x | L | M | T |
| 4 | L | S | x | L | B | 16 | S | x | x | M | W | 28 | L | L | x | x | T |
| 5 | x | S | M | M | B | 17 | L | S | S | S | E | 29 | L | x | x | x | T |
| 6 | L | S | x | L | B | 18 | L | S | M | S | E | 30 | x | x | x | x | T |
| 7 | x | S | x | L | B | 19 | S | x | S | L | T | 31 | L | L | x | x | T |
| 8 | S | x | x | S | W | 20 | S | x | M | L | T | 32 | L | L | x | x | T |
| 9 | S | x | x | S | W | 21 | x | L | x | L | T | 33 | x | L | x | x | T |
| 10 | S | S | x | S | W | 22 | x | x | x | L | T | 34 | x | x | x | x | T |
| 11 | S | x | L | M | W | 23 | x | L | L | S | T | 35 | L | x | x | x | T |
| 12 | S | x | S | S | W | 24 | x | L | x | L | T | 36 | x | L | x | x | T |

**Table 6.** Core of the attributes

| $U$ | $a$ | $b$ | $c$ | $d$ | $m$ | $U$ | $a$ | $b$ | $c$ | $d$ | $m$ | $U$ | $a$ | $b$ | $c$ | $d$ | $m$ |
|---|---|---|---|---|---|---|---|---|---|---|---|---|---|---|---|---|---|
| 1 | S | S | S | M | B | 13 | S | L | - | M | W | 25 | L | - | S | M | T |
| 2 | - | S | M | M | B | 14 | S | - | M | S | W | 26 | L | - | L | S | T |
| 3 | - | S | L | L | B | 15 | S | L | - | M | W | 27 | L | - | L | M | T |
| 4 | L | S | - | L | B | 16 | S | - | - | M | W | 28 | L | L | - | - | T |
| 5 | - | S | M | M | B | 17 | L | S | S | S | E | 29 | L | - | - | - | T |
| 6 | L | S | - | L | B | 18 | L | S | M | S | E | 30 | - | - | - | - | T |
| 7 | - | S | - | L | B | 19 | S | - | S | L | T | 31 | L | L | - | x - | T |
| 8 | S | - | - | S | W | 20 | S | - | M | L | T | 32 | L | L | - | - | T |
| 9 | S | - | - | S | W | 21 | - | L | - | L | T | 33 | - | L | - | - | T |
| 10 | S | S | - | S | W | 22 | - | - | - | L | T | 34 | - | - | - | - | T |
| 11 | S | - | L | M | W | 23 | - | L | L | S | T | 35 | L | - | - | - | T |
| 12 | S | - | S | S | W | 24 | - | L | - | L | T | 36 | - | L | - | - | T |

**Table 7.** Final deinterlacing system ($x$ = don't care)

| $U$ | $a$ | $b$ | $c$ | $d$ | $m$ | $U$ | $a$ | $b$ | $c$ | $d$ | $m$ | $U$ | $a$ | $b$ | $c$ | $d$ | $m$ |
|---|---|---|---|---|---|---|---|---|---|---|---|---|---|---|---|---|---|
| 1 | S | S | S | M | B | 8 | S | x | L | M | W | 15 | x | L | S | L | T |
| 2 | x | S | M | M | B | 9 | S | L | x | M | W | 16 | x | L | L | S | T |
| 3 | x | S | L | L | B | 10 | L | S | S | S | E | 17 | x | L | x | L | T |
| 4 | L | S | x | L | B | 11 | L | S | M | S | E | 18 | L | x | S | M | T |
| 5 | S | S | x | S | W | 12 | S | x | S | L | T | 19 | L | x | L | S | T |
| 6 | S | x | S | S | W | 13 | S | x | M | L | T | 20 | L | x | L | M | T |
| 7 | S | S | x | S | W | 14 | S | L | x | L | T | 21 | L | L | x | x | T |

$$\begin{array}{rl} \text{if } (a_L b_S d_S (c_S \vee c_M)) & \longrightarrow m_E \\ \text{else if } (b_S (d_M (a_S c_S \vee c_M)) \vee (d_L (a_L \vee c_L)) & \longrightarrow m_B \\ \text{else if } (a_S (d_S (b_S \vee c_S \vee c_M)) \vee (d_M (b_L \vee c_L)) & \longrightarrow m_W \\ \text{else} & \longrightarrow m_T \end{array} \tag{21}$$

## 5 Experimental Results

In this section, we compare the objective and subjective quality as well as the computational CPU time for the different proposed interpolation methods. We conducted an extensive simulation to test the performance of our algorithm. We ran our experiments on four "real-world" HDTV sequences with a field size of $1920 \times 1080i$: Mobcal, Parkrun, Shields, and Stockholm as shown in Figure 4. These sequences are different from the sequences employed to highlight which sequences were used for information acquisition.

As a measure of objective dissimilarity between a deinterlaced image and the original one, we use the peak signal-to-noise ratio (PSNR) in decibels (dB), as follows:

$$PSNR(Img, Org) = 10 \log_{10} \frac{S^2}{MSE(Img, Org)} \tag{22}$$

This similarity measure relies on another measure, namely the mean square error (MSE):

$$MSE(Img, Org) = \frac{\sum_{i=1}^{N} \sum_{j=1}^{M} (Org(i,j) - Img(i,j))^2}{N \times M} \tag{23}$$

where $Org$ is the original image, $Img$ is the deinterlaced image of size $N \times M$ and $S$ is the maximum possible intensity value (with m-bit integer values, S will be $2^m - 1$).

For the objective performance evaluation, the chosen video sequences were inputted into the four conventional algorithms (**Bob**, **Weave**, **ELA** and **STELA**) as well as into the new algorithm. Tables 8 and 9 portray the results of the deinterlacing methods for the selected sequences in terms of PSNR and computational CPU time, respectively . The results point out that the proposed algorithm yields the second or third best performance in terms of PSNR. Moreover, the proposed algorithm only requires 84.65% of CPU time than that of **ELA**. Especially, it shows nearly the same objective performance when compared to the **STELA** method in terms of PSNR, even though it has only about 80.78% of CPU time.

Fig. 4 compares the visual performance of our proposed algorithm with several major conventional methods. We can observe that these conventional methods have the following main shortcomings in contrast to the presented method:

1. **Bob** exhibits no motion artifacts and has minimal computational requirements. However, the input vertical resolution is halved before the image is interpolated, thus reducing the detail in the progressive image.

**Table 8.** Results of different interpolation methods for four 1920 × 1080$i$ sequences (PSNR in dB)

| Method | Mobcal | Parkrun | Shields | Stockholm |
|---|---|---|---|---|
| Bob | 28.463 | 21.131 | 24.369 | 26.586 |
| Weave | 25.624 | 19.031 | 21.743 | 24.223 |
| ELA | 27.978 | 21.296 | 24.436 | 26.762 |
| STELA | 28.472 | 21.268 | 24.499 | 26.774 |
| Proposed | 28.395 | 21.262 | 24.458 | 26.763 |

**Table 9.** Results of different interpolation methods for four 1920 × 1080$i$ sequences (CPU time)

| Method | Mobcal | Parkrun | Shields | Stockholm |
|---|---|---|---|---|
| Bob | 0.6523 | 0.7958 | 0.7196 | 0.6490 |
| Weave | 0.5243 | 0.6967 | 0.6113 | 0.5778 |
| ELA | 0.9687 | 0.9758 | 0.9664 | 0.9059 |
| STELA | 1.000 | 1.000 | 1.000 | 1.000 |
| Proposed | 0.7656 | 0.8669 | 0.8188 | 0.7799 |

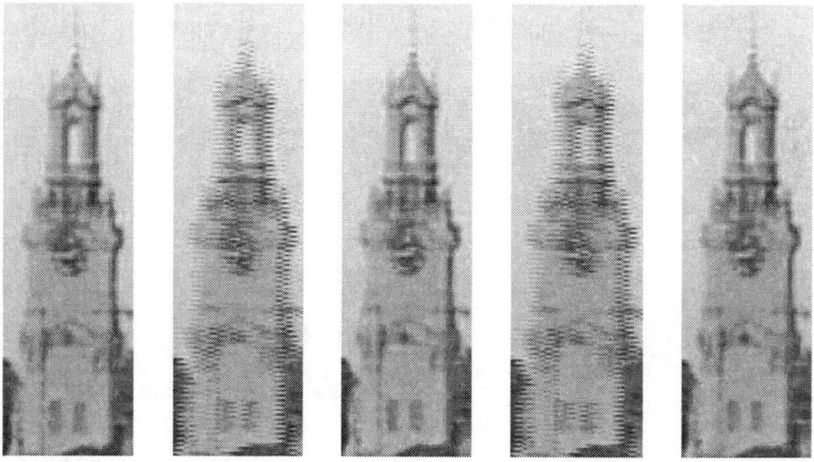

(a) **Bob** (b) **Weave** (c) **ELA** (d) **STELA** (e) **RSD**

**Fig. 4.** Subjective quality comparison of the $45^{th}$ Stockholm sequence

2. **Weave** results in no degradation of the static images. However, the edges exhibit significant serrations, which is an unacceptable artifact in a broadcast or professional television environment.

3. The **ELA** algorithm provides good performance. It can eliminate the blurring effect of bilinear interpolation and bears both sharp and straight edges. However, due to misleading edge directions, interpolation errors often become larger in areas of high-frequency components. In addition, some defects may occur when an object exists only in the same parity field.
4. **STELA** can estimate the motion vector to be zero in the static region so that it can reconstruct the missing pixel perfectly, results in no degradation. However, it gradually reduces the vertical detail as the temporal frequencies increase. The vertical detail the previous field is combined with the temporally shifted current field, indicating that some motion blur occurred. Despite of all this, STELA gives the best quality among the four conventional methods.

From the experimental results it is observed that the proposed algorithm has good objective and subjective qualities for different sequences also keeping a low computational CPU time required to achieve the real-time processing.

## 6 Conclusions

In this chapter, we proposed a RST-based deinterlacing method. Using rough set theory, it is now possible to cope with a deinterlacing system having ambiguous decisions given by a decision maker. Our proposed information acquisition model selects the most suitable deinterlacing method among four deinterlacing procedures and it successively builds the approximations of the deinterlacing sequence by evaluating the four methods in each condition.

Decision making and interpolation results are presented. The results of computer simulations demonstrate that the proposed method outperforms a number of schemes in literature.

## References


1. Seidner, D.: IEEE Trans. Image Processing 14, 1876–1889 (2005)
2. Jerri, A.: Proceedings of the IEEE 65, 1565–1595 (1977)
3. Janssen, J., Stessen, J., de With, P.: An advanced sampling rate conversion technique for video and graphics signals. In: International Conference on Image Processing and its Applications, pp. 771–775 (1997)
4. Chen, M., Huang, C., Hsu, C.: IEEE Trans. Consumer Electronics 50, 1202–1208 (2004)
5. Oh, H., Kim, Y., Jung, Y., Morales, A., Ko, S.: IEEE International Conference on Consumer Electronics, pp. 52–53 (2000)
6. Bellers, E., de Haan, G.: Advanced de-interlacing techniques. In: Proceedings of ProRisc/IEEE Workshop on Circuits, Systems and Signal Processing (1996)
7. Swan, P.: Method and apparatus for providing interlaced video on a progressive display. U.S. Patent 5–864–369 (1999)
8. Doyle, T.: Interlaced to sequential conversion for EDTV applications. In: Proceedings of the 2nd International Workshop on Signal Processing of HDTV, pp. 412–430 (1990)

9. Pawlak, Z.: Rough Sets - Theoretical Aspects of Reasoning about Data. Kluwer Academic, Dordrecht (1991)
10. Zhang, X., Zhang, F., Zhao, Y.: Generalization of RST in ordered information table. In: Yeung, D.S., Liu, Z.-Q., Wang, X.-Z., Yan, H. (eds.) ICMLC 2005. LNCS (LNAI), vol. 3930, pp. 2027–2032. Springer, Heidelberg (2006)
11. Sugihara, K., Tanaka, H.: Rough set approach to information systems with interval decision values in evaluation problems. In: Bello, R., Falcon, R., Pedrycz, W., Kacprzyk, J. (eds.) Granular Computing: At the Junction of Rough Sets and Fuzzy Sets, Springer, Heidelberg (2007)
12. Pan, L., Zheng, H., Nahavandi, S.: The application of rough set and Kohonen network to feature selection for object extraction. In: Proceedings of ICMLC 2003, pp. 1185–1189 (2003)
13. Grzymala-Busse, J.: LERS - A system for learning from examples based on rough sets. In: Slowinski, R. (ed.) Intelligent Decision Support. Handbook of Applications an Advances of the Rough Set Theory, Kluwer Academic, Dordrecht (1992)
14. Hu, X.: Using rough set theory and database operations to construct a good ensemble of classifiers for data mining applications. In: Proceedings of the ICDM 2001, pp. 233–240 (2001)
15. Wu, X., Wang, Q.: Application of rough set attributes reduction in quality evaluation of dissertation. In: Proceedings of ICGC 2006, pp. 562–565 (2006)
16. Peng, Y., Liu, G., Lin, T., Geng, H.: Application of rough set theory in network fault diagnosis. In: Proceedings of ICITA 2005, pp. 556–559 (2005)
17. Kusiak, A.: IEEE Trans. Electronics Packaging Manufacturing 24, 44–50 (2001)
18. Agrawal, A., Agarwal, A.: Rough logic for building a landmine classifier. In: Proceedings of ICNSC 2005, pp. 855–860 (2005)
19. Torres, L.: Application of rough sets in power system control center data mining. In: Proceedings of PESW 2002, pp. 627–631 (2002)
20. Jeon, G., Jeong, J.: A fuzzy interpolation method using intra and inter field information. In: Proceedings of ICEIC 2006 (2006)

# Part II: Fuzzy and Rough Sets in Machine Learning and Data Mining

# Learning Membership Functions for an Associative Fuzzy Neural Network

Yanet Rodríguez, Rafael Falcón, Alain Varela, and María M. García

Computer Science Department
Central University of Las Villas
Carretera Camajuani km 5 1/2 Santa Clara, Cuba
yrsarabia@uclv.edu.cu

**Summary.** Some novel heuristic methods for automatically building triangular, trapezoidal, Gaussian and sigmoid membership functions are introduced, providing a way to model linear attributes as linguistic variables. The utilization of such functions in five different fashions in the context of an Associative Fuzzy Neural Network outperformed two existing methods. Also, these heuristic methods are suitable for being applied to other knowledge representation formalisms that use fuzzy sets.

**Keywords:** membership functions, fuzzy sets, associative fuzzy neural network, heuristic methods.

## 1 Introduction

Fuzzy logic has proved to be an essential methodology for dealing with uncertain and imprecise environments. We have witnessed for some decades an increasing utilization of fuzzy sets in nearly every area of knowledge processing, including classification and clustering tasks.

But a key problem emerges: how to construct the fuzzy sets that are needed to carry out these activities? One first thought would be to rely on human experts so as to have them provide the required information. This, nevertheless, is not always possible. There might not be experts available or they might hesitate about the most convenient way of modeling a linear attribute in a cases base, for example. In such situations we are compelled to devise and properly attempt some machine learning techniques [1] that aim to automate the process of building the fuzzy sets for the linear attributes we are dealing with.

Lots of procedures for building membership functions (MFs) have been devised and can be easily found in literature, ranging from very simple heuristics that start slicing the domain of values of a linear variable to the extensive use of evolutionary algorithms and well-known numerical procedures such as interpolation.

It is our goal, then, to introduce four heuristic methods for building some kinds of MFs from available data and subsequently define five ways in which linear attributes are to be modeled by using them. The underlying principle

R. Bello et al. (Eds.): Granular Computing, STUDFUZZ 224, pp. 151–161, 2008.
springerlink.com 

behind these methods is the consideration of the existing relationships between data, yielding more "reliable" functions in the sense of their interpretability.

The feasibility of the so-built membership functions will be properly tested by means of an associative[1] fuzzy neural network named Fuzzy-SIAC which was introduced in [2] and is capable of fashioning linear attributes as fuzzy sets. Such fuzzy neural network [3] plays the role of the inference engine. Two other existing methods for creating membership functions were also taken into consideration for comparison purposes with our proposed approach.

The chapter is structured as follows: a brief look at several existing approaches for building membership functions is presented in the next section, whereas Section 3 elaborates on the description of our proposed approach. The main characteristics of the associative fuzzy neural network used as the inference engine in our study are depicted in section 4. Later on, section 5 is devoted to thoroughly display and discuss the achieved experimental results. Conclusions and future work are finally outlined.

## 2 Building Membership Functions: The State of the Art

In order to provide the user with an overall idea on how research about automatic building of membership functions has been conducted so far, we will enumerate several methods found in the literature that utilize different approaches and explain them in some detail. A comparison between such procedures highlighting their advantages and drawbacks is beyond the scope of this chapter.

### 2.1 By Using a Discretization Method and a Simple Heuristic

Several authors from the University of Otago, New Zealand, headed by Zhou proposed a method for automatically building trapezoidal MFs [4]. To begin with, they applied the Chi2 discretization algorithm [5] so as to automatically determine the number and width of the MFs.

Four-point trapezoidal functions are used that cause each input value to belong to a maximum of two of them, the membership degrees for which will always add up to one. A degree of overlap between adjacent functions of either a 25% or 50% is accomplished by a quite straightforward strategy.

### 2.2 By Using the Measurement Theory and Interpolation

A little earlier but more complex (and efficient) method to construct membership functions from training data is thoroughly depicted in [6].

The authors from the Massachusetts Institute of Technology (MIT) outline a methodology that makes intensive use of a mathematically axiomatic

[1] This kind of neural networks makes no distinction between input and output neurons and their weights are computed once and left unchanged. The Interactive Activation and Competition (IAC), Hopfield and Brain-State-in-a-Box networks are examples of associative networks.

method known as "measurement theory", which offers a suitable framework for constructing a membership function in cases where the membership is based on subjective preferences. Given a finite set of membership values by a human expert, the remaining values shall be obtained by means of interpolation.

Further, constrained interpolation must be used to ensure the interpolated results remain a membership function, i.e. they are monotonic and convex and bounded in [0, 1]. The paper clearly explains that using least-squares or cubic splines to interpolate leads to a non-convex result and hence the authors introduce several modifications to the constrained interpolation method using Bernstein polynomials [7] to fit to this problem.

### 2.3 By Means of Mathematical Morphology

Another different outlook for generating a fuzzy partition for a numerical attribute is propounded by Marsala and Bouchon-Meunier [8]. The algorithm named FPMM (Fuzzy Partition using Mathematical Morphology) assumes the attribute's values are associated with a class value and is executed during each step of the construction of a decision tree.

The chief thought ruling the algorithm is to employ operators coming from the mathematical morphology setting [9], namely basic operators (erosion and dilatation) as well as compound operators (opening and closure). The procedure is based on several rewriting systems which are represented as transductions. Each of these rewriting systems is based on a mathematical morphology operator.

Two algorithms are defined that reduce and enlarge an arbitrary sequence of letters, respectively. A composition of both approaches is encouraged so as to obtain two general operators. Finally, a procedure that smoothes a word induced by a training set is implemented and, from that word, the FPMM approach determines the fuzzy partition which holds nothing but trapezoidal membership functions.

### 2.4 By Means of Evolutionary Algorithms

Researchers have found in Evolutionary Algorithms (EA) an invaluable potential for optimizing both continuous and discrete functions. Owing to their distributed nature which allows them to explore several prospective solutions simultaneously, they are capable of surveying multi-dimensional, non-linear, non differentiable search spaces and locating the optimum in a reasonable number of iterations with few or none additional overhead imposed to the problem.

In this section we will confine ourselves to display two reported studies on the use of Genetic Algorithms to dynamically construct membership functions from training data.

#### The Bacterial Algorithm in Presence of Fuzzy Rules

Trapezoidal MFs are used in [10] for being general enough and widely used. They are embedded in fuzzy rules which describe a fuzzy system. The purpose of the research work conducted was to profit from an approach derived from Genetic Algorithms (GA) named "Bacterial Algorithm" [11] so as to find the initial membership functions and subsequently to adapt their parameters.

The Bacterial Algorithm (BA) is nothing but a GA but with a modified mutation operator called "bacterial mutation", emulating a biological phenomenon of microbial evolution. The simplicity of the algorithm as well as its ability to reach lower error values in a shorter time became it extremely appealing for the authors. The trapezoidal MFs are encoded in each gene of every chromosome of the BA.

The outlined procedure relies on an existing rule base associated to a certain fuzzy system. At first, all membership functions in the chromosome are randomly initialized. Afterwards, the bacterial mutation operator is applied to a randomly part of the chromosome and the parameters of the MFs are changed. The best individual transfers the mutated region into the other individuals. This cycle is repeated for the remaining parts until all parts of the chromosome have been mutated and tested.

### Building and Tuning Membership Functions with GA

Another approach profiting from the parallel distributed nature of the Genetic Algorithms for dynamically building and tuning beta and triangular membership functions is deeply elaborated in [12].

The MFs are initially generated by way of the application of some discretization algorithm which, however, remains unclear in the paper. Once the set of disjoint intervals has been reckoned, the MFs are constructed by setting the middle point of the interval as the point reaching the highest membership degree. Thus the triangular and beta functions preserve their symmetry.

As to the remaining parameters, they are computed and afterwards carefully tuned so as to meet the constraints regarding the degree of overlap between two adjacent functions. A user-driven parameter specifying the desired degree of overlap is taken into consideration when computing the fitness function of the GA, which in turn is fashioned as a multi-objective optimization (MOO) problem. The constituent parts of the overall fitness function represent the properties that either the bell or the triangular function is to satisfy concerning the degree of overlap that must be kept. Those individual properties are aggregated and properly weighed into a single fitness function. A Dynamic Weighted Aggregation (DWA) scheme is employed in which the weights vary as the number of generations of the GA increases.

For a thorough description of the GA parameters utilized in the experiments, go over the material [12]. The experimental results clearly demonstrate that the proposed approach outperforms an existing method.

## 3 A New Approach to Create Membership Functions

Formally, the process by which individuals from some universal set X are determined to be either members or non-members of a crisp set can be defined by a characteristic or discriminative function [13]. This function can be generalized such that the values assigned to the elements of X fall within a specified range and are referred to as the membership degrees of these elements in X (fuzzy sets approach).

Basically, we propose the computation of a suitable MF to model a linear attribute in two stages. The first one deals with getting the linguistic terms, which is accomplished through partitioning the universe of the linear attribute (linguistic variable) into several disjoint intervals. During the second stage, a MF is built for every linguistic term.

### 3.1 Getting the Linguistic Terms

Several methods that aim to split a continuous variable are available. We chose `CAIM` [14] as a discretization method (for classification problems) and `K-means` [15] as a clustering algorithm (intended to cope with multi-objective problems). Unlike the "Equal Width" or "Equal Frequency" discretization methods that build up a partition based on shallow considerations having little or nothing to do with the actual data the cases base holds, all of these methods allow for several deeper criteria such as the Class-Attribute Interdependence for making up the partition.

This, in turn, enables the fact that "better" MFs can be attained, in the sense that the amount of intervals eliciting from the discretization stage is fairly manageable by an external user and thus the overall interpretability of the yielded linguistic terms is enhanced. By gazing at the approaches listed in the prior section, one realizes that they all undergo the same drawback: low or null interpretability. It is very difficult -if not impossible- for the user to suitably assign a meaning to the MFs bore by those methodologies. From this standpoint, the bid approach contributes to provide a fitting number of intervals from which to build the membership functions in a more formal and readable way.

Let $X$ be the set of values that appear in the case base (set of training examples) for a linear attribute $x$. We apply one of the above methods in order to model it as a linguistic variable, associating a linguistic term to each resultant group $G_i$.

There will be as many MFs as groups were obtained. The j-th group is represented by $[A_j, B_j]$, where $A_j$ and $B_j$ are the lower and upper boundaries, respectively.

### 3.2 Building the Membership Functions

Here we have a discrete or continuous ordered universe $Y$ and the MF $j$ for the linguistic term $T_j$ corresponding to the j-th fuzzy set should be achieved. The support [13] of this function will be the set of all points $y \in Y$ in $[A_j, B_j]$ having a membership degree greater than zero ($\mu_j(y) > 0$). In this section we explain some heuristic methods that automatically build different sorts of MFs from training examples. Specifically, the triangular, trapezoidal, Gaussian and sigmoid membership functions have been considered [3]. We are not going to delve into them, for they share a widespread use and their parameters are straightforward to understand.

A triangular MF is specified by three parameters $a$, $b$ and $c$, where $b$ is the central vertex and was computed as follows:

$$b_j = \frac{y_i * \beta_i + y_{i+1} * \frac{\beta_i+\beta_{i+1}}{2} + \ldots + y_{k-1} * \frac{\beta_{i-2}+\beta_{i-1}}{2} + y_k * \beta_{k-1}}{\beta_i + \frac{\beta_i+\beta_{i+1}}{2} + \ldots + \frac{\beta_{i-2}+\beta_{i-1}}{2} + \beta_{k-1}} \quad (1)$$

Where:

$j$: denotes the j-th interval $[A_j, B_j]$,
$i$: the first data index in $[A_j, B_j]$,
$k$: the last data index in $[A_j, B_j]$,
$y_i$ : value of i-th data in $[A_j, B_j]$ and
$\beta_i$: a similarity measure between $y_i$ and $y_{i+1}$

Hong proposed a heuristic method [13] for determining such three parameters, where the similarity between adjacent data is calculated from their difference. The previous expression is a generalization of Hong's expression for $b_j$ assuming $b_i = 1\ \forall y \in Y$. Thus, $b_j$ becomes the mean of all values in $[A_j, B_j]$. The same idea used by Hong can be applied to the remaining parameters $a_j$ and $c_j$, which are obtained by interpolation.

A trapezoidal MF is described by four parameters $a$, $b$, $c$ and $d$. Notice that this kind of function is reduced to a triangular one when $b = c$. For this reason, we will take advantage of the previous procedure for computing the triangular MF parameters. Let $a_j = A_j$ and $d_j = B_j$. Let $M_j = b_j$, which is calculated from interval $[A_j, B_j]$. Afterwards, the same procedure shall be applied to the new intervals $[A_j, M_j]$ and $[M_j, B_j]$ so as to reckoning $b_j$ and $c_j$, respectively.

A Gaussian MF is specified by two parameters $c$ and $\sigma$ where $c_j$ represents its center and $\sigma_j$ stands for its width. The previous ideas regarding the triangular function's parameters fit, hence $c_j$ is also calculated as the mean of all points in $[A_j, B_j]$. The $\sigma_j$ parameter relates to $c_j$ as shown in (2), preserving the function's symmetry.

$$\sigma_j = 2 * v_{min},\ v_{min} = min\left(|A_j - c_j|, |B_j - c_j|\right) \quad (2)$$

Finally, a sigmoid MF is fully described by parameters $c$ and $\alpha$. Depending on the sign of $\alpha$ , this function is inherently positive (open right) or negative (open left) and therefore suitable for representing concepts such as "very thick" (linguistic term $T_k$, 0 ¡ j ¡ k) or "very thin" (linguistic term $T_0$). The parameter $c$ is computed likewise as for the Gaussian function while (3) displays how to calculate $\alpha$.

$$\alpha = \ln(0.25) * |v - c| \quad (3)$$

$$v = \begin{cases} A_j & \text{if } min\left(|c - A_j|, |c - B_j|\right) = |c - A_j| \\ B_j & \text{otherwise} \end{cases} \tag{4}$$

It is worthwhile remarking that the way the MFs are built leaves no room for a degree of overlap between adjacent MFs, since every parameter necessary to de¯ne it is never computed outside the interval representing the linguistic term from which the MF stands.

## 4 The Associative Fuzzy Neural Network

In order to test the feasibility of the novel methods, the Fuzzy-SIAC (Fuzzy Simple Interactive Activation and Competition) described in [2] was chosen as the inference engine.

It keeps a resemblance with the IAC network [16] in the sense that the neurons are organized into clusters (each cluster represents an attribute) but there is no competition between the neurons belonging to the same cluster. An ANFIS-like [3] preprocessing layer was added as shown in Figure 1.

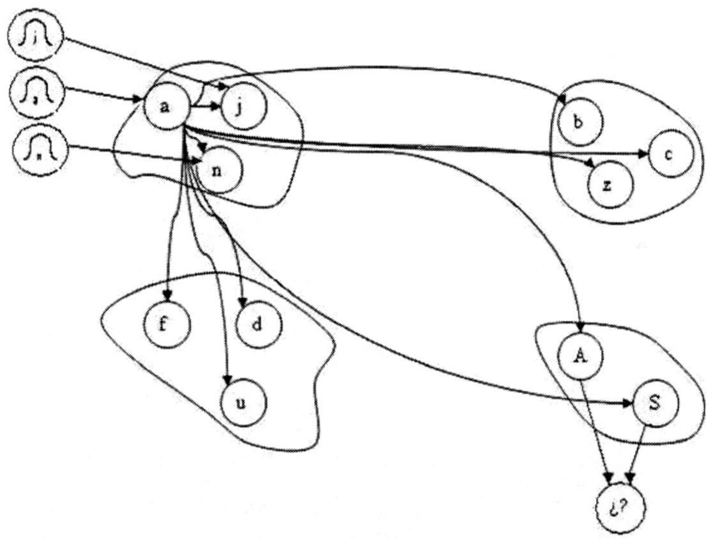


**Fig. 1.** The topology of the Fuzzy-SIAC network

A weight $w_{ij}$ denoting the strength of the connection between neurons $i$ and $j$ belonging to groups $I$ and $J$ respectively ($I \neq J$) labels every arc of the neural network. One way to measure such strength is by counting how many times the values represented by neurons i and j simultaneously appear throughout the cases base. Another choice for computing $w_{ij}$ would be the Pearson's correlation coefficient [17].

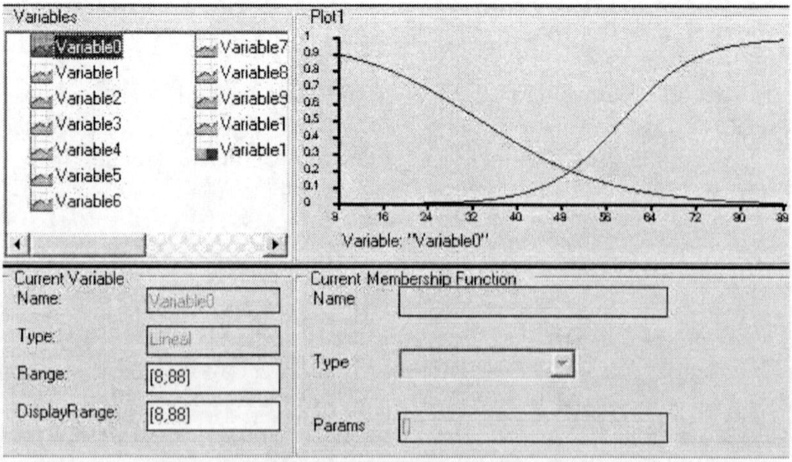


**Fig. 2.** A linear attribute shaped through variant 5

## 5 Experimental Results and Discussion

Several experiments were carried out in order to assess the feasibility of the new heuristic methods. They all utilize Fuzzy-SIAC (outlined in Section 3) as their inference engine.

Five fuzzification alternatives for a linear attribute were defined from the above heuristic methods:

1. Use triangular functions alone for representing all of the linguistic terms.
2. The triangular MFs are restricted to the first and last linguistic term whereas the remaining linguistic terms are modeled via trapezoidal MFs.
3. Use trapezoidal functions alone for representing all of the linguistic terms.
4. Use Gaussian functions alone for representing all of the linguistic terms. This variant takes advantage of the smoothness of the Gaussian function.
5. Gaussian and sigmoidal functions are intermingled, using the sigmoid functions to fashion the first and last linguistic terms as pictured in Figure 2.

Nineteen well-known international databases from the UCI Machine Learning Repository [18] were selected in order to properly validate the incidence degree the fuzzy modeling of the linear attributes might have over the network's performance. Table 1 displays the classification accuracy achieved by the ANN whose weights were computed by "relative frequency" and in which a 10-fold cross-validation procedure was applied.

The aforementioned Hong's heuristic for triangular MFs (column 1) and the Zhou's method for trapezoidal MFs [4](column 2) were chosen as benchmarks. The third column exhibits the higher performance accomplished by any of the five fuzzification alternatives previously reported.

The encouraging and clearly superior outcomes reported at Table 1 were appropriately confirmed by non-parametric statistic tests. First of all, the Friedman

**Table 1.** The Fuzzy-SIAC's performance reached over 19 cases bases by using three different heuristics for automatically building membership functions

| Nr. | Cases base | Hong's | Zhou's | Proposed methods |
|---|---|---|---|---|
| 1 | Ionosphere | 66.69 | 78.43 | **84.99** |
| 2 | Iris | 69.33 | **92.00** | 72.67 |
| 3 | Sonar | 58.05 | 78.93 | **80.23** |
| 4 | Liver-disorders | 58.87 | 59.75 | **65.23** |
| 5 | Pima Indians | 34.76 | 68.10 | **75.14** |
| 6 | Credit-app | 78.73 | 85.51 | **85.83** |
| 7 | Heart-disease | 51.02 | 52.72 | **57.17** |
| 8 | Hepatitis | 69.04 | **69.63** | 69.00 |
| 9 | Wine | 59.52 | 86.02 | **93.91** |
| 10 | Echocardiogram | 69.56 | **70.98** | 70.27 |
| 11 | Horse colic | 75.56 | 77.16 | **77.44** |
| 12 | Waveform | 33.14 | 76.02 | **78.54** |
| 13 | Thyroid-disease | 24.55 | 43.27 | **55.49** |
| 14 | Glass | 88.33 | **93.35** | 89.95 |
| 15 | Vehicle | 37.86 | 61.35 | **62.90** |
| 16 | Page blocks | 73.77 | 87.66 | **90.06** |
| 17 | WPBC | 27.81 | 40.55 | **76.30** |
| 18 | Sick-euthyroid | 17.67 | 22.92 | **34.71** |
| 19 | Labor | 79.33 | 89.00 | **91.33** |
| | AVERAGE | 56.50 | 70.17 | **74.27** |

test yielded a significance level of 0.000 for 3-related samples, thus uncovering significant differences among the various approaches compared. Subsequently, the Wilcoxon test also reported meaningful differences when comparing our approach with Hong's (sig. level = 0.000) and Zhou's (sig. level = 0.007), supporting the proposed methods. It was also evidenced that none of the five fuzzification alternatives excelled over the rest, for no meaningful differences were spotted among them.

As Zhou's computational method uses `Chi`$^2$ as its underlying discretization algorithm to obtain the linguistic terms, it happened that, for some cases bases, the amount of intervals generated as output was nearly the number of real values the attribute held. No doubt the serious implication it brought about the network's overall performance. That's why the choice of `CAIM` as the discretization algorithm lying beneath the novel heuristic methods accomplished a far better outcome and a sense of "reliability" regarding both the number of linguistic terms eventually yielded and the intervals themselves, for `CAIM` returns the lowest possible amount of intervals and maximizes the class-attribute interdependence as well.

It is also worth stressing the slight improvements attained when applying the novel heuristic methods to a real application: the anticancer drug design outlined in [19]. A classification accuracy of 78.28% was reached in contrast to that supplied under expert's criteria that claimed to use sigmoid and bell functions (77.86%).

## 6 Conclusions and Future Work

In this chapter, we employed fuzzy sets in the selection of the representative values (linguistic terms) for linear attributes as a preliminary step to define the topology for an Associative Fuzzy Neural Network.

The way the fuzzy sets have been obtained is clearly outlined: by making a partition of the domain of the linguistic variable into several disjoint intervals and, later on, defining a fuzzy set via its MF for each of the intervals. Heuristic methods for automatically making up triangular, trapezoidal, Gaussian and sigmoid functions from data were introduced. The MFs gotten this way are easy to understand by the experts in the application domain and outperformed Hong's and Zhou's methods when tested with the fuzzy ANN.

As a future work, we are testing an incipient tuning algorithm for the parameters of the proposed MFs. The method utilizes supervised learning in order to improve the performance achieved by the network and shares some key points with the tuning algorithms associated with the NEFCLASS, ANFIS and other neuro-fuzzy systems [3].

## References

1. Mitchell, T.: Machine Learning. McGraw-Hill Science/Engineering/Math (1997)
2. Rodríguez, Y., et al.: Extending a Hybrid CBR-ANN Model by Modeling Predictive Attributes using Fuzzy Sets. Lecture Notes on Artificial Intelligence, vol. 4140. Springer, Heidelberg (2007)
3. Nauck, D., Klawonn, F., Kruse, R.: Foundations of Neuro-Fuzzy Systems. John Wiley & Sons Ltd., Chichester (1997)
4. Zhou, Q., Purvis, M., Kasabov, N.: A membership function selection method for fuzzy neural networks. In: Proceedings of the International Conference on Neural Information Processing and Intelligent Systems, pp. 785–788. Springer, Singapore (1997)
5. Liu, H., Setiono, R.: Chi2: Feature selection and discretization of numeric attributes. In: Proceedings of the IEEE 7th International Conference on Tools with AI (1997)
6. Chen, J., Otto, K.: Fuzzy Sets and Systems 73, 313–327 (1995)
7. McAllister, D., Roulier, J.: ACM Transactions on Mathematical Software 7(3), 331–347 (1981)
8. Marsala, C., Bouchon-Meunier, B.: Fuzzy Partitioning Using Mathematical Morphology in a Learning Scheme. In: Proceedings of the Fifth IEEE International Conference on Fuzzy Systems (1996)
9. Serra, J.: Image analysis and mathematical morphology. Academic Press, New York (1982)
10. Botzheim, J., Hamori, B., Kóczy, L.: In: Reusch, B. (ed.) Computational Intelligence. Theory and Applications. LNCS, vol. 2206, pp. 218–227. Springer, Heidelberg (2001)

11. Salmeri, M., Re, M., Petrongari, E., Cardarilli, G.: A novel bacterial algorithm to extract the rule base from a training set. In: Proceedings of the Ninth IEEE International Conference on Fuzzy Systems (2000)
12. Piñeiro, P., Arco, L., Garcia, M.: Algoritmos genéticos en la construcción de funciones de pertenencia. In: Revista Iberoamericana de Inteligencia Artificial (AEPIA), vol. 18(2) (2003)
13. Hong, J., Xizhao, W.: Fuzzy Sets and Systems 99, 283–290 (1998)
14. Kurgan, L., Cios, K.: IEEE Transactions on Knowledge and Data Engineering 16, 145–153 (2004)
15. Jang, J., Sun, C., Mizutani, E.: Neuro-Fuzzy and Soft Computing. Prentice-Hall, Englewood Cliffs (1998)
16. Gledhill, J.: Neuralbase: A neural network system for case based retrieval in the Help Desk diagnosis domain. Master Thesis, Royal Melbourne Institute of Technology University, Melbourne, Australia (1995)
17. Garcia, M., Rodríguez, Y., Bello, R.: Usando conjuntos borrosos para implementar un modelo para sistemas basados en casos interpretativos. In: Proceedings of the International Joint Conference, 7th Ibero-American Conference, 15th Brazilian Symposium on AI, Springer, Heidelberg (2000)
18. Murphy, P., Aha, D.: UCI repository of machine learning databases. University of California-Irvine, Department of Information and Computer Science (1994)
19. Rodríguez, Y.: Sistema computacional para la determinación de propiedades anticancerígenas en el diseño de un fármaco. In: IV Congreso Internacional de Informática Médica de La Habana (Informática 2003)

# An Incremental Clustering Method and Its Application in Online Fuzzy Modeling

Boris Martínez[1], Francisco Herrera[1], Jesús Fernández[1], and Erick Marichal[2]

[1] Faculty of Electrical Engineering, Central University of Las Villas (UCLV)
Carretera Camajuaní Km. 5.5, Santa Clara, Cuba
{boris,herrera}@uclv.edu.cu

[2] University of Informatics Sciences (UCI), Faculty 2
Carretera San Antonio de Los Baños Km. 2.5, La Habana, Cuba
emarichal@uci.cu

**Summary.** Clustering techniques for the generation of fuzzy models have been used and have shown promising results in many applications involving complex data. This chapter proposes a new incremental clustering technique to improve the discovery of local structures in the obtained fuzzy models. This clustering method is evaluated on two data sets and the results are compared with the results of other clustering methods. The proposed clustering approach is applied for nonlinear Takagi–Sugeno (TS) fuzzy modeling. This incremental clustering procedure that generates clusters that are used to form the fuzzy rule antecedent part in online mode is used as a first stage of the learning process.

**Keywords:** Online learning, evolving/incremental clustering, fuzzy system, Takagi–Sugeno fuzzy model.

## 1 Introduction

Many real-world problems are changing, non-linear processes that require fast, adapting non-linear systems capable of following the process dynamics. Therefore, there are demands for effective approaches to design self-developing systems which at the same time should be flexible and robust. Recently, several algorithms for online learning with self-constructing structure have been reported [1] [2] [3] [4] [5] [6].

During the past few years, significant attention has been given to data-driven techniques for the generation of flexible models and among these techniques are fuzzy systems. It is well known that fuzzy systems are universal approximators [7], i.e., they can approximate any nonlinear continuous function to any prescribed accuracy if sufficient fuzzy rules are provided. The Takagi–Sugeno (TS) fuzzy model [8] has become a powerful practical engineering tool for complex systems modeling because of its capability to describe a highly nonlinear system using a small number of rules.

R. Bello et al. (Eds.): Granular Computing, STUDFUZZ 224, pp. 163–178, 2008.
springerlink.com 

Fuzzy modeling involves structure and parameter identification. Most methods for structure identification are based on data clustering. Clustering algorithms can be divided into two classes, offline and online. Although a great number of clustering algorithms have been proposed, the majority of them process the data offline, hence, the temporal structure is ignored [9]. Online clustering algorithms should be adaptive in the sense that up-to-date clusters are offered at any time, taking new data items into consideration as soon as they arrive. For continuous online learning of the TS fuzzy model, an online clustering method responsible for the model structure (rule base) learning is needed [4]. Also, according to the incremental/evolving fuzzy systems paradigm, the structure (rules/clusters) of the fuzzy system is not fixed, it gradually evolves (can expand or shrink), and an incremental clustering method is needed [1] [4] [6] [10] [11].

The incremental/evolving fuzzy learning allows the system to infer rules continually as new data become available without forgetting the previously learned ones and without referring at any time to the previously used data. Hence, the system becomes self-adaptive and the acquired knowledge becomes self-corrective. As new data arrive, new rules may be created and existing ones modified allowing the system to evolve over time. Despite its importance in real time applications, incremental/evolving learning remains a topic at its very earlier stages [6] [12]. There are several real-world applications where data become available over time. In such applications, it is important to devise learning mechanisms to induce new knowledge without 'catastrophic forgetting' and/or to refine the existing knowledge. The whole problem is then summarized in how to accommodate new data in an incremental way while keeping the system under use [6].

In this chapter, a brief analysis of two clustering methods is made: the Evolving Clustering Method (ECM) [13], an online incremental algorithm used in Dynamic Evolving Neural–Fuzzy Inference System (DENFIS) [1]; and the Agglomerative Clustering Method (AddC) [9], an online agglomerative clustering algorithm for non-stationary data. Also, a new incremental approach based on these methods, the Evolving–Agglomerative Clustering Method (eACM), is proposed.

In the fuzzy modeling approach used in this work, after the number of rules is decided by using eACM, the consequent parameters are tuned by using a recursive least squares method. It is important to note that learning could start without prior information and only a few data samples. Furthermore, the clusters adapt their radius to the spatial information brought in by new data samples, while the cluster history is taking into account. This allows the updating of the zone of influence of the fuzzy rules. These features make the approach potentially useful in adaptive control, robotic, diagnostic systems and as a tool for knowledge acquisition from data [4].

The rest of the chapter is organized as follows. The clustering algorithms are discussed in Section 2. Section 3 gives a description of TS fuzzy modeling. Section 4 presents the simulation results and Section 5 draws the concluding remarks.

## 2 Clustering Algorithms

The clustering techniques working in online mode are fundamentally used when it is needed to modify a dynamic process model in real time, or when there are restrictions such as time, computational cost, etc. The most popular online clustering algorithms are the one-pass methods (or single-pass methods) [14]. Such clustering approach needs only one data pass, it handles one data point at a time and then, discards it. This kind of method reduces the mathematical operations and accelerates the clustering process, which is ideal for real time operations.

This chapter focuses on the application of a single-pass clustering method in fuzzy modeling. In the following, a review of three one-pass clustering algorithms (ECM, AddC and eACM) is made. These methods are distance-based. In this chapter, the distance between vectors $\boldsymbol{x}$ and $\boldsymbol{y}$ means a normalized Euclidean distance defined as follows:

$$\|\boldsymbol{x}-\boldsymbol{y}\| = \frac{\left(\sum_{i=1}^{n} |x_i - y_i|^2\right)^{1/2}}{n^{1/2}}, \qquad \boldsymbol{x},\boldsymbol{y} \in \mathbb{R}^n \,. \tag{1}$$

### 2.1 Evolving Clustering Method (ECM)

The ECM is a fast algorithm for dynamic clustering of data. In any cluster, the maximum distance between a sample point which belongs to the cluster and the corresponding cluster center, is less than or equal to a threshold value, $Rthr$. This clustering parameter would affect the number of clusters to be created. Each cluster $C_j$ is characterized by the center, $\boldsymbol{Cc_j}$, and the radius, $Ru_j$.

In the clustering process, new clusters will be created or some created clusters will be updated. When the sample $\boldsymbol{z_i}$ can not belong to any existing clusters, a new cluster is created, its cluster center, $\boldsymbol{Cc}$, is located at the sample point $\boldsymbol{z_i}$ and its cluster radius, $Ru$, is set to zero. In the second case, a sample point $\boldsymbol{z_i}$ will be included in a created cluster $C_a$. The cluster $C_a$ is updated by moving its center, $\boldsymbol{Cc_a}$, and increasing its radius value, $Ru_a$. The new center $\boldsymbol{Cc_{anew}}$ is located on the line connecting input vector $\boldsymbol{z_i}$ and the old cluster center $\boldsymbol{Cc_a}$, so that the distance from the new center $\boldsymbol{Cc_{anew}}$ to the sample point $\boldsymbol{z_i}$ is equal to the new radius $Ru_{anew}$. A cluster will not be updated any more when its cluster radius, $Ru$, has reached the threshold value $Rthr$. If it is regarded that the current sample belongs to an existing cluster $C_j$ ($\|\boldsymbol{z_i}-\boldsymbol{Cc_j}\| \leq Ru_j$), neither a new cluster is created nor any cluster is updated. Note that the algorithm does not keep any information of passed samples. A more detailed description of ECM is given in [1] and [13].

### 2.2 Agglomerative Clustering Method (AddC)

The AddC is a clustering algorithm which minimizes the global distortion. The basic idea is that each point of data can belong to a new cluster. Thus, a new cluster is placed on each and every new point. This implies that a cluster must be allocated at the cost of the existing clusters. This is done by merging the two closest clusters into one. The solution is affected by the maximum number of clusters, $Kmax$. This parameter must be user-defined and small perturbations in either the parameter or the data can result in drastically different solutions. On the other hand, the resulting algorithm, does not neglect small clusters. If a small cluster is distinct enough, it will not be lost by being merged into an existing cluster. Each cluster $C_j$ is characterized by the center, $\boldsymbol{Cc_j}$, and a weight, $W_j$, which represents the number of points of the cluster.

The algorithm is simple and fast and it can be summarized in the following three steps: For each data point arriving;

1. Move the closest cluster towards the point.
2. Merge the two closest clusters. This results in the creation of a redundant cluster.
3. Set the redundant cluster equal to the data point.

Three criteria are addressed at each time step, minimization of the within cluster variance, maximization of the distances among the clusters and adaptation to temporal changes in the distribution of the data. In the first step, the within cluster variance is minimized by updating the representation of the closest cluster. The second step maximizes the distances between the clusters by merging the two clusters with the minimum distance (not considering their weight). Finally, temporal changes in the distribution of the data are anticipated by treating each new point as an indication to a potential new cluster. For a more detailed description of AddC see [9].

### 2.3 Evolving–Agglomerative Clustering Method (eACM)

The proposed method is an online incremental clustering algorithm without any optimizing process for dynamically estimating the number of clusters and finding their centers in a data set. This method uses two clustering parameters: a threshold radius, *Rthr*, and a threshold similarity value, *Sthr*. The basic idea is that each data sample can belong to a new cluster and the maximum distance between a data point which belongs to the cluster and the corresponding cluster center is limited. Since, this work look for an incremental clustering algorithm to improve the discovery of local structures in the obtained fuzzy models, a similarity measure to merge two similar clusters and a weight to update a cluster center are used. This is done to favor the membership to concrete classes to obtain the convexity of fuzzy membership functions [15]. Each cluster $C_j$ is characterized by the center, $\boldsymbol{Cc_j}$, and a weight, $W_j$.

The eACM algorithm is described as follows:

1. First data point $\boldsymbol{z_1}$ is assigned to the first cluster, $C_1$, whose cluster center, $\boldsymbol{Cc_1}$, is that data point. Set $W_1 = 1$ and $K = 1$ (number of clusters).
2. Get the current data point $\boldsymbol{z_i}$.
3. Compute the distance between this data point, $\boldsymbol{z_i}$, and each cluster center $\boldsymbol{Cc_j}$.

$$d(i,j) = \|\boldsymbol{z_i} - \boldsymbol{Cc_j}\| \quad j = 1, \ldots, K \,. \tag{2}$$

4. Find the cluster $C_m$ which is closest to data point $\boldsymbol{z_i}$,

$$d(i,m) = \min_j d(i,j) = \min_j \|\boldsymbol{z_i} - \boldsymbol{Cc_j}\| \quad j = 1, \ldots, K \,. \tag{3}$$

5. If $d(i,m) \le Rthr$, the cluster $C_m$ is updated as follows:

$$\begin{aligned} \boldsymbol{Cc_m} &= \boldsymbol{Cc_m} + \frac{\boldsymbol{z_i} - \boldsymbol{Cc_m}}{W_m + 1} \,, \\ W_m &= W_m + 1 \,. \end{aligned} \tag{4}$$

6. Compute the distance between each pair of cluster centers $\boldsymbol{Cc_\alpha}$ and $\boldsymbol{Cc_\beta}$.

$$D(\alpha,\beta) = \|\boldsymbol{Cc_\alpha} - \boldsymbol{Cc_\beta}\|, \quad \alpha \neq \beta \,. \tag{5}$$

7. Find the clusters $C_\gamma$ and $C_\delta$ with the minimum distance between its centers.

$$D(\gamma,\delta) = \min_{\alpha,\beta,\alpha\neq\beta} D(\alpha,\beta) = \min_{\alpha,\beta,\alpha\neq\beta} \|\boldsymbol{Cc_\alpha} - \boldsymbol{Cc_\beta}\| \,. \tag{6}$$

8. If $D(\gamma,\delta) \le Sthr$, the two redundant clusters, $C_\gamma$ and $C_\delta$, are merged by computing their weighted average location and cumulative number of points. Finally, decrease the number of clusters.

$$\begin{aligned} \boldsymbol{Cc_\gamma} &= \frac{\boldsymbol{Cc_\gamma} W_\gamma + \boldsymbol{Cc_\delta} W_\delta}{W_\gamma + W_\delta} \,, \\ W_\gamma &= W_\gamma + W_\delta \,, \\ K &= K - 1 \,. \end{aligned} \tag{7}$$

9. Initialize a new cluster with the last data point $\boldsymbol{z_i}$.

$$\begin{aligned} K &= K + 1 \,, \\ \boldsymbol{Cc_K} &= \boldsymbol{z_i} \,, \\ W_K &= 1 \,. \end{aligned} \tag{8}$$

10. While there remains data to be clustered, go to step 2.

At the end of the process, if some cluster centers are distinct enough but only a few points are close to it (e.g., distant noises, outliers), these clusters (revealed by a very small weight) can be removed. In this case, the next step is added:

11. (Post processing) - Remove all clusters with a negligible weight (counter).

$$\forall j,\ j = 1, \ldots, K, \quad \text{if } W_j < \varepsilon, \text{ eliminate } C_j \text{ and set } K = K - 1 \,. \tag{9}$$

The eACM has several properties that makes it promissory for system identification and knowledge acquisition from data, such as: the procedure begins without previous information on the characteristics of the data and with only one sample data; the class dimension is limited; it is simple, fast and the computational load is low still in the case of great amounts of data; it is able to detect classes with great differences in size; it does not require the previous determination of the number of classes; and the algorithm follows new data while preserving the existing structure.

Nevertheless, this proposed algorithm can only detect hyper-spherical classes, the final result depends on the order in which the data are presented and a previous definition of two parameters ($Rthr$ and $Sthr$) is needed.

### 2.4 Quantitative Analysis of Clustering Algorithms

To quantitatively analyze the performance of the proposed algorithm, two examples were used: a randomly generated Gaussian mixtures [9] and the Box–Jenkins data set [16]. For the purpose of comparative analysis, the following four clustering methods are applied on the same data sets:

- AddC, agglomerative clustering method [9], (one pass)
- ECM, evolving clustering method [1] [13], (incremental, one pass)
- ISC, incremental supervised clustering [6] [17], (incremental, one pass)
- eACM, evolving agglomerative clustering (incremental,one pass)

After the data was clustered by the different methods, several indexes were measured. Taking the distance between each example point, $\boldsymbol{z_i}$, and the closest cluster center, $\boldsymbol{Cc_j}$, the index $J$ is defined by the following equation:

$$J = \sum_{i=1}^{S} \min_{j} \|\boldsymbol{z_i} - \boldsymbol{Cc_j}\| , \tag{10}$$

where $S$ is the size of the data set. The global distortion was calculated as follows:

$$J_G = \frac{1}{S} J = \frac{1}{S} \sum_{i=1}^{S} \min_{j} \|\boldsymbol{z_i} - \boldsymbol{Cc_j}\| , \tag{11}$$

While, the global distortion provides a measure of the average performance, the local distortion provides a good measure of the quality of the representation of each individual cluster. Hence, the local distortion is determined as follows:

$$J_L = \sum_{i=1}^{K} \frac{1}{S_j} \left( \sum_{i, \boldsymbol{z}_i \in C_j}^{S_j} \min_{j} \|\boldsymbol{z_i} - \boldsymbol{Cc_j}\| \right) , \tag{12}$$

where $K$ is the number of clusters generated, $C_j$ is the $j$–th cluster and $S_j$ the number of points in $C_j$. The distortion of each point is the distance between the point and its most representative centroid, normalized by the size of its originating cluster. This ensures that the affect each cluster has on the performance measure is relatively equal, even small clusters influence the final result [9].

**Example 1: Gaussian Mixtures Data Set**

The number of Gaussian mixtures generated was six. Each of them had a randomly generated number of points and shape. The maximum distances between an example point which belongs to a cluster and the corresponding cluster center ($MaxDist$) and the values of the indices ($J$, $J_G$, $J_L$), defined by (10), (11) and (12), are measured for comparison and shown in Table 1. The graphical results are shown in Fig. 1.

The ECM algorithm performs relatively unsatisfactory. This is because neither a new cluster is created nor any cluster is updated if it is regarded that the

**Table 1.** Results obtained by using different clustering methods for clustering the Gaussian mixtures data set into 6 clusters

| Method | AddC | ECM | ISC | eACM |
|---|---|---|---|---|
| $J$ | 268.34 | 463.62 | 241.70 | 209.15 |
| $J_G$ | 0.0822 | 0.1420 | 0.0740 | 0.0641 |
| $J_L$ | 0.4606 | 0.9158 | 0.4219 | 0.3851 |
| $maxDist$ | 0.2679 | 0.3564 | 0.2181 | 0.2102 |

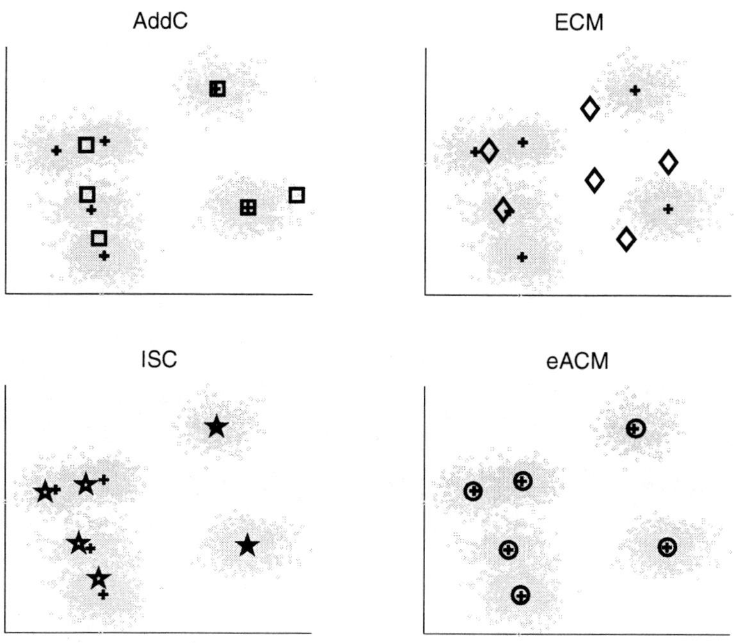


**Fig. 1.** Results of clustering the Gaussian mixture by several clustering methods: data (○), sources (+), centers (◊ - ECM, □ - AddC, △ - ISC, ◯ - eACM)

current sample belongs to an existing cluster. On the other hand, the other algorithms continuously update its cluster centers.

The eACM algorithm succeeds in approaching other results in the minimization of the indices. The proposed method also minimizes the *maxDist* parameter and its centers are close to the sources.

**Example 2: Box–Jenkins' Gas Furnace Data Set**

The gas furnace time series [16] is a well- known bench-mark data set and has been frequently used by many researches in the area of neural networks and fuzzy system for control, prediction and adaptive learning [1] [3] [13]. The example consists of 296 input–output samples recorded with the sampling period of 9 s. The gas combustion process has one input variable, methane gas flow, and one output variable, the carbon dioxide ($CO_2$) concentration. The instantaneous value of the output at the moment (t) can be regarded as being influenced by the methane flow at a time moment (t-4) and the carbon dioxide $CO_2$ produced in the furnace at a time moment (t-1).

In this example, each partition has 10 clusters. The results are shown in Table 2 and Figure 2. The index values for eACM simulation are comparable with the index values produced by other methods. Note in Table 1 and Table 2 that eACM obtains minimum values of *MaxDist* for clustering, which indicates

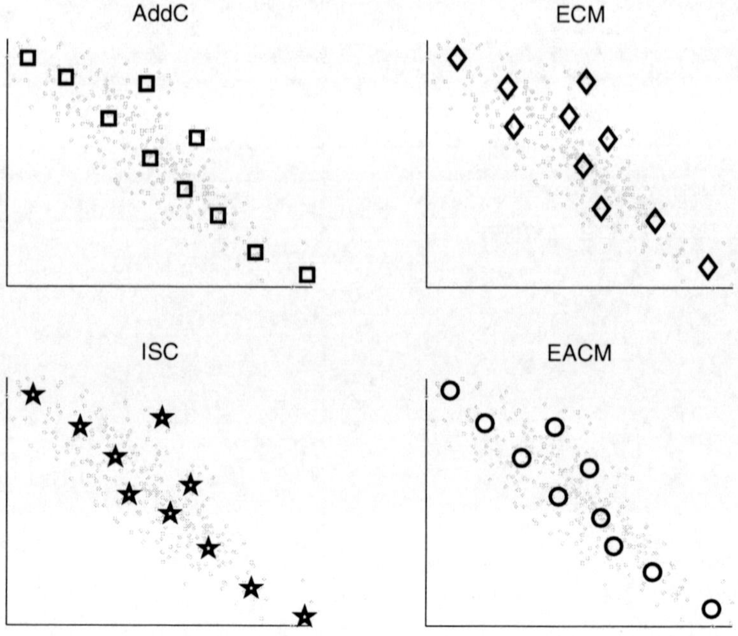


**Fig. 2.** Results of clustering the Box–Jenkins data set by several clustering methods: data (○), sources (+), centers (◇ - ECM, □ - AddC, △ - ISC, ◯ - eACM)

**Table 2.** Results obtained by using different clustering methods for clustering the Box–Jenkins data set into 10 clusters

| Method | AddC | ECM | ISC | eACM |
|---|---|---|---|---|
| $J$ | 25.593 | 27.368 | 25.592 | 25.043 |
| $J_G$ | 0.0876 | 0.0973 | 0.0876 | 0.0858 |
| $J_L$ | 0.8851 | 0.9435 | 0.8656 | 0.8927 |
| $maxDist$ | 0.1873 | 0.2057 | 0.1796 | 0.1787 |

that this method partitions the data set more uniformly than other methods [1]. Looking at the results from a different point of view, it can be stated that if all these clustering methods obtained the same value of $MaxDist$, eACM would result in less number of clusters.

## 3 Takagi–Sugeno Fuzzy Modeling

The aim of this section is to describe a computationally efficient and accurate algorithm for online Takagi–Sugeno (TS) fuzzy model generation. This algorithm is based on the DENFIS learning approach, which is executed in online mode [1]. The approach combines evolving–Agglomerative Clustering Method (eACM) for the structure identification of the rule base, and least-squares (LS) procedures for consequent parameters determination.

### 3.1 Takagi-Sugeno Fuzzy System

Our online dynamic fuzzy system uses the well-known Takagi–Sugeno inference engine [8]. Such fuzzy system is composed of $N$ fuzzy rules indicated as follows:

$$\begin{aligned} \Re_i : \textbf{if } x_1 \textbf{ is } A_{i1} \textbf{ and } \ldots \textbf{ and } x_r \textbf{ is } A_{ir} \\ \textbf{then } y_i = a_{i0} + a_{i1}x_1 + \ldots + a_{ir}x_r,\ i = 1, ..., N \ , \end{aligned} \tag{13}$$

where $x_j$, $j = 1, \ldots, r$, are input variables defined over universes of discourse $X_j$ and $A_{ij}$ are fuzzy sets defined by their fuzzy membership functions $\mu_{Aij} : Xj \rightarrow [0, 1]$. In the consequent parts, $y_i$ is rule output and $a_{ij}$ are scalars.

For an input vector $\boldsymbol{x} = [x_1, x_2, \ldots, x_r]^T$, each of the consequent functions can be expressed as follows:

$$y_i = \boldsymbol{a_i}^T \boldsymbol{x_e}, \quad \boldsymbol{x_e} = \left[1, \boldsymbol{x}^T\right]^T . \tag{14}$$

The result of inference, the output of the system $y$, is the weighted average of each rule output $y_i$, indicated as follows:

$$y = \frac{\sum_{i=1}^{N} w_i y_i}{\sum_{i=1}^{N} w_i} = \frac{\sum_{i=1}^{N} w_i \boldsymbol{a_i}^T \boldsymbol{x_e}}{\sum_{i=1}^{N} w_i} , \tag{15}$$

$$w_i = \prod_{j=1}^{r} \mu_{Aij}(x_j), \quad i = 1, \ldots, N \;, \tag{16}$$

is the firing strength of the rule $i$. Equation (15) can be rewritten in the form:

$$y = \sum_{i=1}^{N} \tau_i y_i = \sum_{i=1}^{N} \tau_i \boldsymbol{a}_i^T \boldsymbol{x_e} \;, \tag{17}$$

$$\tau_i = \frac{w_i}{\sum_{i=1}^{N} w_i} \;, \tag{18}$$

where $\tau_i$ represents normalized firing strength of the $i$–th rule. TS fuzzy rule-based model, as a set of local models, enables application of a linear LS method since this algorithm requires a model that is linear in the parameters [18].

Finally, all fuzzy membership functions are Gaussian type functions because, in practice, partitions of this type are recommended when Tagaki–Sugeno consequents are used [19]. Gaussian membership functions depend on two parameters as given by the following equation:

$$\mu(x_d; c_d, \sigma) = exp\left(-\left(\frac{x_d - c_d}{2\sigma}\right)^2\right) \;, \tag{19}$$

where $c_d$ is the value of the cluster center on the $x_d$ dimension, $\sigma$ is proportional to $R$, where $R$ is the distance from cluster center to the farthest sample that belongs to this cluster, i.e., the radius/zone of influences of the cluster/rule.

### 3.2 Algorithm for Structure Identification and Parameters Determination

The online learning algorithm consists of two main parts: structure identification and parameters determination. The object of structure identification is to select fuzzy rules by input-output clustering. In online identification, there is always new data coming, and the clusters should be changed according to the new data. If the data do not belong to an existing cluster, a new cluster is created. If the new cluster is too near to a previously existing cluster then the old cluster is updated. The appearance of a new cluster indicates a region of the data space that has not been covered by the existing clusters (rules). This could be a new operating mode of the plant or reaction to a new disturbance. A new rule is generated only if there is significant new information present in the data. This step uses eACM algorithm.

After online clustering is applied to adjust the centers and widths of membership functions using (19), the linear functions in the consequence parts are created and updated using a linear least squares estimator. For this, the $k$–th element on the main diagonal of the diagonal matrix $\boldsymbol{T_i}\,(i = 1, \ldots, K)$ is formed using the values of the normalized firing strength, obtained from (18). Hence, a matrix composition $\boldsymbol{X'}$ can be formed [20]:

$$\boldsymbol{X'} = [(\boldsymbol{T}_1\boldsymbol{X}_e), (\boldsymbol{T}_2\boldsymbol{X}_e), \ldots, (\boldsymbol{T}_K\boldsymbol{X}_e)] \ , \tag{20}$$

where matrix $\boldsymbol{X}_e = [\mathbf{1}, \boldsymbol{X}]$ is formed by the following rows $\boldsymbol{x}_e^T(k) = [1, \boldsymbol{x}^T(k)]$.

The least-square estimator formula

$$\boldsymbol{a} = \left(\boldsymbol{X'}^T\boldsymbol{X'}\right)^{-1}\boldsymbol{X'}^T\boldsymbol{Y} \ , \tag{21}$$

is used to obtain the initial matrix of consequent parameters $\boldsymbol{a}=[\boldsymbol{a}_1^T \ \boldsymbol{a}_2^T \ldots \boldsymbol{a}_K^T]^T$. This matrix is calculated with a learning data set that is composed of $m$ data pairs.

Equation (21) can be rewritten as follows:

$$\begin{aligned} \boldsymbol{P} &= \left(\boldsymbol{X'}^T\boldsymbol{X'}\right)^{-1} \ , \\ \boldsymbol{a} &= \boldsymbol{P}\boldsymbol{X'}^T\boldsymbol{Y} \ . \end{aligned} \tag{22}$$

In this chapter, a Recursive Least Squares (RLS) estimator with a forgetting factor is used. Let the $k$–th row vector of a matrix $\boldsymbol{X'}$ is denoted as $\boldsymbol{x'}^T(k)$ and the $k$–th element of $\boldsymbol{Y}$ is denoted as $\boldsymbol{y}(k)$. Then $\boldsymbol{a}$ can be calculated iteratively as follows:

$$\begin{aligned} \boldsymbol{P}(k+1) &= \frac{1}{\lambda}\left(\boldsymbol{P}(k) - \frac{\boldsymbol{P}(k)\boldsymbol{x'}(k+1)\boldsymbol{x'}^T(k+1)\boldsymbol{P}(k)}{\lambda + \boldsymbol{x'}^T(k+1)\boldsymbol{P}(k)\boldsymbol{x'}(k+1)}\right) \ , \\ \boldsymbol{a}(k+1) &= \boldsymbol{a}(k) + \boldsymbol{P}(k+1)\boldsymbol{x'}(k+1)\left(\boldsymbol{y}(k+1) - \boldsymbol{x'}^T(k+1)\boldsymbol{a}(k)\right) \ , \end{aligned} \tag{23}$$

where $\lambda$ is a constant forgetting factor which typical values between 0.8 and 1. The initial values, $\boldsymbol{P}(0)$ and $\boldsymbol{a}(0)$, are calculated using (22). The equations in (23) have an intuitive interpretation: the new parameter vector is equal to the old parameter vector plus a correcting term based on the new data $\boldsymbol{x'}^T(k)$.

The recursive procedure for online learning of TS models used in this chapter, includes the following stages.

1. Initialization of the fuzzy model. For this:
   (a) Take the first $m$ data samples from the data set.
   (b) Apply the eACM algorithm to obtain cluster centers.
   (c) Create the antecedents with (19) and use (22) to obtain the initial values of $\boldsymbol{P}$ and $\boldsymbol{a}$.
2. Read the next data sample.
3. Recursive update of the cluster centers by using eACM algorithm.
4. Possible modification of the rule base. A new fuzzy rule is created if there is significant new information present in the clusters created by the clustering algorithm. For this, the following rule is used:
   **if** the cluster weight is not negligible ($W_i > \varepsilon$), **then** a new rule is created.
5. Upgrade of the antecedent parameters by using (19)
6. Recursive calculation of the consequent parameters by using (23)
7. Prediction of the output for the next time step by the TS fuzzy model.

The execution of the algorithm continues for the next time step from stage 2.

It should be noted that in [4] it is stated that using the potential instead of the distance to a certain rule center only [1] for forming the rule base results in rules that are more informative and a more compact rule base. The reason is that the spatial information and history are part of the decision whether to upgrade or modify the rule base. Here, for the same objective, the weights in conjunction with the distance are used. Also, the proposed online incremental clustering approach ensures an evolving rule base by dynamically upgrading and modifying it while inheriting the bulk of the rules (N-1 of the rules are preserved even when a modification or an upgrade take place).

## 4 Experimental Results

The developed identification method is applied in two benchmark examples: Box–Jenkins' identification problem and the Mackey–Glass chaotic time series. These data sets are frequently used as problems in the system identification area. MATLAB software is used for computation and analysis.

### 4.1 Box–Jenkins' Gas–Furnace Identification Problem

Box–Jenkins' gas furnace data [16] consists of 292 consecutive data pairs of the methane flow at a time moment (t-4) and the produced $CO_2$ at a time moment (t-1) as input variables, with the produced $CO_2$ at the moment (t) as an output variable.

The first 15 samples ($m = 15$) are used to obtain initial fuzzy model, while the remaining data are used for online TS learning. With the aim of achieving a fair comparison with other available models, a two rule fuzzy model is obtained. Table 3 compares characteristics of our model with a number of models taken from [3], using mean square error ($MSE$) as error index. An $MSE = 0.1544$ was achieved with $\lambda = 0.95$ for online adaptive identification. The $MSE$ is 0.1648 for non adaptive identification by using all the data ($m = 292$) at stage 1 of the online learning procedure.

Figure 3 illustrates the evolution of the parameters of the two fuzzy rules.

### 4.2 Mackey–Glass Time Series Data Set

In this example, the data set is generated from the Mackey–Glass differential delay equation defined by:

$$\frac{\mathrm{d}x}{\mathrm{d}t} = \frac{0.2\,x(t-\tau)}{1 + x^{10}(t-\tau)} - 0.1\,x(t)\,, \tag{24}$$

where $\tau = 17$ and the initial condition is $x(0) = 1.2$. The aim is using the past values of $x$ to predict some future value of $x$. The task is to predict value $x(t+85)$ from the input vectors $[x(t-18)\ x(t-12)\ x(t-6)\ x(t)]$ (same as in [1] [4]).

The following experiment was conducted: 3000 data points, for $t = 201 : 3200$, are extracted from the time series and used as training data; and 500 data points,

**Table 3.** Box–Jenkins' Problem: Comparison of structure and accuracy

| Models | Inputs | Rules | MSE |
|---|---|---|---|
| Box and Jenkins (1970) | 6 | - | 0.202 |
| Wang and Langari (1995) | 6 | 2 | 0.066 |
| Wang and Langari (1996) | 2 | 5 | 0.172 |
| Kim et.al. (1997) | 6 | 2 | 0.055 |
| Lin et.al. (1997) | 4 | 12 | 0.157 |
| Chen et.al. (1998) | 2 | 3 | 0.268 |
| Wang and Rong (1999) | 2 | 29 | 0.137 |
| Lo and Yang (1999) | 6 | 2 | 0.062 |
| Kang et.al. (2000) | 2 | 5 | 0.161 |
| Kukolj and Levi (2004) | 2 | 2 | 0.129 |
| This model | 2 | 2 | 0.154/0.165 |

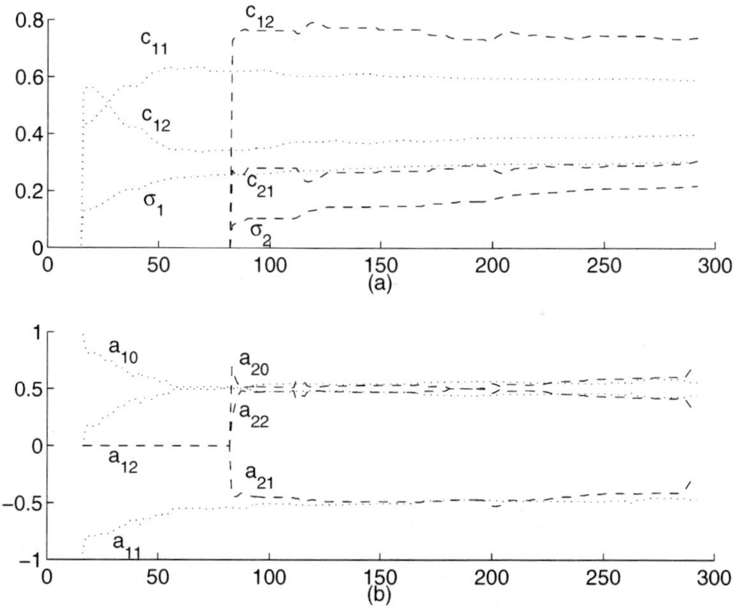


**Fig. 3.** Evolution of parameters: (**a**) antecedent part; (**b**) consequent part

for $t = 5001 : 5500$, are used as testing (validation) data. The learning mechanism is always active, even for the testing data.

To evaluate the performance of the models, the Non-Dimensional Error Index ($NDEI$) is used. This error index is defined as the ratio of the root mean square error ($RMSE$) over the standard deviation of the target data.

**Table 4.** Mackey–Glass' Time Series: Comparison of structure and accuracy

| Models | Rules, nodes or units | NDEI |
|---|---|---|
| EFuNN | 1125 | 0.094 |
| Neural gas | 1000 | 0.062 |
| DENFIS | 883 | 0.042 |
| EFuNN | 193 | 0.401 |
| ESOM | 114 | 0.320 |
| RAN | 113 | 0.373 |
| eTS | 113 | 0.095 |
| DENFIS | 58 | 0.276 |
| This model | 25 | 0.223 |

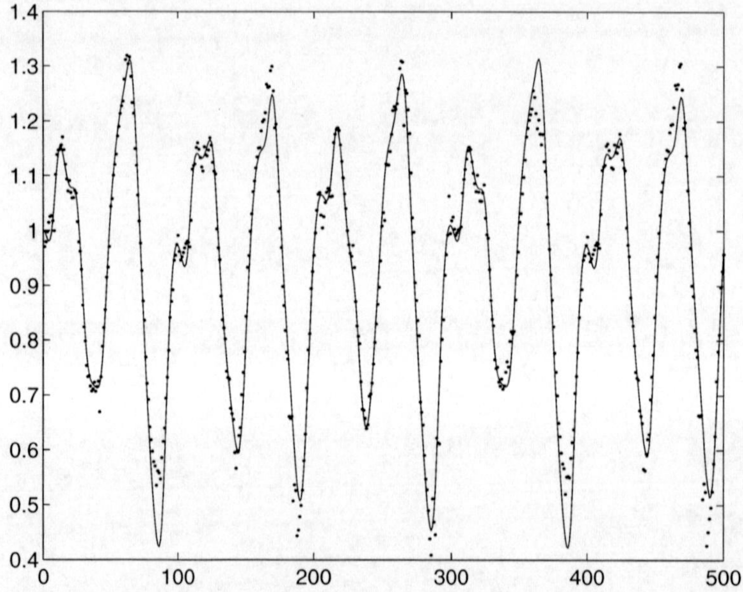


**Fig. 4.** Prediction (85 steps ahead) of the Mackey–Glass chaotic time series, real data (–), model prediction (·)

$$NDEI = \frac{RMSE}{std\,(y(t))} \; . \tag{25}$$

For the purpose of a comparative analysis, the results of some existing online learning models applied on the same problem are taken from [1] and [4]. The results summarized in Table 4 and Fig. 4 show that our approach can yield a more compact model, and hence a more transparent rule base than the similar fuzzy and neuro-fuzzy approaches with comparable $NDEI$.

## 5 Conclusions

This chapter presents an approach for online Takagi–Sugeno fuzzy model generation. This approach relies on a new incremental clustering algorithm conceived for this purpose. Quantitative analysis of the new algorithm performance in clustering simulated data, demonstrated its superior performance in minimizing the maximum distances between an example point which belongs to a cluster and the corresponding cluster center ($MaxDist$), and a superior or comparable performance in minimizing the distortions (local and global) to existing clustering algorithms.

The proposed learning approach combines this clustering algorithm with a recursive least squares procedure. The experiments showed a good performance of the proposed method when compared with other learning algorithms working in online mode. Their characteristics make it potentially useful for applications in adaptive control, real time applications, robotic, diagnosis systems, etc, as well as a tool for knowledge acquisition.

Further directions for research include: improvement of this online learning method; applying to real problems of adaptive process control, complex process identification and control.

## Acknowledgments

This research study has been partially supported by the Ministry of Higher Education (MES) of the Republic of Cuba under Project 6.111 "Application of intelligent techniques in biotechnological processes".

## References

1. Kasabov, N., Song, Q.: IEEE Trans. Fuzzy Syst. 10(2), 144–154 (2002)
2. Victor, J., Dourado, A.: Evolving Takagi-Sugeno fuzzy models. Adaptive Computation Group–CISUC, Coimbra Portugal (2003)
3. Kukolj, D., Levi, E.: IEEE Trans. Syst. Man, Cybern. -Part B. 34(1), 272–282 (2004)
4. Angelov, P., Filev, D.: IEEE Trans. Syst. Man, Cybern. -Part B. 34(1), 484–498 (2004)
5. Yu, W., Ferreyra, A.: On-line clustering for nonlinear system identification using fuzzy neural networks. In: 2005 IEEE International Conference on Fuzzy Systems, Reno USA, pp. 678–683 (2005)
6. Bouchachia, A., Mittermeir, R.: Soft Comput. 11(2), 193–207 (2007)
7. Wang, L.X.: Adaptive fuzzy systems and control, 2nd edn. Prentice Hall, Englewood Cliffs (1997)
8. Takagi, T., Sugeno, M.: IEEE Trans. Syst. Man, Cybern. 15(1), 116–132 (1985)
9. Guedalia, I., London, M., Werman, M.: Neural-Comput. 11(2), 521–540 (1999)
10. Angelov, P., Zhou, X.-W.: Evolving fuzzy systems from data streams in real-time. In: EFS 2006. 2006 International Symposium on Evolving Fuzzy Systems, Ambleside Lake District UK, pp. 26–32 (2006)

11. Angelov, P., Kasabov, N.: IEEE SMC eNewsLetter (June 1–13, 2006)
12. Angelov, P., Filev, D., Kasabov, N., Cordon, O.: Evolving fuzzy systems. In: EFS 2006. Proc. of the 2006 International Symposium on Evolving Fuzzy Systems, pp. 7–9. IEEE Press, Los Alamitos (2006)
13. Song, Q., Kasabov, N.: A novel on-line evolving clustering method and its applications. In: Fifth Biannual Conference on Artificial Neural Networks and Expert Systems, pp. 87–92 (2001)
14. Martínez, B., Herrera, F., Fernández, J.: Métodos de agrupamiento clásico para el modelado difuso en línea. In: International Convention FIE 2006, Santiago de Cuba Cuba (2006)
15. Díez, J., Navarro, J., Sala, A.: Revista Iberoamericana de Automática e Informática Industrial 1(2), 32–41 (2004)
16. Box, G., Jenkins, G.: Time series analysis, forecasting and control. Holden Day, San Francisco USA (1970)
17. Bouchachia, A.: Incremental rule learning using incremental clustering. In: IPMU 2004. 10th Conference on Information Processing and Management of Uncertainty in Knowledge-Based Systems, Perugia Italy (2004)
18. Passino, K., Yurkovich, S.: Fuzzy control. Addison-Wesley, Menlo Park CA (1998)
19. Sala, A.: Validación y aproximación funcional en sistemas de control basados en lógica borrosa. Algoritmos de inferencia con garantía de consistencia. PhD Thesis, Universidad Politécnica de Valencia, Valencia Spain (1998)
20. Setnes, M., Babuska, R., Verburger, H.B.: IEEE Transactions on Systems, Man, and Cybernetics - Part C 28, 165–169 (1998)

# Fuzzy Approach of Synonymy and Polysemy for Information Retrieval

Andrés Soto[1], José A. Olivas[2], and Manuel E. Prieto[2]

[1] Department of Computer Science, Universidad Autónoma del Carmen
CP 24160, Ciudad del Carmen, Campeche, México
soto_andres@yahoo.com

[2] SMILe Research Group (Soft Management of Internet e-Laboratory)
Department of Computer Science, Universidad de Castilla La Mancha
Paseo de la Universidad 4, 13071-Ciudad Real, Spain
JoseAngel.Olivas@uclm.es, Manuel.Prieto@uclm.es

**Summary.** Development of methods for Information Retrieval based on conceptual aspects is vital to reduce the quantity of unimportant documents retrieved by the search engines. In this chapter, a method for expanding user queries is presented, such that for each term in the original query, all of its synonyms by a certain meaning with maximum concept frequency are introduced. To measure the degree of concept presence in a document (or even in a document collection), a concept frequency formula is introduced. New fuzzy formulas are also introduced to calculate the synonymy degree between terms to manage with concepts (meanings). With them, even though a certain term does not appear in a document, some degree of its presence could be estimated based on its degree of synonymy with terms that do appear in the document. A polysemy index is also introduced in order to simplify the treatment of weak and strong words.

**Keywords:** web information retrieval, fuzzy set, query expansion, vector space model, synonymy, polysemy.

## 1 Introduction

Information Retrieval (IR) has changed considerably since Calvin Mooers coined the term at MIT in 1948–50. In the last years, with the expansion of the World Wide Web, the amount of information on it have increased enormously thus nowadays Web Information Retrieval System (WIRS) represent one of the main targets of IR.

Although Internet is still the newest information medium, it is the fastest growing medium of all times. According to Lyman and Varian[1], between 1999 and 2002 new stored information grew about 30% a year. In 2002, the Web used to contain about 170 terabytes of information on its surface (i.e. fixed web

[1] Lyman, Peter and Hal R. Varian, "How Much Information", University of California, Berkeley, 2003. http://www.sims.berkeley.edu/how-much-info-2003

R. Bello et al. (Eds.): Granular Computing, STUDFUZZ 224, pp. 179–198, 2008.
springerlink.com 

pages), which means seventeen times the size of the USA Library of Congress print collections in volume. As Population Reference Bureau[2] registered a world population of around 6.3 billions in 2003, then almost 800 MB of recorded information was produced per person each year those years. Around January 2003, SearchEngineWatch.com[3] reported 319 million searches per day at the major search engines. Whois.Net, the Domain-Based Research Services[4], reported an increase of 30% in the number of domains registered from 32 millions in 2003 to 95 millions in June 2006.

Google, Yahoo, AOL Search and MSN Search are some of the most important Web search engines today. They are able to retrieve millions of page references in less than a second. Therefore they have a high level of efficiency. Unfortunately, most of the information retrieved could be considered irrelevant. For that reason, efficacy level could be considered poor since a user could receive millions of documents for her/his query but just few of them are useful.

Efficacy and relevance level strongly depend on the fact that most crawlers just look for words or terms without considering their meaning. Crawlers often use the Vector Space Model (VSM) [1] to keep documents indexed by the terms contained in them. Terms are weighted by their frequency in the documents, thus more frequent terms are considered more important. Similarity between a query and a document is considered a function of the matching degree between the terms in the query and the terms in the document, according to the term frequency. Page ranking usually consider that document relevance directly depend on the number of links connected with the document page. Therefore search systems work based on word matching instead of concept matching. Therefore, search methods should change from only considering lexicographical aspects to considering conceptual ones too [2] [3].

Taking into account the huge quantity of information in the Web nowadays, its incredible growing rate and the limited capacities of persons to even look at all of them, it is vital to reduce as much as possible the quantity of unimportant documents retrieved by search engines, while keeping all of the important ones. Then, the problem could be postulated as:

1. To retrieve only the important documents, according to personal concerns and
2. If there are, even though, too many important documents, just retrieve the most important ones.

Therefore, documents should be categorized some how by their level of importance according to some user preferences. After that, summit to the user just the most important ones.

[2] World Population Data Sheet. 2005, Population Reference Bureau (PRB). http://www.prb.org/

[3] Searches Per Day 2006. Danny Sullivan, Editor-In-Chief. http://searchenginewatch.com/reports/article.php/2156461

[4] Whois.Net: Domain-Based Research Services 2006. http://www.whois.net

E-Learning systems are close related with WIRS. E-Learning students have to look at large quantities of documents, so it is convenient to give them tools to narrow down the available resources based on the student's background knowledge, learning objectives and pedagogical approaches [4].

Nowadays, Soft Computing (SC) techniques are playing an important role [5] [6] to improve web search results. Different approaches and solutions have been proposed in the last years.

Clustering could be considered as the unsupervised learning process of organizing objects into groups whose members are similar in some way, such that non trivial relations and structures can be revealed. Document clustering methods have been widely applied in IR, supported by the hypothesis that documents relevant to a given query should be more similar to each other than to irrelevant documents, so they would be clustered together [7].

User models and tools to customize the information space are needed to take into account the user preferences. Some approaches focus their studies in the definition of models for representing documents based on extensions of the original Vector Space Model. Other approaches lead to the construction of flexible adaptive sites, based on user profiles and user behavior patterns, using data mining techniques [8] [9] [10] [11]. Others incorporate the multi-agent paradigm, such that agents could search the Web based on the user preferences and needs [12].

Fuzzy measures have been used to retrieve and classify information [4] [13]. Some search systems, in addition, use fuzzy association rules to expand user queries by finding new terms [13] [14].

On the other hand, systems based on term interrelations stored in ontologies (not fuzzy) such as WordNet, a semantic net of word groups [15] can be found as well. In [16] a system based on WordNet is proposed in which vector elements have three values to identify the corresponding tree in the net and the sense used in the document. This kind of systems require a special matching mechanism, like the ontology matching algorithm proposed in [17] to compare the words with the associated concepts.

Using WordNet also, a disambiguation method based on training the system with corpus of documents is proposed in [18]. Other corpus-based search system that uses the probability that certain concepts co-occur together for disambiguating meanings is proposed in [19]. Another approach to sense disambiguation by studying the local context of the words and comparing them to the habitual context of each one of the word senses is proposed in [20]. This system requires the usual context words to be stored in a repository.

The concept of relative synonymy to define a model of concept-based vectors is introduced in [21][22]. In this model, a term can be represented by a conceptual vector which is obtained by the linear combination of the definitions of the whole set of concepts. This system requires a concept repository.

Soft Management of Internet e-Laboratory (SMILe) research group [23] at Castilla La Mancha University is deeply involved in the development of Information Retrieval methods for the World Wide Web based on conceptual characteristics of the information contained in documents. Several models and tools have been developed by the members of the group, such as FIS-CRM (Fuzzy Interrelations and Synonymy Conceptual Representation Model) and the FISS Metasearcher [24][25], FzMail: a tool for organizing documents such as for example e-mail messages [26], the agents-based meta-search engine architecture GUMSe [27] or T-DiCoR for Three-Dimensional Representation of Conceptual Fuzzy Relations [28].

FIS-CRM (Fuzzy Interrelations and Synonymy Conceptual Representation Model) [24][25] [27] [29]is a methodology oriented towards processing the concepts contained in any kind of document, which can be considered an extension of the Vector Space Model (VSM), that uses the information stored in a fuzzy synonymy dictionary and fuzzy thematic ontologies. The dictionary provides the synonymy degree between pairs of synonyms and the ontologies bring the generality degree (hypernym, hyponym) between words. The generality degree value is calculated by the method proposed in [30]. The synonymy dictionary used in FIS-CRM was developed by S. Fernandez [31] [32]. It is an automatic implementation using Prolog of Blecua's Spanish dictionary of synonyms and antonyms [33] which includes around 27 thousands words.

In this chapter, new formulas based on those developed in FIS-CRM will be introduced. Therefore, it would be convenient to explain it with some extend. FIS-CRM approach is kept but a new version of the formulas is introduced in order to manage with synonymy and polysemy. With these new fuzzy formulas, the whole process of concept matching is simplified. As in FIS-CRM, although a certain term does not appear in a document, some degree of its presence could be estimated based on its degree of synonymy based on terms that do appear in the document. To measure the degree of concept presence in a document (or even in a document collection), a concept frequency formula is introduced. Finally, a method for expanding user queries is also presented, such that for each term in the original query, all of its synonyms by a certain meaning with maximum concept frequency are presented.

Unlike FIS-CRM, in this chapter, WordNet [34] will be used as storage of synonymy relations for English language. WordNet is a large lexical English database developed under the direction of George A. Miller. Nouns, verbs, adjectives and adverbs are grouped into sets of cognitive synonyms (synsets), each expressing a distinct concept. Synsets are interlinked by means of conceptual-semantic and lexical relations. WordNet distinguishes between nouns, verbs, adjectives and adverbs because they follow different grammatical rules. Every synset contains a group of synonymous words or collocations (a collocation is a sequence of words that go together to form a specific meaning, such as "car pool"); different senses of a word are in different synsets. WordNet also provides general definitions. The meaning of the synsets is further clarified with short defining glosses, which includes definitions and/or example sentences. One of WordNet purposes is to support automatic text analysis and artificial intelligence applications.

As of 2006, the WordNet database contains about 150,000 words organized in over 115,000 synsets for a total of 207,000 word-sense pairs; in compressed form, it is about 12 megabytes in size. The database and software tools have been released under a BSD style license and can be downloaded and used freely. It also includes an ANSI Prolog version of the WORDNET database.

## 2 Information Retrieval Methods Based on Concepts

The fundamental basis of FIS-CRM is to share the occurrences of a contained concept among the fuzzy synonyms that represent the same concept and to give a weight to those words which represent a more general concept than the contained word does. FIS-CRM constructs a vector space based on the number of occurrences of the terms contained in a set of documents. Afterwards, it readjusts the vector weights in order to represent concept occurrences, using for this purpose the information stored in the dictionary and ontologies. The readjusting process involves sharing the occurrences of a concept among the synonyms which converge to the concept and give a weight to the words that represent a more general concept than the contained ones. In this way, FIS-CRM readjusts the VSM vector weights in order to represent concept occurrences, using for this purpose the information stored in the dictionary and the ontologies.

Synonymy is usually conceived as a relation between expressions with identical or similar meaning. From the ancient times a controversy has existed about how to consider synonymy: whether as an identity relation between language expressions or as a similarity relation. In FIS-CRM, synonymy is understood as a gradual, fuzzy relation between terms as in [31] [32].

Fuzzy sets were introduced in 1965 by L. A. Zadeh [35]. A fuzzy set is a set without a crisp, clearly defined boundary. In classical set theory, an element either belongs or not to the set according to a crisp condition. In fuzzy set theory, elements could have only a partial degree of membership to the set. The membership function defines for each point in the input space its degree of membership, a number between 0 and 1. The input space is sometimes referred to as the universe of discourse.

Jaccard's coefficient is used in FIS-CRM to calculate the synonymy degree between two terms [27]. The method assumes that the set of synonyms of every sense of each word is available stored in a synonymy dictionary [32]. Given two sets $X$ and $Y$, their similarity is measured by (1).

$$sm(X,Y) = \frac{|X \cap Y|}{|X \cup Y|} \tag{1}$$

On the other hand, let us consider two words $w_1$ and $w_2$ with $m_i$ and $m_j$ possible meanings respectively, where $1 \leq i \leq M_1$ and $1 \leq j \leq M_2$. Then $S(w, m_i)$ represents the set of synonyms provided by the dictionary for every entry $w$ in the concrete meaning $m_i$. Then, the degree of synonymy $SD$ between two words $w_1$ and $w_2$ by the meaning $m_1$ is defined in (2).

$$SD(w_1, m_1, m_2) = \max_{1 \leq j \leq M_2} sm(S(w_1, m_1), S(w_2, m_j)) \tag{2}$$

$S(w, m)$ represents the set of synonyms provided by the dictionary for a word $w$ with meaning $m$ and $M_2$ is the number of meanings $m_j$ associated with word $w_2$.

A concept in FIS-CRM [27] [36] is not an absolute concept that has a meaning itself, i.e. there is not any kind of concept definition set or concept index. In FIS-CRM a concept is dynamically managed by means of the semantic areas of different words. Every word has a semantic area. The semantic area of a pair word sense is defined by the set of synonyms of that pair. The width of the semantic area of a word is intrinsic to the semantic shades of that word. Obviously, it can not be measured but if two overlapping semantic areas are compared, it could be assume that the one whose number of synonyms is larger should have a larger semantic area. The semantic area of a weak word (i.e. a word with several meanings) is the union of the semantic areas of each of its senses.

For example, if a term $t_1$ in a document is related to another more general term $t_2$ by means of a generality interrelation, the semantic area $SA_1$ of the first one will be included in the semantic area $SA_2$ of the second one. In this case, $SA_1$ is included in $SA_2$ with a membership degree equal to the generality degree between both terms, $GD(t_1, t_2)$.

$$GD(t_1, t_2) = \frac{O(t_1 \wedge t_2)}{O(t_1)} \tag{3}$$

Where $O(t_1)$ is the number of occurrences of $t_1$ and $O(t_1 \wedge t_2)$ is the number of co-occurrences of $t_1 \wedge t_2$.

In this case, it is considered that $t_1$ occurs once and the number of occurrences of the concept referred by $t_2$ is equal to $GD(t_1, t_2)$.

Considering a concept, obtained from the occurrences of various synonyms, as a fuzzy set, it is possible to define the membership degree of each one of the words that form the concept to the concept itself. Assuming that $m$ words (synonyms each other) co-occur in a document, the membership degree $(t_i, C)$ of each term $t_i$ to the concept $C$, which they converge to, is defined by (4).

$$\mu(t_i, C) = \min_{j=1}^{m} \{SD(t_i, t_j)\} \tag{4}$$

Once this value is defined, it is possible to define the number $N$ of occurrences of a concept $C$ (formed by the co-occurrence of $m$ synonyms) in a query or document by (5), in which $w_i$ is the weight of the term $t_i$ in the document, that in this case and in order to simplify, is the number of occurrences of the term $t_i$. The vector with the term weights is called the VSM vector.

$$N = \sum_{i=1}^{m} w_i \times \mu(t_i, C) \tag{5}$$

After obtaining the weights $w_i$, FIS-CRM proceed to readjust the weights by sharing the number of occurrences of each concept among the words of the synonyms set whose semantic area is more representative to that concept, obtaining

FIS-CRM vectors based on concept occurrences. Thus, a word may have a weight in the new vector even if it is not contained in it, as long as the referenced concept underlies the document.

The main handicap of the sharing process in FIS-CRM is managing with weak words (words with several meanings). The sense of disambiguation of weak words is implicitly carried out by the sharing process. Three situations are distinguished depending on the implication of weak or strong words (words with only one meaning). So, there are three types of synonymy sharing:

1. Readjustment occurrences among strong words: when one or more strong synonyms co-occur in a document or query.
2. Readjustment occurrences among strong and weak words: when one or more strong synonyms co-occur with one or more weak synonyms.
3. Readjustment occurrences among weak words: when one or more weak synonyms co-occur, without any strong synonym.

**Readjustment occurrences among strong words.** Let us consider a piece of the VSM vector $(w_i)$, $1 \leq i \leq m$, containing several occurrences of $m$ strong synonyms, where $w_i$ reflects the number of occurrences of the term $t_i$ (see Table 1). Let us assume that these synonyms converge to a concept $C$ whose most suitable set of synonyms is formed by $n$ strong terms.

**Table 1.** Readjustment among strong words

| Terms | $t_1$ | $t_2$ | ... | $t_m$ | $t_{m\ 1}$ | ... | $t_n$ |
|---|---|---|---|---|---|---|---|
| VSM vector | $w_1$ | $w_2$ | ... | $w_m$ | 0 | ... | 0 |
| FIS-CRM vector | $w'_1$ | $w'_2$ | ... | $w'_m$ | $w'_{m\ 1}$ | ... | $w'_n$ |

Then, the FIS-CRM vector $(w'_i)$, $1 \leq i \leq n$, would be obtained by (6) where $w'_i$ is the readjusted weight of the term $t_i$.

$$w'_i = N \times \mu(t_i, C) \times \sqrt{\frac{1}{\sum_{i=1}^{n} \mu(t_i, C)^2}} \qquad (6)$$

For example, let us suppose that the terms A and B are synonyms, which co-occur in a document with 2 and 3 occurrences respectively. And let us also suppose that the most suitable synonyms set they converge to contains the words C, D and E. Let us assume that the synonymy degrees among these terms are defined as shown in Table 2.

Then, the VSM vector will be like the one below. In this case, the number of occurrences $N$ of the concept formed by the co-occurrence of A and B is 4.5, obtained by the expression (5). The corresponding FIS-CRM vector is shown below in Table 3.

**Table 2.** Example of synonymy degrees

| Terms | A | B | C | D | E |
|---|---|---|---|---|---|
| A | | 0.9 | 0.8 | 0.7 | 0.6 |
| B | | | 0.7 | 0.8 | 0.9 |
| C | | | | 0.5 | 0.6 |
| D | | | | | 0.9 |
| E | | | | | |

**Table 3.** Example of readjustment among strong words

| Terms | A | B | C | D | E |
|---|---|---|---|---|---|
| VSM vector | 2 | 3 | 0 | 0 | 0 |
| FIS-CRM vector | 2.35 | 2.35 | 1.83 | 1.83 | 1.56 |

**Readjustment occurrences among strong and weak words.** This type of adjustment is carried out when one or several weak synonyms co-occur in a document. Let us consider a piece of the VSM vector of a document containing $m$ weak synonymous words, where $w_i$ is the number of occurrences of the term $t_i$. In Table 4, the first $m$ terms are the weak ones contained in the document. The next $f = n - m$ terms are the strong ones contained in the document. The last $g = p - n$ terms are the strong terms of the set of synonyms not contained in the document. In this case, the number $N$ of concept occurrences, which the $n$ synonyms converge to, is shared among the strong synonyms, from $t_{m+1}$ to $t_p$.

**Table 4.** Readjustment among weak and strong words

| Terms | $t_1$ | ... | $t_m$ | $t_{m\ 1}$ | ... | $t_n$ | $t_{n\ 1}$ | ... | $t_p$ |
|---|---|---|---|---|---|---|---|---|---|
| VSM vector | $w_1$ | ... | $w_m$ | $w_{m\ 1}$ | ... | $w_n$ | 0 | ... | 0 |
| FIS-CRM vector | 0 | ... | 0 | $w'_{m\ 1}$ | ... | $w'_n$ | $w'_{n\ 1}$ | ... | $w'_p$ |

It is important to point out that in order to calculate the number $N$, when managing the synonymy degree between two weak words, we must take into consideration the number that identifies the sense obtained by the disambiguation process. In the case of the synonymy degree between a strong word and a weak word it is implicitly disambiguated taking the value $SD(strong, weak)$ as in (2).

The weights of the strong synonyms (from $t_{m+1}$ to $t_p$) of the corresponding FIS-CRM vector are calculated by (6), assigning weight 0 (zero) to the first $m$ terms (the weak ones). Then, the occurrences are shared only among the strong synonyms, leaving the weak terms without any weight.

**Table 5.** Readjustment among strong words

| Terms | $t_1$ | $t_2$ | ... | $t_m$ | $t_{m\ 1}$ | ... | $t_n$ |
|---|---|---|---|---|---|---|---|
| VSM vector | $w_1$ | $w_2$ | ... | $w_m$ | 0 | ... | 0 |
| FIS-CRM vector | $w'_1$ | $w'_2$ | ... | $w'_m$ | $w'_{m\ 1}$ | ... | $w'_n$ |

**Readjustment occurrences among weak words.** This type of sharing is carried out when several weak synonyms co-occur and they do not have strong synonyms to share the occurrences of the concept they converge to. As in the previous cases, let us consider a piece of the VSM vector $(w_i)$, $1 \leq i \leq m$, shown in Table 5, containing several occurrences of $m$ weak synonyms, where $w_i$ reflects the number of occurrences of the term $t_i$. And let us consider the set of $n$ synonyms of the right disambiguated sense (all of them are weak terms). In this case, the number of occurrences of the concept to which the $m$ weak terms converge is shared among all the synonyms of its right set of synonyms. In this case we should take the same considerations as the ones explained in the previous section about the identification of the number of the senses involved.

## 3 Fuzzy Model for Synonymy and Polysemy

The approach of considering synonymy as an equivalence relation completely differs from that which considers it as a gradual relation. The latter one is closer to the behavior of synonymy in dictionaries, where it is possible to find synonyms which are not equivalent. For example, auto and automobile share a common meaning: "a motor vehicle with four wheels; usually propelled by an internal combustion engine" [15]. But automobile has another meaning: as a verb, it means "to travel in an automobile". Therefore, auto and automobile are not equivalent terms, but similar ones. In what follows, synonymy will be considered as an asymmetric relation.

Let $V$ be a set of terms which belongs to a particular dictionary and $M$ the set of meanings associated to the terms in $V$. Therefore, each term in $V$ has one or more meanings in $M$ and each meaning in $M$ has one or more terms associated in $V$.

Let *meaning* be a binary crisp relation such that $meaning\,(t,m) = 1$ if and only if there is a $t$ in $V$, $m$ in $M$ such that $m$ represents a meaning of term $t$.

$$meaning : V \times M \rightarrow \{0,1\} \tag{7}$$

Let $M(t)$ be the set of different meanings associated with a certain term $t$.

$$M(t) = \{m \in M \mid meaning\,(t,m) = 1\} \tag{8}$$

Polysemy is the capacity for a word or term to have multiple meanings. Terms with only one meaning are considered *strong*, while terms with several meanings are considered *weak.* Romero [26] considers that "the main handicap of the sharing process is managing weak words" and consider three situations depending on the implication of weak or strong words. In the above example, *auto* is a strong word, while *automobile* is weaker than *auto*, and *car* is even weaker because it has five meanings.

In order to manage all those situations in just one way rather than case by case, an index $I_p(t)$ will be defined in (9) to represent the polysemy degree of term $t$. Therefore a strong term will have zero degree of polysemy, while weak terms will increase their degree of polysemy as increases their number of meanings. Let us denote with $N_m(t)$ the number of meanings associated with the term $t$, that is, the number of elements of the set $M(t)$ defined above.

$$I_p : V \rightarrow [0,1] \text{ where } I_p(t) = 1 - \frac{1}{N_m(t)} \tag{9}$$

Obviously, if t is a strong term (i.e. a term with only one meaning), then $N_m(t) = 1$ and $I_p(t) = 0$ which means that $t$ is not polysemous, i.e. its polysemy degree is null (zero). On the other hand, the greater the number of meanings of $t$, the greater the polysemy degree and the closer the index $I_p(t)$ to 1. Therefore the polysemy index $I_p(t)$ is a measure of the term weakness. At the same time, $(1 - I_p(t))$ could be interpreted as a measure of the strength of the term $t$. Therefore $I_p(auto) = 0$, $I_p(automobile) = 0.5$ and $I_p(car) = 0.8$.

Let us define a fuzzy relation $S$ (see (10)) between two terms $t_1, t_2 \in V$ such that $S(t_1, t_2)$ expresses the degree of synonymy between the two terms.

$$S(t_1, t_2) = \frac{|M(t_1) \cap M(t_2)|}{|M(t_1)|} \tag{10}$$

Therefore:

1. If $M(t_1) \cap M(t_2) = \emptyset$ there are no synonyms between them. This implies $|M(t_1) \cap M(t_2)| = 0$ so that $S(t_1, t_2) = 0$, so the degree of synonymy between them is zero, i.e. there is no synonymy.
2. If $M(t_1) \subseteq M(t_2)$ then $t_2$ includes all meanings of $t_1$. Therefore $M(t_1) \cap M(t_2) = M(t_1)$ so that $|M(t_1) \cap M(t_2)| = M(t_1)$ so that $S(t_1, t_2) = 1$, thus $t_1$ is a "full" synonym of $t_2$ (with the maximum degree).
3. In other cases, when $t_1$ do not share some meanings with $t_2$, then $0 < |M(t_1) \cap M(t_2)| \leq |M(t_1)|$ so that $0 < S(t_1, t_2) \leq 1$, so the degree of synonymy varies.

That way, the degree of synonymy between auto and automobile will be 1, which means that the concept *auto* totally corresponds with the concept *automobile.* But, in the other way, the degree of synonymy between *automobile* and *auto* is just 0.5 because *automobile* just corresponds with *auto* in half of the meanings.

Let us denote $T(m)$ as the set of terms that share a meaning $m$:

$$T(m) = \{t \in V \mid meaning\,(t, m) = 1\} \tag{11}$$

Then, for all $m \in M$ and $t_1, t_2 \in T(m)$ so that $S(t_1, t_2) > 0$. Therefore, if the term $t_2$ appears in a particular document but the term $t_1$ does not, some degree of presence of term $t_1$ could be calculated for that particular document, considering the degree of synonymy between them.

Let us suppose, for example, that the term "matching" appears 20 times in a document holding 320 terms. According to WordNet [34], "matching" has two possible meanings:

1. intentionally matched ($m_1$)
2. being two identical ($m_2$)

The meaning $m_1$ is shared by two terms, $T_1$ = {matching, coordinated}, while $m_2$ is shared by four terms, $T_2$ = {matching, duplicate, twin, twinned}. Therefore, all of them share some degree of synonymy.

## 4 Adjusting the Vector Space Model

Consider $D$ as a collection of documents such that each document $D_j$ is composed by terms from the vocabulary $V$.

$$D = \{D_1, D_2, D_3, \ldots, D_{nd}\} \tag{12}$$

Term frequency $tf$ (see formula (13)) is a well known measure [37] [38] [39] [40] of the importance of a term $t_i$ in $V$ within a document $D_j$ where $n_{ij}$ (resp. $n_{kj}$) is the number of occurrences of term $t_i$ (resp. $t_k$) in the document $D_j$ and $n_{*j}$ is the number of terms in the same document.

$$tf_{ij} = \frac{n_{ij}}{\sum_k n_{kj}} = \frac{n_{ij}}{n_{*j}} \tag{13}$$

This measure is one of the most referenced in Information Retrieval, but it considers all terms in the same way, independently of their meaning. Therefore, it would be interesting to have a formula which allows measuring the importance within a document not only of a term but of a meaning.

According to the previous example about "matching", let us suppose two situations (see Table 6):

1. "matching" appears 20 times in the document while the other synonyms do not appear in it.
2. "matching" appears 20 times, "coordinated" appears 15 and the other synonyms do not appear.

Let us define a coefficient $R_j(m)$ which could be interpreted as a measure of the use of a meaning $m$ in $M$ in a document $D_j$ based on the number of occurrences of the terms associated with that meaning.

$$R_j(m) = \sum_{t_i \in T(m)} (n_{ij}\,(1 - I_p(t))) = \sum_{t_i \in T(m)} \frac{n_{ij}}{N_m(t_i)} \tag{14}$$

**Table 6.** Calculating term frequency

| Term | Number of occurrences | | Term frequency | |
|---|---|---|---|---|
| | a) | b) | a) | b) |
| matching | 20 | 20 | 0.063 | 0.063 |
| coordinated | 0 | 15 | 0 | 0.047 |
| duplicate | 0 | 0 | 0 | 0 |
| twin | 0 | 0 | 0 | 0 |
| twinned | 0 | 0 | 0 | 0 |

$R_j(m)$ relates to the number of occurrences of the different terms (synonyms) associated with a certain meaning $m$, according with their respective polysemy degree. Thus, for strong terms (i.e. $N_m(t_i) = 1$ so $I_p(t_i) = 0$), any time the term occurs, it should be interpreted as a reference to its only meaning, so the total number of occurrences of $t_i$ is added to $R_j(m)$. On the contrary, if the term is weak (i.e. $N_m(t_i) > 1$ therefore $0 < I_p(t_i) < 1$), then just a proportional part to the polysemy degree (weakness) of the term is added; then, as weaker the term, the lesser its contribution to $R_j(m)$.

On the other hand, it is easy to observe that if a term $t_i$ has different meanings (i.e. $N_m(t_i) > 1$), then the number of occurrences of $t_i$ will influence proportionally the corresponding values $R_j(m_i)$ for each one of the meanings $m_i$ of $t_i$ ( i.e. $m_i$ in $M(t_i)$).

Unfortunately, defined in that way, the value $R_j(m)$ could be difficult to analyze without knowing the corresponding value of $R_j$ for the other meanings. Therefore, coefficient $Cf_j(m)$ is defined in 15 such that $0 < Cf_j(m) < 1$.

$$Cf_j(m) = \frac{R_j(m)}{\sum_k n_{kj}} = \frac{\sum_{t_i \in T(m)} (n_{ij}\ (1 - I_p(t_i)))}{n_{*j}} \tag{15}$$

That way, it is easy to compare the relative importance of the different meanings in a document $D_j$. As can be seen, the coefficient $Cf_j(m)$ resembles the term frequency one, consequently it will be called the *concept frequency* of meaning $m$ in the document $D_j$.

**Table 7.** Estimating concept frequency

| Meaning | Situation a) | | Situation b) | |
|---|---|---|---|---|
| | $R_j(m)$ | $Cf_j(m)$ | $R_j(m)$ | $Cf_j(m)$ |
| $m_1$ | 10 | 0.03 | 15 | 0.047 |
| $m_2$ | 10 | 0.03 | 10 | 0.03 |

Based on Table 6 the corresponding values of $R_j(m)$ and $Cf_j(m)$ for the meanings $m_1$ and $m_2$ are shown in Table 7. In situations a) and b), the terms "duplicate", "twin" and "twinned" are influenced by the concept frequency of $m_2$. In situation b), the meaning $m_1$ is more influenced than $m_2$ because synonyms "matching" and "coordinated" do appear in the document.

It is easy to calculate the concept frequency for a meaning $m$ for two documents $D_1$ and $D_2$ and compare them by some distance. A popular measure of similarity is the cosine of the angle between two vectors $\mathbf{X_a}$ and $\mathbf{X_b}$. The cosine measure is given by the following expression:

$$s(\mathbf{X_a}, \mathbf{X_b}) = \frac{\mathbf{X_a^T} \cdot \mathbf{X_b}}{\parallel \mathbf{X_a} \parallel_2 \cdot \parallel \mathbf{X_b} \parallel_2} \tag{16}$$

Cosine measure is the most popular measure for text documents [42]. As the angle between the vectors shortens and the two vectors get closer the cosine angle approaches 1, meaning that the similarity of whatever is represented by the vectors increases.

By calculating $Cf_1(m)$ and $Cf_2(m)$ for all the meanings $m$ in $M$, two vectors $\mathbf{Cf_1^M}$ and $\mathbf{Cf_2^M}$ are obtained. For those vectors, a fuzzy relation $similar^M$ between two documents is defined by:

$$similar^M(D_1, D_2) = s(Cf_1^M, Cf_2^M) \tag{17}$$

A new model for document clustering is proposed in [43] [44] to manage with conceptual aspects. To measure the presence degree of a concept in a document collection, the above concept frequency formulas are used. A fuzzy hierarchical clustering algorithm is used to determine an initial clustering. Then an improved soft clustering algorithm is applied.

Two different datasets which are widely known were used to evaluate the effectiveness of the clustering method.

1. SMART collection[5] contains 1400 CRANFIELD documents from aeronautical systems papers, 1033 MEDLINE documents from medical journals and 1460 CISI documents from information retrieval papers.
2. REUTERS data set consists of 21578 articles from the Reuters news service in 1987 [45].

Some of the most relevant evaluation measures [46] were used to compare and analyse the performance of the clustering method.

1. Measures of the document representation method:
   a) Mean Similarity (MS): Average of similarity of each element with the rest of the set.
   b) Number of Outliers (NO): An outlier is an object that is quite different from the majority of the objects in a collection.
2. Measures of the clustering results:
   a) Internal quality measures that depend on the representation

[5] ftp://ftp.cs.cornell.edu/pub/smart

   i. Cluster Self Similarity (CSS): the average similarity between the documents in a cluster
   ii. Size of Noise Cluster (SNC): number of elements unclassified in the hierarchical structure.
b) External quality measures based on a known categorization.
   i. F-Measure [47]: combines the precision (p) and recall (r) values from IR [48] [49].

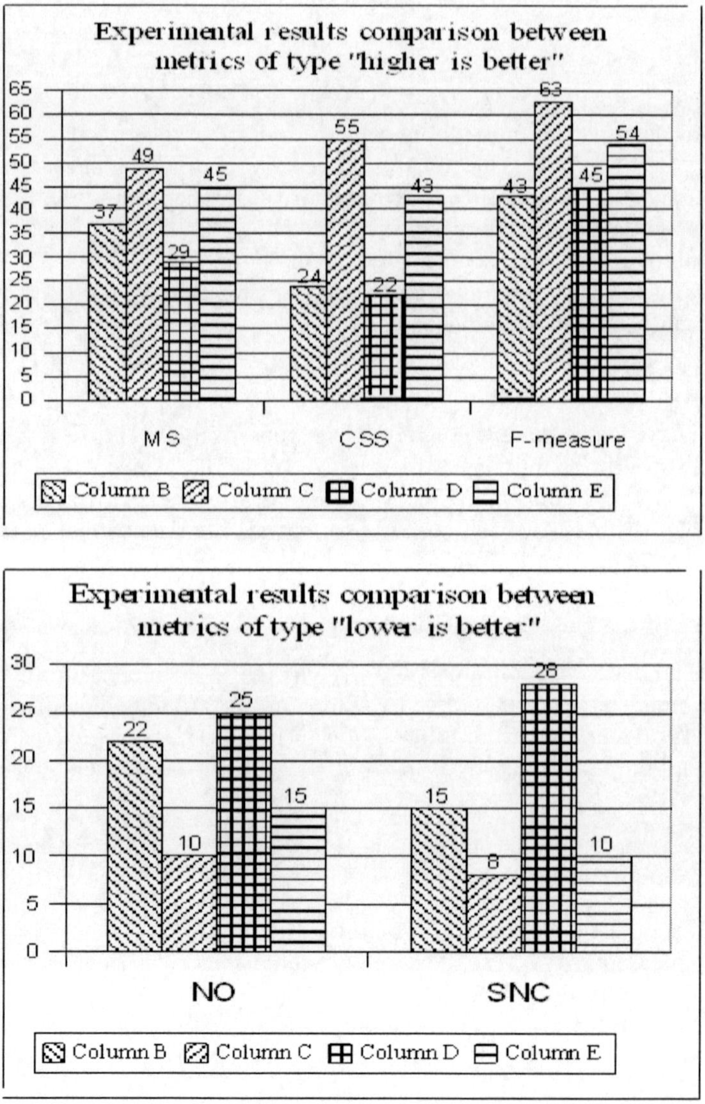


**Fig. 1.** Experimental results comparison

The F-measure of cluster $j$ and class $i$ is given by:

$$F(i,j) = \frac{2 \times r_{ij} \times p_{ij}}{r_{ij} + p_{ij}} \tag{18}$$

For an entire cluster hierarchy the F-measure of any class is the maximum value obtained at any node in the tree. An overall value for the F-measure is computed by taking the weighted average of all values for the F-measure as follows, where $n$ is the number of documents and the maximum is calculated over all clusters at all levels:

$$F = \sum \frac{n_i}{n} \times \max\{F(i,j)\} \tag{19}$$

The results obtained by this model are compared with those obtained by the classical methods [48], such as the tf-idf representation method [45] and the fuzzy c-means clustering algorithm [49]. The experimental results are shown in Table 8, expressed in percentages. In the first part of the table are grouped the results corresponding to metrics of type "higher is better". Then, in the second part, are grouped the results corresponding to metrics of type "lower is better". Fig. 1 pictures the results.

**Table 8.** Experimental Results

| | TF-IDF & FCM | Hybrid Model | TF-IDF & FCM | Hybrid Model |
|---|---|---|---|---|
| Metric | SMART | SMART | REUTERS | REUTERS |
| MS | 37 | 49 | 29 | 45 |
| CSS | 24 | 55 | 22 | 43 |
| F-measure | 43 | 63 | 45 | 54 |
| | | | | |
| NO | 22 | 10 | 25 | 15 |
| SNC | 15 | 8 | 28 | 10 |

## 5 Query Expansion

Iterative searching is a natural approach to improve relevance level by using collection frequency weights. Usually it is supposed that, some information is obtained about which documents are relevant and which others are not by an initial search. The information thus obtained can be used to modify the original query by adding new terms, selected from relevant documents to construct the new queries. This process is known as query expansion [41].

Our approach is to use initially a collection of documents provided by the user as relevant, maybe from an initial search as told before or from the files the user keeps on hard disk or by links provided by some Web tool as Yahoo Search MyWeb Beta or Google Bookmarks for IE Toolbar Version 4. Based on those documents, concept frequencies can be calculated easily ordered as shown below.

Previous defined expressions (13)–(15) could be easily extended for a whole collection of documents $D$. Therefore, term frequency of $t_i$ in $V$ for $D$ would be defined in (20), where $n_{i*}$ (resp. $n_{k*}$) is the number of occurrences of term $t_i$ (resp. $t_k$) in the whole collection of documents $D$ and $n_{**}$ is the number of terms in the whole collection.

$$tf_{i*} = \frac{n_{i*}}{\sum_k n_{k*}} = \frac{n_{i*}}{n_{**}} \tag{20}$$

Expression (14)will be redefined and a measure $RD(m)$ of the use of a meaning $m$ in $M$ in the whole collection $D$ shall be introduced.

$$RD(m) = \sum_{t_i \in T(m)} (n_{i*}(1 - I_p(t_i))) \tag{21}$$

And then, the concept frequency coefficient $Cf_j(m)$ is also redefined for the whole collection $D$.

$$CfD(m) = \frac{RD(m)}{\sum_k n_{k*}} = \frac{\sum_{t_i \in T(m)} (n_{i*}(1 - I_p(t_i)))}{n_{**}} \tag{22}$$

Once concept frequencies are calculated, the corresponding meanings could be considered ordered also. Therefore, the meaning $m$ of a term $t$ with the maximum concept frequency coefficient $CfD(m)$ for collection $D$ will be denoted as $\max_m(t)$.

$$\max_m(t) = \max_{m_i \in M(t)} (m_i) \tag{23}$$

$$m_i \geq m_j \Longleftrightarrow CfD(m_i) \geq CfD(m_j)$$

Then, when the user makes a query $Q$, it is expanded to a new query $Q_e$ defined in this way:

$$Q_e = \bigcup_{t \in Q} T(\max_m(t)) \tag{24}$$

Such that for each term $t$ in $Q$, all the terms associated with $t$ by a maximum meaning will be included in $Q_e$, which are all the synonyms of $t$ by a meaning $m$ which has maximum concept frequency. According to the example shown in Table 7 situation b), let us suppose that the meaning $m_1$ has the maximum concept frequency coefficient $CfD(m_1)$, therefore if "matching" is used in a query $Q$, then all the terms in $T_1$ (i.e. "matching", "coordinated") will be included in $Q_e$.

## 6 Conclusions

Nowadays, search engines are able to retrieve efficiently millions of page references in less than a second, but unfortunately, with a low level of efficacy because users receive millions of useless documents, irrelevant for their query. The low level of efficacy strongly depend on the fact that most crawlers just look for words or terms in the documents without considering their meaning.

Development of new methods for Web Information Retrieval based on conceptual characteristics of the information is vital to reduce the quantity of unimportant documents retrieved by today search engines. Our research group is deeply involved in the development of IR methods for WWW based on conceptual characteristics of the information contained in documents. This chapter could be considered as another attempt in that direction. Based on that, this model is a logical extension and complement of the FIS-CRM model.

Both models are oriented to measure the presence of concepts in documents by using fuzzy interpretations of synonymy. In both cases, synonymy is considered a similarity relation, not an equivalent relation. Based on this formulas, even though a certain term does not appear in a document, it is possible to estimate some degree of its presence according to the degree of synonymy shared with terms that do appear in the document.

FIS-CRM uses a Spanish dictionary, which include about 27 thousands words and several thematic ontologies. Our approach uses an English dictionary, WordNet, which contains about 150,000 words organized in over 115,000 synonymy sets for a total of 207,000 word-sense pairs.

A concept in FIS-CRM is not an absolute concept that has a meaning itself, i.e. there is not any kind of concept definition set or concept index. In FIS-CRM a concept is dynamically managed by means of the semantic areas of different words. In our approach, a concept or meaning is the definition of a term that appears in a dictionary, in this case, WordNet. Those meanings define the synsets of WordNet and are used by our approach to manage with the weak words. A polysemy index was defined, which help to share the term occurrences between the different sense. The main handicap of the sharing process in FIS-CRM is managing with weak words (words with several meanings). Three situations are distinguished depending on the implication of weak or strong words. In the approach presented in this chapter, the introduction of the polysemy index extremely simplifies the management of weak and strong words, incorporating all the three cases mentioned in only one formula.

With the concept frequency coefficient, it is possible to measure how similar are two or more documents depending on their use of some concept. In this approach, this coefficient could also be used to order a document collection in relation with the use made by the different documents of some concept. This way it is possible to elaborate a user profile in order to help him/her to expand the queries, based on his/her previous search history and interests.

**Acknowledgements.** This project has been partially supported by SCAIWEB PAC06-0059 project, JCCM, Spain.

## References


1. Salton, G., Wong, A., Yang, C.: Communications of the ACM 18(11), 613–620 (1975)
2. Ricarte, I., Gomide, F.: A reference model for intelligent information search. In: Proceedings of the BISC Int. Workshop on Fuzzy Logic and the Internet, pp. 80–85 (2001)
3. Baeza-Yates, R., Ribeiro, B.: Modern Information Retrieval. Addison-Wesley-Longman, ACM Press, New York (1999)
4. Mendes, M., Sacks, L.: Evaluating fuzzy clustering for relevance-based information access. In: FUZZ-IEEE 2003. Proc. of the IEEE International Conference on Fuzzy Systems (2003)
5. Pasi, G.: Mathware and Soft Computing 9, 107–121 (2002)
6. Herrera-Viedma, E., Pasi, G.: Fuzzy approaches to access information on the Web: Recent developments and research trends. In: Proceedings of the Third Conference of the EUSFLAT, pp. 25–31 (2003)
7. Zamir, O., Etzioni, O.: Grouper: A dynamic clustering interface to web search results. In: Proceedings of the WWW8 (1999)
8. Martin-Bautista, M., Vila, M., Kraft, D., Chen, J., Cruz, J.: Journal of Soft Computing 6(5), 365–372 (2002)
9. Perkovitz, M., Etzioni, O.: Artificial Intelligence 118, 245–275 (2000)
10. Tang, Y., Zhang, Y.: Personalized library search agents using data mining techniques. In: Proceedings of the BISC Int. Workshop on Fuzzy Logic and the Internet, pp. 119–124 (2001)
11. Cooley, R., Mobashe, B., Srivastaba, J.: Grouping web page references into transactions for mining world wide web browsing patterns. Technical report TR 97-021, University of Minnesota, Minneapolis (1997)
12. Hamdi, M.: MASACAD: A multi-agent approach to information customization for the purpose of academic advising of students. In: Applied Soft Computing Article, Elsevier B.V. Science Direct (in Press, 2006)
13. Lin, H., Wang, L., Chen, S.: Expert Systems with Applications 31(2), 397–405 (2006)
14. Delgado, M., Martin-Bautista, M., Sanchez, D., Serrano, J., Vila, M.: Association rules and fuzzy associations rules to find new query terms. In: Proc. of the Third Conference of the EUSFLAT, pp. 49–53 (2003)
15. Miller, G.: Communications of the ACM 11, 39–41 (1995)
16. Gonzalo, J., Verdejo, F., Chugur, I., Cigarran, J.: Indexing with W*ordNet* synsets can improve retrieval. In: Proc. of the COLING/ACL Work. on usage of W*ordNet* in natural language processing systems (1998)
17. Kiryakov, A., Simov, K.: Ontologically supported semantic matching. In: Proceedings of NODALIDA 1999: Nordic Conference on Computational Linguistics, Trondheim (1999)
18. Loupy, C., El-Bèze, M.: Managing synonymy and polysemy in a document retrieval system using WordNet. In: Proceedings of the LREC 2002: Workshop on Linguistic Knowledge Acquisition and Representation (2002)
19. Whaley, J.: An application of word sense disambiguation to information retrieval. Technical Report PCS-TR99-352, Dartmouth College on Computer Science (1999)
20. Leacock, C., Chodorow, M.: Combining local context and Wordnet similarity for word sense disambiguation. In: WordNet, an Electronic Lexical Database, pp. 285–303. MIT Press, Cambridge (1998)

21. Lafourcade, M., Prince, V.: Relative Synonymy and conceptual vectors. In: Proceedings of the Sixth Natural Language Processing Pacific Rim Symposium, vol. 202, pp. 127–134 (2001)
22. Lafourcade, M.: Conceptual vectors and fuzzy templates for discriminating hyperonymy (is-a) and meronymy (part-of) relations. In: Konstantas, D., Léonard, M., Pigneur, Y., Patel, S. (eds.) OOIS 2003. LNCS, vol. 2817, pp. 19–29. Springer, Heidelberg (2003)
23. Olivas, J., de la Mata, J., Serrano-Guerrero, J., Garcés, P., Romero, F.: Desarrollo de motores inteligentes de búsqueda en Internet en el marco del grupo de investigación SMILe-ORETO. In: Olivas, J., Sobrino, A. (eds.) Recuperación de información textual, Text Information Retrieval, Universidad de Santiago de Compostela, pp. 89–102 (2006)
24. Garcés, P., Olivas, J., Romero, F.: FIS-CRM: A Representation Model Based on Fuzzy Interrelations for Internet Search. In: Proceedings of ICAI 2002, pp. 219–224 (2002)
25. Olivas, J., Garces, P., Romero, F.: Int. Journal of Approx. Reasoning 34(2-3), 201–219 (2003)
26. Romero, F., Olivas, J., Garces, P., Jimenez, L.: FzMail: A Fuzzy Tool for Organizing E-Mail. In: The 2003 International Conference on Artificial Intelligence ICAI 2003, Las Vegas, USA (2003)
27. de la Mata, J., Olivas, J., Serrano-Guerrero, J.: Overview of an Agent Based Search Engine Architecture. In: ICAI 2004. Proceedings of the International Conference on Artificial Intelligence, Las Vegas, USA, pp. 62–67 (2004)
28. Olivas, J., Rios, S.: In: Larsen, H.L., Pasi, G., Ortiz-Arroyo, D., Andreasen, T., Christiansen, H. (eds.) FQAS 2006. LNCS (LNAI), vol. 4027, pp. 681–690. Springer, Heidelberg (2006)
29. Garcés, P., Olivas, J., Romero, F.: Journal of the American Society for Information Science and Technology JASIST 57(4), 564–576 (2006)
30. Widyantoro, D., Yen, J.: Incorporating fuzzy ontology of term relations in a search engine. In: Proceedings of the BISC Int. Workshop on Fuzzy Logic and the Internet, pp. 155–160 (2001)
31. Fernandez, S.: Una contribuciń al procesamiento automt́ico de la sinonimia utilizando Prolog. Ph.D. thesis, Santiago de Compostela University, Spain (2001)
32. Fernandez, S., Grana, J., Sobrino, A.: A Spanish e-dictionary of synonyms as a fuzzy tool for information retrieval. In: JOTRI 2002. Actas de las I Jornadas de Tratamiento y Recuperación de Información, León, Spain (2002)
33. Blecua, J.: Diccionario avanzado de sinónimos y antónimos de la Lengua Española. Diccionarios de lengua española Vox, Barcelona 647 (1997)
34. WordNet, An Electronic Lexical Database. The MIT Press, Cambridge, MA (1998)
35. Zadeh, L.: Information and Control 8, 338–353 (1965)
36. Olivas, J., Garcés, P., de la Mata, J., Romero, F., Serrano-Guerrero, J.: Conceptual Soft-Computing based Web search: FISCRM, FISS Metasearcher and GUMSe Architecture. In: Nikravesh, M., Kacprzyk, J., Zadeh, L. (eds.) Forging the New Frontiers: Fuzzy Pioneers II. Studies in Fuzziness and Soft Computing. Springer, Heidelberg (2007)
37. Sparck, J.: Journal of Documentation 28,11–21 (1972)
38. Sparck, J., Walker, S., Robertson, S.: Information Processing and Management 36, 779–808 (2000)
39. Robertson, S.: Journal of Documentation 60, 503–520 (2004)

40. Robertson, S., Sparck, J.: Simple, proven approaches to text retrieval. Technical Report 356, University of Cambridge Computer Laboratory (2006)
41. Ghosh, J.: Scalable clustering in the handbook of data mining. In: Nong, Y. (ed.), vol. 10, pp. 247–278. Lawrence Erlbaum Assoc., Mahwah (2003)
42. Romero, F., Soto, A., Olivas, J.: Fuzzy clustering based on concept measuring in documents. In: Proceedings of the EUROFUSE workshop New Trend. Fuzzy Preference Modeling (2007)
43. Salton, G., McGill, M.: Introduction to Modern Information Retrieval. McGraw-Hill, New York (1983)
44. Yang, Y.: Journal of Information Retrieval 1(1–2), 67–88 (1999)
45. Larsen, B., Aone, C.: Fast and effective text mining using linear-time document clustering. In: Proceedings of the KDD 1999 (1999)
46. Van Rijsbergen, C.: Information Retrieval, 2nd edn. Buttersworth, London (1989)
47. Kowalski, G.: Information Retrieval Systems - Theory and Implementation. Kluwer Academic Publishers, Dordrecht (1997)
48. Barrett, R., Selker, T.: AIM: A new approach for meeting information needs. Technical report, IBM Research (1995)
49. Pedrycz, W.: Pattern Recognition Letters 17, 625–631 (1996)

# Rough Set Theory Measures for Quality Assessment of a Training Set

Yailé Caballero[1], Rafael Bello[2], Leticia Arco[2], Yennely Márquez[1], Pedro León[1], María M. García[2], and Gladys Casas[2]

[1] Computing Department, University of Camagüey, Cuba
yailec@yahoo.com, yaile.caballero@reduc.edu.cu

[2] Computer Science Department, Central University of Las Villas, Cuba
{rbellop,leticiaa,gladita,mmgarcia}@uclv.edu.cu

**Summary.** The accelerated growth of the information volumes on processes, phenomena and reports brings about an increasing interest in the possibility of discovering knowledge from data sets. This is a challenging task because in many cases it deals with extremely large, inherently not structured and fuzzy data, plus the presence of uncertainty. Therefore it is required to know a priori the quality of future procedures without using any additional information. In this chapter we propose new measures to evaluate the quality of training sets used by supervised learning classifiers. Our training set assessment relied on measures furnished by rough set theory. Our experimental results involved three classifiers (k-NN, C-4.5 and MLP) from international databases. New training sets are built taking into account the results of the measures and the accuracy obtained by the classifiers, aiming to infer the accuracy that the classifiers would obtain by using a new training set. This is possible using a rule generator (C4.5) and a function estimation algorithm (k-NN).

**Keywords:** Rough set theory, measures, quality assessment, machine learning, knowledge generation.

## 1 Introduction

The likelihood of discovering knowledge from data sets has reached its top interest level nowadays because of the speedy growth of digital information. Machine learning (ML) studies the learning problem in the context of machines, i.e. how machines are able to acquire the knowledge that allows them to solve particular problems [11]. ML is intended to automate the learning process in such a way that knowledge can be found with a minimum of human dependency. A system is able to learn either by obtaining new information or by modifying the knowledge it currently holds so as to make it more effective. The outcome of learning is to outfit the machine (or man) with novel knowledge that enables to embrace (provide solutions to) a wider range of problems as well as to achieve either more accurate or cheap solutions or at least, to simplify the knowledge stored.

Automatically processing large amounts of data to find useful knowledge is the primary target of knowledge discovery from databases (KDD) and it can be

R. Bello et al. (Eds.): Granular Computing, STUDFUZZ 224, pp. 199–210, 2008.
springerlink.com 

defined as a non-trivial procedure of identifying valid, novel, potentially useful and eventually comprehensible patterns from data [8].

ML and KDD have a very strong relation. Both of them acknowledge the importance of induction as a way of thinking. These techniques have a broad application range.

Lots of Artificial Intelligence procedures are applicable to such sort of problems. Rough sets can be considered sets with fuzzy boundaries, sets that cannot be precisely characterized using the available set of attributes [12][13]. The basic concept of RST is the notion of approximation space.

Therefore, the goal of our research is to define and apply measures to evaluate the quality of decision systems by using RST.

The quality of the discovered knowledge mainly depends on two contributing factors: the training set and the learning method being used. Frequently, what is assessed is the quality of knowledge coming out from the application of some learning method, and this evaluation involves the control set, i.e., post-learning assessment. From the data available, different learning methods are utilized so as to ascertain which one yields a more fitting knowledge. Accordingly, being able to estimate the data quality before engaging into a learning process is a relevant aspect for saving time and computational resources.

This is the aim of the present study, which introduces some measures based on the rough set theory (RST), enabling us to estimate the quality of the training sets for classifiers' learning. Section 2 examines the problem to be solved; Section 3 is devoted to the essentials of RST and its associated measures as well as the novel measures proposed; Section 4 elaborates on the results achieved with our study for three classifiers: the k-NN method, the C-4.5 algorithm, and the Multilayer Perceptron (MLP). The knowledge generation issue for C4.5 and k-NN classifiers is the key point of Section 5 whereas the chapter's conclusions are finally outlined in Section 6.

## 2 A Look into the Problem

The problem of classifiers' learning requires that the examples be pairs of the type $(X, c)$; where $X$ stands for the set of features characterizing the object, and c stands for the object's class. Building a classifier implies finding out a function f, such that $c = f(X)$. The learning algorithm generates a function h that approximates f. The methods for doing so are divided into inductive and lazy ones. Examples of inductive methods are the algorithm C-4.5 algorithm [15] and the Multilayer Perceptron [16], while k-NN [7] is a typical method for lazy learning [11]. The k-NN classifier [7] employs distance functions to make predictions out of stored instances. The classifier's input is a vector q with unknown value for the decision class whereas the output is a prediction for its class. The error in classifying each instance of the training set is referred to as Leaving One Outside Classification Error (LOOCE). The aim of the k-NN classifier is to lower the LOOCE coefficient, whose calculation depends on whether the class values are continuous or discrete. The C4.5 algorithm is an extension of the

ID3 algorithm proposed by Quinlan in 1993 [15], which does allow the features to dwell on a continuous domain. It falls within a subset of classifiers widely known as "decision-making trees", which are those trees whose internal nodes are labelled as attributes; the protuberant branches designate boolean conditions relying on the attributes' values, whereas the leaves denote the categories or decision classes. Such algorithms provide a practical method for approaching concepts and functions that carry discrete values [3][4]. Neural network models are specified by their topology (structure, type of link), the features of the nodes (neuron's model), and the learning rule (weight computation method). In a MLP neural network, the network?s topology arranges neurons by layers hence setting up links from the front layers up to the rear layers. The neural model is an $S$-fashioned function (despite the fact that other continuous functions with fixed boundaries have been devised) and the learning method is the powerful, well-known backpropagation algorithm. As previously pointed out, the study of the relation between the training set, the efficiency, and the performance achieved during the learning process is conducted through the trial-and-error method on an experimental basis. In other words, successive training processes take place and later, their outcomes are validated. Yet, regardless of the $h$-function shaping method, the degree to which the latter approximates $f$ largely depends on the information held by the training set, being a crucial standpoint for knowledge extraction. Hence, researching on such data is of the utmost importance. Both the learning method and the procedures improving such sets rest on the original information held on them. Appraising the quality of a training set may become the way for making decisions on how to develop the learning stage. In other words, given a training set (TS), a function g is sought that comprises an indicator that deems the knowledge drawing capabilities of the TS which is a key point in building and running the classifier. There is no abundant literature on methods for the a priori assessment of training sets. Example are outlined in [6][10]. The present chapter recommends a solution to the aforementioned matter which borrows measures from rough set theory (RST), one of the most widespread conceptual frameworks for data analysis [9][14].

## 3 Rough Set Theory

Rough set theory [14], introduced by Z. Pawlak (1982), has often proved to be an excellent mathematical tool for the analysis of a vague description of objects. The adjective "vague", referring to the quality of the information, means either inconsistency or ambiguity caused by the granularity of the information in a knowledge system. The rough set philosophy is based on the assumption that with every object of the universe there is associated a certain amount of information (data, knowledge) expressed by means of some attributes used for its description. Objects having the same description are indiscernible with respect to the available information. The indiscernibility relation modeling the indiscernibility of objects thus constitutes a mathematical foundations of RST; it induces a partition of the universe into clusters of indiscernible objects, called "elementary sets" that can be used to build knowledge about a real or abstract world. The use of the indiscernibility relation results in information granulation [12][14]. In this section we recall some basic notions related to rough sets and the extension of RST via similarity relations. Also, we mention some measures of closeness of concepts and measures comprising entire decision systems. Finally, we propose some new measures for decision systems using RST.

### 3.1 Basic Concepts of Rough Set Theory

An information system is a pair $IS=(U,A)$, where $U$ is a non-empty, finite set called "the universe" and $A$ is a non-empty, finite set of attributes. Elements of $U$ are called objects. A decision system is a pair $DS=(U, A\bigcup\{d\})$, where $d\in A$ is the decision attribute.

The essentials of RST are the lower and upper approximations of a subset $X\subseteq U$. They were originally introduced with reference to an indiscernibility relation $R$. Let $R$ be a binary relation defined on $U$ which represents indiscernibility. By $R(x)$ we denote the set of objects which are indiscernible to $x$.

In classic RST, $R$ is defined as an equivalence relation (reflexive, symmetric and transitive). $R$ induces a partition of $U$ into equivalence classes corresponding to $R(x)$, $x\in U$. This classic approach to RST is extended by accepting objects which are not indiscernible but sufficiently close or similar that can be grouped into the same class [14]. The aim is to construct a similarity relation $R'$ from the indiscernibility relation $R$ by relaxing the original conditions for indiscernibility. This relaxation can be performed in many ways, thus giving many possible definitions for similarity. However, this similarity relation $R'$ must satisfy some minimal requirements. $R$ being an indiscernibility relation (equivalence relation) defined on $U$, $R'$ is a similarity relation extending $R$ iff $\forall x \in U$, $R(x)\subseteq R'(x)$ and $\forall x \in U$, $\forall y\in R'(x)$, $R(y)\subseteq R'(x)$, where $R'(x)$ is the similarity class of $x$, i.e. $R'(x)=\{y\in U\colon yR'x\}$. $R'$ is reflexive, any similarity class can be seen as a grouping of indiscernibility classes and $R'$ induces a covering of $U$ [18].

Notice that $R'$ is not imposed to be symmetric, even if most definitions of similarity usually involve symmetry. Notice also that $R'$ is not imposed to be transitive. Unlike non-symmetry, non-transitivity has been often assumed for similarity. This clearly shows that an object may belong to different similarity classes simultaneously. It means that the covering induced by $R'$ on $U$ may not be a partition. The requirement of any similarity relation is reflexivity. $R'$ can always be seen as an extension of the trivial indiscernibility relation $R$ defined by $R(x)=\{x\}$, $\forall x\in U$.

The approximation of a set $X\in U$, using an indiscernibility relation $R$, has been introduced as a couple of sets called $R$-lower and $R$-upper approximations of $X$. We consider here a more general definition of approximations which can handle any reflexive $R'$. The $R'$-lower and $R'$-upper approximations of $X$ are defined in [17].

When a similarity relation is used instead of the indiscernibility relation, other concepts and properties of RST (approximation measures, reduction and dependency) remain unchanged.

### 3.2 RST-Based Measures for Decision Systems

RST offers several gauges for the analysis of decision systems. Accuracy and quality of approximation and quality of classification measures are three representatives of these techniques. Rough membership function is an important function to develop new measures for the analysis of decision systems.

The accuracy of approximation of a rough set $X$ (denoted by $\alpha(X)$) measures the amount of roughness for a given set. If $\alpha(X)$=1, $X$ is crisp (exact) with respect to a set of attributes, if $\alpha(X)$¡1, $X$ is rough (vague) with respect to it [17].

The quality of approximation coefficient is denoted by $\gamma$ $(X)$ and expresses the percentage of objects which can be correctly classified into class $X$. Moreover, $0\leq\alpha(X)\leq\gamma\ (X) \leq 1$, and $\gamma\ (X)$ =0 if $\alpha(X)$=0, while $\gamma(X)$=1 if $\alpha(X)$=1 [17]. Let $C_1$, ...,$C_m$, $C_m$ the decision classes of the decision system *DS*.

The Quality of Classification coefficient describes the inexactness of the approximated classifications. It mean the percentage of objects which can be correctly classified in the system. If this coefficient equals 1, the decision system is consistent, otherwise it is inconsistent [17], (see (1)).

$$\Gamma(DS) = \frac{\sum \mid R'_{lower}(C_i) \mid}{\mid U \mid} \tag{1}$$

Both the accuracy and quality of approximation are associated to the respective class of a decision system; but in most cases, it is necessary to appraise the accuracy and quality of the entire decision system. Thus, two new measures to calculate the accuracy of classification were proposed in [1].

Right now we will introduce the generalized versions of both the accuracy and quality of approximation. A distinctive, common feature in each case is the presence of a weight per class, which can either be fixed by expert criteria or computed via some heuristic method.

**Generalized Accuracy of Classification.** This expression computes the weighted mean of the accuracy per class. The experts can either determine the weight per class by following some particular criterion or they can use heuristics to define the importance of each class

$$A(DS)_{Generalized} = \frac{\sum(\alpha(C_i) \cdot w(C_i))}{\sum w(C_i)} \tag{2}$$

**Generalized Quality of Classification.** The following expression computes the weighted mean of the quality of approximation per class.

$$\Gamma(DS)_{Generalized} = \frac{\sum(\gamma(C_i) \cdot w(C_i))}{\sum w(C_i)} \tag{3}$$

In both expressions (2) and (3), $w(C_i)$ is a value between 0 and 1 representing the weight of class $C_i$.

During the experimental stage it came up that the classifiers' correlations experienced several differences depending on which measure was being employed: whether those corresponding to the decision system or the ones describing datasets in terms of their classes. To avoid this undesirable effect, the generalized approximation ratio was proposed in [1].

**Generalized Approximation Ratio.** This measure involves parameters used for general description and by classes in the decision system, without making explicit a distinction among classes [2]

$$T(DS) = \frac{\sum | R'_{lower}(C_i) |}{\sum | R'_{upper}(C_i) |} \tag{4}$$

## 4 A Study on the Estimation Capability of the RST Measures

For this study we used 25 international datasets: (Balance-Scale, Balloons, Breast-Cancer Wisconsin, Bupa (Liver Disorders), Credit, Dermatology, E Coli, Exactly, Hayes-Roth, Heart-Disease (Hungarian), Iris, LED, Lung Cancer, M of N, Monks-1, Mushroom (Agaricus-Lepiota), Pima Indians Diabetes, Promoter Gene Sequence, Tic-Tac-Toe, House-Votes, Wine Recognition, Yeast) online at http://www.ics.uci.edu/˜mlearn/MLRepository.html

The procedure followed is described below:

1. For each set of samples, a 10-fold cross validation was applied to avoid the superposition of the training sets.
2. The Quality of Classification (QC), Generalized Accuracy of Classification (GAC), Generalized Quality of Classification (GQC), Generalized Approximation Ratio (GAR) were calculated for each training set.
3. Later on, the accuracy of the MLP, k-NN and C-4.5 classifiers was computed. The classification was calculated by applying the algorithms found in the Weka[1] environment.

The results obtained show that:

There is correspondence between each classifier's results and the new RST-reliant measures because high values of such gauges correspond to high values of the classifiers likewise as for the low values. In order to support the observations about the accomplished results, a statistical processing was realized. Pearson?s correlation method was calculated and the coefficients obtained reached values close to 1 in most of the cases with a ?bilateral signification? less than 0.01; such outcomes allowed to conclude that there exists an underlying linear correlation between the RST measures and the classifiers' performance. See table 1 for further details. That is, the RST measures can be used to estimate the quality of the training set in order to use them in a subsequent learning process.

The correlation is significant at level 0.01 (bilateral).

[1] Weka - Machine Learning Software in Java http://sourceforge.net/projects/weka/

**Table 1.** Correlations found between the classifiers and RST measures

| RST measure | Pearson Correlation | Classifiers | | |
|---|---|---|---|---|
| | | k-NN | MLP | C-4.5 |
| QC | Coefficient | 0.943 | 0.954 | 0.981 |
| | Signification | 0.000 | 0.000 | 0.000 |
| WAC | Coefficient | 0.919 | 0.946 | 0.944 |
| | Signification | 0.001 | 0.000 | 0.000 |
| GAC | Coefficient | 0.940 | 0.958 | 0.978 |
| | Signification | 0.001 | 0.000 | 0.000 |
| GQC | Coefficient | 0.876 | 0.931 | 0.887 |
| | Signification | 0.004 | 0.001 | 0.003 |
| GAR | Coefficient | 0.956 | 0.971 | 0.967 |
| | Signification | 0.000 | 0.000 | 0.000 |

# 5 Generating Knowledge from the Clues Provided by the RST-Based Gauges

Moreover, we can use these results to infer useful knowledge in order decide which supervised classifier (k-NN, C4.5 or MLP) is the more convenient one for a specific training set. It is also useful to evaluate qualitatively the result of the overall accuracy that this classifier might obtain using the new training set. The C4.5 rule generator has been utilized to do that. Once the most fitting classifier for the specific training set is identified, it is possible to guess the expected accuracy value using a k-nearest neighbor method.

## 5.1 Building the Dataset

Six new datasets were created (as shown in Figs. 1 and 2); two of them correspond to each supervised classifier (one dataset carrying the numerical accuracy value and the other holding the discretized accuracy value).

These datasets will hold 250 cases, 4 predictive features, represented by the measures: Quality of Classification, Generalized Accuracy of Classification, Generalized Quality of Classification and Generalized Approximation Ratio, all of them ranging from 0 to 1. The objective feature (class) for the three datasets with the discretized accuracy is represented by the following labels:

A ⟶ "Not applicable, very low accuracy"
B ⟶ "Applicable, low accuracy"
C ⟶ "Applicable, medium accuracy"

D⟶“Applicable, high accuracy”
E ⟶“Applicable, very high accuracy”

The MDL algorithm [19] was used for the discretization process.

### 5.2 Machine Learning Techniques to Rule Generation

The process of supervised classification is carried out using the C4.5 algorithm, so as to run a qualitative evaluation of the expected accuracy of the studied classifiers for a new, unknown training set. The input consisted of the three datasets depicted in Figure 1 wherein the classifiers' performance was properly discretized in order to turn it into the class attribute.

A set of classification rules is obtained for each classifier (k-NN, MLP, C4.5) using the C.45 algorithm. Such knowledge bases allow inferring the performance of each classifier according to the decision classes (A, B, C, D, E).

In this chapter we display the set of classification rules obtained for the k-NN classifier.

Instances: 250
Attributes: 5 (QC, GAR, GAC, GQC, Class)
Class = (A, B, C, D, E)
Evaluation mode: 10-fold cross-validation

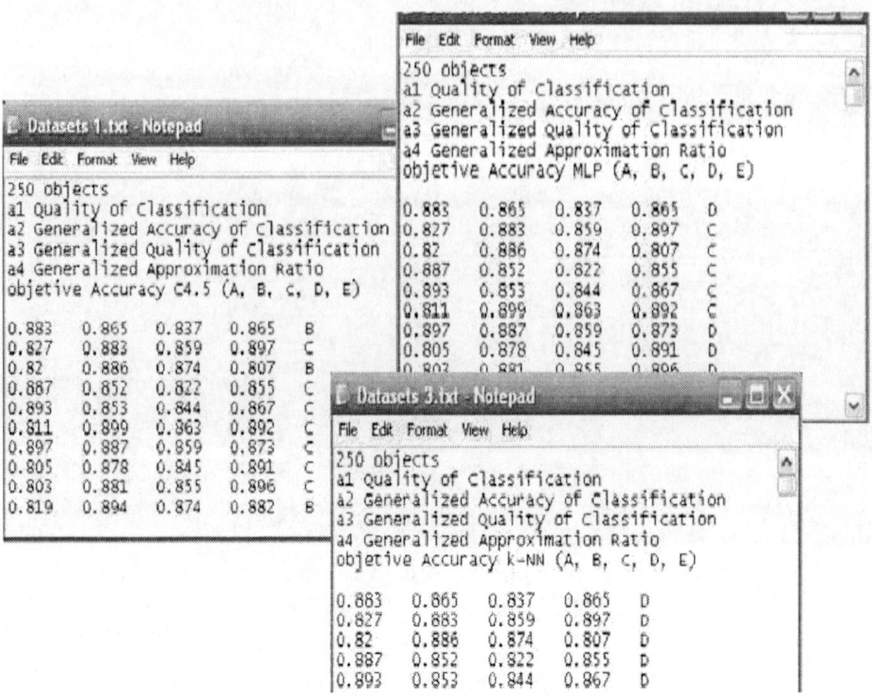


**Fig. 1.** Datasets with the discretized accuracy value

=== Classifier model (full training set)===
C4.5 pruned tree

---

GAR ≤ 0.807
— GQC ≤ 0.593: A (28.0)
— GQC > 0.593
— — QC ≤ 0.827
— — — QC ≤ 0.737
— — — — QC ≤ 0.591: B (13.0)
— — — — QC > 0.591
— — — — — QC ≤ 0.705: A (14.0)
— — — — — QC > 0.705: B (13.0)
— — — QC > 0.737
— — — — GAR ≤ 0.659: C (28.0)
— — — — GAR > 0.659
— — — — — GAC ≤ 0.781: B (11.0/2.0)
— — — — — GAC > 0.781: A (7.0/3.0)
— — QC > 0.827: A (24.0/3.0) GAR > 0.807
— GAR ≤ 0.938 — — QC ≤ 0.798
— — — QC ≤ 0.768: E (2.0)
— — — QC > 0.768: D (19.0)
— — QC > 0.798 — — — GQC ≤ 0.889
— — — — GAC ≤ 0.845
— — — — — QC ≤ 0.811: E (3.0/1.0)
— — — — — QC > 0.811: B (3.0/1.0)
— — — — GAC > 0.845: C (30.0/2.0)
— — — GQC > 0.889
— — — — GAC ≤ 0.936: E (3.0/1.0)
— — — — GAC > 0.936: D (4.0/1.0)
— GAR > 0.938
— — GQC ≤ 0.992
— — — GAC ≤ 0.985
— — — — QC ≤ 0.991
— — — — — GQC ≤ 0.985: D (2.0)
— — — — — GQC > 0.985: E (6.0/1.0)
— — — — QC > 0.991: D (5.0)
— — — GAC > 0.985: C (2.0)
— — GQC > 0.992: D (23.0/11.0)

Number of Leaves : 20
Tree size: 39

The overall performance measures using the C4.5 method to predict the accuracy of the classifiers are shown in Table 2; these are the results of applying also a 10-fold cross validation [5].

**Table 2.** Overall performance measures using C4.5

| Overall performance measures | Classifiers Performance Values | | |
|---|---|---|---|
| | k-NN | MLP | C4.5 |
| Correctly Classified Instances (%) | 95.906 | 95.313 | 96.491 |
| Kappa statistic | 0.9090 | 0.8970 | 0.9230 |
| Mean absolute error | 0.0450 | 0.0500 | 0.0509 |
| Root mean squared error | 0.1640 | 0.2030 | 0.1850 |
| Relative absolute error (%) | 9.8110 | 10.976 | 10.943 |
| Root relative squared error (%) | 34.574 | 42.744 | 38.950 |

### 5.3 Appraising the Performance of the Fittest Classifier

Once the most appropriate classifier has been found for a specific training set, the accuracy value is estimated using the datasets created with the numerical value of the classifier's performance (see Fig. 2), by means of the k-NN method.

The closest neighbor is then chosen, yielding an approximate value of the selected classifier's performance. Table 3 portrays the results of the k-NN method's performance for the estimation of the currently studied classifier's effectiveness.

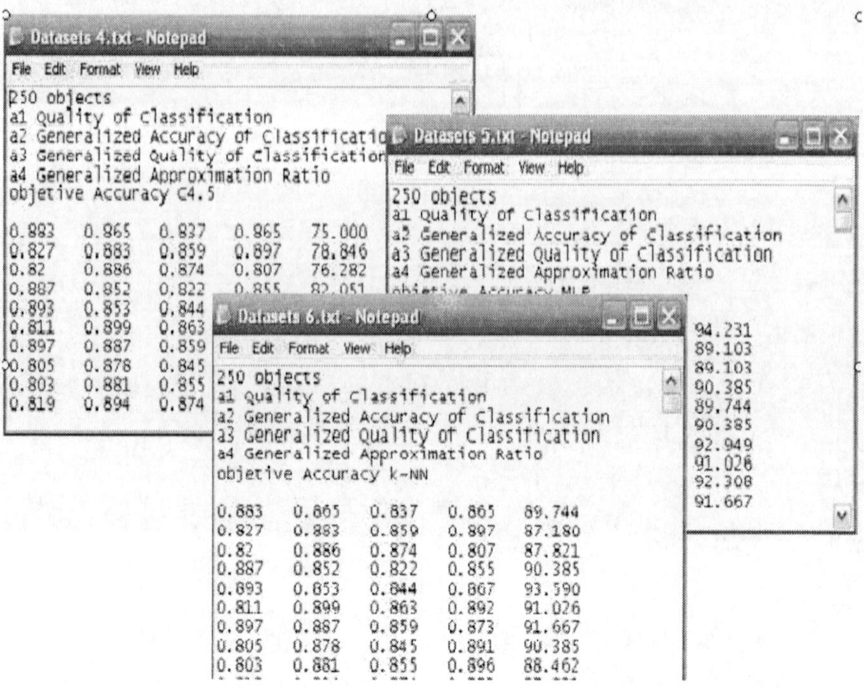


**Fig. 2.** Datasets created with the numerical accuracy value

**Table 3.** Overall performance measures using k-NN

| Overall performance measures | Classifiers Performance Values | | |
|---|---|---|---|
| | k-NN | MLP | C4.5 |
| Correlation coefficient | 0.969 | 0.914 | 0.913 |
| Mean absolute error | 0.004 | 0.004 | 0.007 |
| Root mean squared error | 0.200 | 0.198 | 0.344 |
| Relative absolute error (%) | 5.915 | 8.840 | 10.413 |
| Root relative squared error (%) | 24.328 | 41.447 | 42.151 |

## 6 Conclusions

In this chapter we have introduced the problem of evaluating the quality of a dataset, with the purpose in mind of using it afterwards as a training set for learning methods, particularly the k-NN and C4.5 algorithms as well as the MLP neural network. We proposed a suit of novel measures to evaluate decision systems as a whole by means of the rough set theory. The results obtained show that there is a meaningful relation between the classifiers' performance and the RST measures, leading to the a priori determination of the quality of future procedures without using any additional information. Machine Learning methods (C4.5) and k-NN) and RST-based measures allow identifying which of the classifiers under consideration is the most suitable for a new training set and appraising the expected behavior it would have for this training set. Gratifying results have been accomplished while generating knowledge from this information.

## References


1. Arco, L., Bello, R., García, M.: On clustering validity measures and the rough set theory. In: Gelbukh, A., Reyes-Garcia, C.A. (eds.) MICAI 2006. LNCS (LNAI), vol. 4293, Springer, Heidelberg (2006)
2. Caballero, Y., Bello, R., Taboada, A., Nowé, A., García, M.: A new measure based in the rough set theory to estimate the training set quality. In: Proc. of the 8th International Symposium on Symbolic and Numeric Algorithms for Scientific Computing (2006)
3. Choubey, S.: A comparison of feature selection algorithms in the context of rough classifiers. In: Proc. of the 5th IEEE International Conference on Fuzzy Systems, pp. 1122–1128 (1996)
4. Chouchoulas, A., Shen, Q.: LNCS (LNAI), vol. 11, pp. 118–127 (1999)
5. Demsar, J.: Journal of Machine Learning Research, 1–30 (2006)
6. Djouadi, A.: Trans. on Pattern Recognition analysis and Machine Learning 12, 92–97 (1990)
7. Garcia, J.: KNN Workshop. Suite para el desarrollo de clasificadores basados en instancias. Bachelor Thesis, Universidad Central de Las Villas, Santa Clara, Cuba (2003)

8. Kodratoff, Y., Ras, Z., Skowron, A.: Knowledge discovery in texts: A definition and applications. In: Raś, Z.W., Skowron, A. (eds.) ISMIS 1999. LNCS, vol. 1609, pp. 16–29. Springer, Heidelberg (1999)
9. Komorowski, J., Pawlak, Z.: Rough Fuzzy Hybridization: A new trend in decision-making. Springer, Heidelberg (1999)
10. Michie, D., Spiegelhalter, D., Taylor, C.: Machine Learning. Neural and Statistical Classification (1994)
11. Mitchell, T.: Machine Learning. McGraw-Hill, New York (1997)
12. Pawlak, Z.: Vagueness and uncertainty: A rough set perspective. In: Studies in Computational Intelligence, vol. 11 (1995)
13. Pawlak, Z.: Rough sets. In: Comm. of ACM, vol. 38 (1995)
14. Pawlak, Z.: International Journal of Computer and Information Sciences 11, 341–356 (1982)
15. Quinlan, J.: C-4.5: Programs for machine learning. San Mateo, California (1993)
16. Rosemblatt, F.: Principles of Neurodynamics (1962)
17. Skowron, A., Stepaniuk, J.: Intelligent systems based on rough set approach. In: Proc of the International Workshop on Rough Sets. State of the Art and Perspectives, pp. 62–64 (1992)
18. Slowinski, R., Vanderpooten, D.: Advances in Machine Intelligence & Soft-Computing 4, 17–33 (1997)
19. Witten, I., Frank, E.: Data Mining: Practical Machine Learning Tools and Techniques. Department of Computer Science, University of Waikato (2005)

# A Machine Learning Investigation of a Beta-Carotenoid Dataset

Kenneth Revett

University of Westminster, Harrow School of Computer Science
London, England HA1 3TP
revettk@westminster.ac.uk

**Summary.** Numerous reports have implicated a diet and/or conditions where levels of carotene/retinol are below minimal daily requirements may pre-dispose individuals to an increased susceptibility to various types of cancer. This study investigates dietary and other factors that may influence plasma levels of these anti-oxidants. A rough sets approach is employed on a clinical dataset to determine the attributes and their values are associated with plamsa levels of carotene/retinol. The resulting classifier produced an accuracy of approximately 90% for both beta-carotene and retinol. The results from this study indicate that age, smoking, and dietary intake of these endogenous anti-oxidants is predictive of plasma levels.

**Keywords:** beta-carotene, cancer, data mining, retinol, rough sets.

## 1 Introduction

Carotenoids are phytochemicals that are found in many leafy vegetables. In particular, a and -carotenes are precursors to vitamin A (also known as retinol). Since the pioneering work of Olson in 1988 on the biological actions of Carotenoids, focusing on their role as anti-oxidants, a number of clinical studies have been undertaken to further investigate their biological role(s) [1]. Vitamin A is involved in boosting the immune system, is a powerful anti-oxidant, and may have an impact on a variety of forms of cancer [2] [3] [4]. Clinical studies have suggested that low dietary intake or low plasma concentrations of retinol, beta-carotene, and/or other carotenoids might be associated with increased risk of developing certain types of cancer [5] [6] [7] [8].

A prospective study of two cohorts of patients with lung cancer has revealed a causal role between levels of carotenoids in their diet and the risk of contracting lung cancer [9]. This study indicated that a-carotene and lycopene intakes were significantly associated with reduced risks of lung cancer. Other carotenoids such as beta-carotene and lutein were negatively associated with increased cancer risks - but the results were not statistically significant. A study investigating dietary carotenoids and colon cancer implicated an inverse relationship between certain carotenoids (as measured by dietary intake) and the risk of colon cancer [10]. A large clinical study examining the relationship between dietary carotenoids

R. Bello et al. (Eds.): Granular Computing, STUDFUZZ 224, pp. 211–227, 2008.
springerlink.com 

and breast cancer found no clear association between intakes of any carotenoids and breast cancer risk in the study population as a whole or when subgroups were defined based on various lifestyle categories such as smoking and alcohol consumption [11]. These studies focused on the effects of dietary intake and cancer risks. Whether there is a direct relationship between dietary intake and plasma levels - which reflect the active concentrations within biological tissues, is a factor not addressed in these studies.

Nierenberg's study on a large sample of patients with nonmelanoma skin cancer investigated the role between dietary consumption, plasma levels of beta-carotene and retinol, in a large cohort of patients with non-melanoma skin cancer [12]. This important study examined the role of carotenoids in a two-stage process: intake rates versus plasma levels and plasma levels versus their impact on disease. In this study, where all subjects were positive for non-melanoma skin cancer, yielded data on factors that influence retinol levels (male sex and oral contraceptive use), dietary carotene levels and female sex were positively correlated with beta-carotene levels, and cigarette smoking and quetelet index were negatively related to beta-carotene consumption. Thus the results from this study indicated that the relationship between dietary intake and plasma levels (at least in cancer patients) is not straightforward. Unfortunately, this study consisted solely of patients already diagnosed with cancer - it would have been interesting to have matched controls.

A study of $\beta$-carotene and lung cancer by Albanes indicates that intervention studies, where patients undergo a regime of controlled dietary supplements in order to control the plasma levels of carotenoids provides very different results from purely epidemiological (or observational) studies [13]. The author highlights the issue of what types of evidence are sufficient to support or refute evidence from clinical studies - which in turn may have an impact on dietary legislation.

The major controversy highlighted in this chapter is the relationship between dietary intake and other life-style factors and the plasma levels of carotenoids. In addition, the study employs a patient cohort with non-cancerous lesions - and hence serve as controls for many forms of cancer (many yield lesions which are often surgically removed as part of their treatment regime). Results from observational studies clearly indicate that a diet high in fruit and vegetables (the principal sources of carotenoids) are less likely to develop lung cancer than those whom consume less fruits and vegetables [13]. Trial intervention studies however do not support the results from observational studies [14]. The data from trial intervention studies indicate the opposite - that high levels are associated with a higher risk of lung cancer. Bendrich provides an explanation for this discrepancy based on enhanced lung function [13]. Increased forced expiratory lung volume could translate into deeper breathing of the carcinogens and other oxidants found in cigarettes. This could result in a greater carcinogen burden in smokers supplemented with beta-carotene (or with higher levels generally) compared to placebo. The hypothesis proposed by Bendrich is supported by a relatively recent study, which suggests that low levels of beta-carotene levels may be an indicator of cellular insult rather than the cause. As Jandacek has colorfully put it, cellular beta-carotene levels may act like a Canary in the coal mine, indicating that there are damaging cellular processes occurring [15]. These reports highlight the critical role of these anti-oxidants as either causative factors or indicators of tissue damage that may ultimately be associated with a variety of cancers [4] [5].

In addition to well-known functions such as dark adaptation and growth, retinoids (a principal naturally occurring anti-oxidant) have an important role in the regulation of cell differentiation and tissue morphogenesis. Following numerous experimental studies on the effects of retinoids on carcinogenesis, their clinical use has already been introduced in the treatment of cancer (acute promyelocytic leukemia) as well as in the chemo-prevention of carcinogenesis of the head and neck region, breast, liver and uterine cervix [8].

Given the importance of this class of chemicals and their potential causative-diagnostic role in carcinogenesis, every effort should be made to study the effects of plasma/tissue levels of these anti-oxidants and their impact on cancer. In particular, beta-carotene (and to a lesser extent retinol) are readily oxidized by free radicals, and the resultant oxidized products may follow one of three pathways: protectant, reactant, or pro-oxidant [15]. The protectant pathway was proposed based on early data which suggested that beta-carotene served as a site of generic cellular oxidation, protecting other more vital cellular components from this destructive process. In this capacity, beta-carotene acted as the terminal electron donor in the catalytic pathway of oxidation, thus terminating this process before extensive cellular damage occurs [15]. This hypothesis is supported by the observation that beta-carotene is more readily oxidized than other unsaturated molecules. The role of beta-carotene as a reactant also requires that beta-carotene is preferentially oxidized, but the bi-products of this reaction are protective. Lastly, through a process of auto-oxidation (in the presence of sufficient oxygen levels such as occurs in the lungs), a chain reaction of oxidation occurs, which may damage cellular membranes. Serum retinol (vitamin A), which is a bi-product of beta-carotene biochemistry, has been reported to suppress cancer in early studies, but recent reports indicate that the same caveats for beta-carotene apply to retinol as well, although the correlation is much weaker for retinol [6] [9]. In addition, it has been reported that retinol levels are buffered much more effectively then beta-carotene, making the relationship between dietary intake and plasma levels difficult to predict. Zhang et al. provide convincing evidence that the relationship between dietary intake and plasma levels of retinol is very poor, yielding a correlation coefficient of 0.08 (with a multivariate adjusted $r^2$) [16]. What is clear is that attempts to manage plasma levels of these antioxidants through dietary consumption and/or vitamin use is a difficult task at best. In addition, the clinical evidence suggests that levels of these anti-oxidants can produce both positive and negative results.

Since clinical studies of the effects of anti-oxidants and carcinogenesis have yielded equivocal results, this work investigated the use of machine learning tools to determine the correlation between clinically measurable variables and plasma levels of beta-carotene and retinol. In particular, the machine learning paradigm of rough sets was applied to a clinically generated dataset containing information on a number of attributes that focused on life style indicators such as smoking, alcohol consumption, vitamin use, and dietary factors. Important questions are addressed regarding how our life-style can influence our health indirectly through processes which may alter the balance of the bodies' redox state through alterations in anti-oxidant levels. In particular, this work investigates the correlation(s) between the attributes contained in this clinical study and the decision outcome - plasma levels of beta-carotene and retinol. The goal is to extract the set of attribute(s) that are quantitatively correlated (either positively or negatively) with the plasma levels of these anti-oxidants. This information would provide clinicians with a suitable set of parameters to focus their clinical efforts on. Lastly, by using rough sets, the resultant classifier generates a set of rules that are in the form of simple 'if ' 'then' rules which can be readily interpreted by a domain expert. This places the results of this study into a context that is directly translatable into a model that can be readily incorporated into a decision support system. In addition, the dataset was investigated by a novel neural network classifier called LTF-C - linear transfer function classifier [17]. The purpose of applying the LTF-C classfier was to strengthen the validity of the results obtained from the rough sets approach as well as to provide a separate measure of the information content of the dataset.

The rest of this chapter is organised as follows: in the next section a brief description of the dataset is presented, followed by a description of the rough sets algorithm, a results section and then a brief conclusion/discussion section.

### 1.1 Dataset Description

This dataset contains 315 observations on 14 variables (including the decision classes). The subjects were patients who had an elective surgical procedure during a three-year period to biopsy or remove a lesion of the lung, colon, breast, skin, ovary or uterus that was found to be non-cancerous. Two of the variables (attributes) consist of plasma levels of beta-carotene and retinol. The dataset was treated as if it contained two decision classes: one containing the beta-carotene levels and all other attributes (except the retinol levels), resulting in 12 attributes and one decision class. The same technique was applied, leaving out the beta-carotene levels (retaining the retinol levels as the decision class.) Therefore, two tables were created - where either beta-carotene or retinol was the decision attribute. This approach to this dataset therefore excludes the possibility of an interaction between the two classic anti-oxidants. Although beta-carotene is a precursor to retinol (vitamin A), the local tissue and/or plasma concentrations are controlled by very different mechanisms that do not yield any easily discernible relationship. The attributes in the dataset are listed in table1. There were no missing values and the attributes consisted of both categorical and continuous data.

The rough sets approach performs best when the attributes (including the decision class) is discretised - this reduces the cardinality of the rule set produced during the rule generation process. There are several different discretisation

**Table 1.** Dataset description used in this study. The data types contain a mixture of ordinal, continuous, or categorical data. The left column is the attribute label and the right column describes the meaning of the attributes and lists allowable values for each attribute.

| Attribute name | Attribute type |
|---|---|
| AGE | Int (years) |
| SEX | (1 = Male,2=female) |
| SMOKSTAT | (1=Never, 2=Former, 3=Current Smoker) |
| QUETELET | Quetelet (weight/(height*2)) |
| VITUSE | Vitamin Use (1=Yes, fairly often, 2=Yes, not often, 3=No) |
| CALORIES | Number of calories consumed per day. |
| FAT | Grams of fat consumed per day |
| FIBER | Grams of fiber consumed per day |
| ALCOHOL | Number of alcoholic drinks consumed per week |
| CHOLESTEROL | Cholesterol consumed (mg per day) |
| BETADIET | Dietary beta-carotene consumed (mcg per day) |
| RETDIET | Dietary retinol consumed (mcg per day) |
| BETAPLASMA | Plasma beta-carotene (ng/ml) |
| RETPLASMA | Plasma Retinol (ng/ml) |

strategies available in the rough sets implementation used in this study (RSES v 2.2.1). The version employed in this study was the minimum description length (MDL/entropy preserving strategy). The final discretisation strategy employed can be evaluated empirically by examining the classification accuracy as a function of discretisation method. For a comprehensive review of discretisation strategies, please consult [18]. The next section describes the machine learning approach typically employed in a rough sets based analysis.

## 2 Rough Sets

Rough set theory was developed and introduced by Z. Pawlak in 1982 as a theoretical framework for extracting knowledge from data [19] [20]. Since its inception, the rough sets approach has been successfully applied to deal with vague or imprecise concepts, extract knowledge from data, and to reason about knowledge derived from the data [21] [22]. This work is another example which demonstrates that rough sets has the capacity to evaluate the importance (information content) of attributes, discover patterns within data, eliminate redundant attributes, and yields the minimum subset of attributes for the purpose of knowledge extraction.

The first step in the process of mining any dataset using rough sets is to transform the data into a decision table. In a decision table (DT), each row consists of an observation (also called an object) and each column is an attribute, one of which is the decision attribute for the observation. The decision table is the starting point for all subsequent work within the rough sets framework. Rough sets works with discrete data, and there are a variety of methods for discretising continuous data. Prior to the discretisation process, any objects with missing values must be handled. Missing values is very often a problem in biomedical datasets and can arise in two different ways. It may be that an omission of a value for one or more subject was intentional - there was no reason to collect that measurement for this particular subject (i.e. 'not applicable' as opposed to 'not recorded'). In the second case, data was not available for a particular subject and therefore was omitted from the table. There are two options available: remove the incomplete records from the DT or try to estimate what the missing value(s) should be. The first method is obviously the simplest, but it may not be feasible to remove records if the DT is small to begin with. Alternatively, data imputation must be employed without unduly biasing the DT. In many cases, an expert with the appropriate domain knowledge may provide assistance in determining what the missing value should be - or else is able to provide feedback on the estimation generated by the data collector. The author's experience suggests that the conditioned mean/mode fill method is most suitable for data imputation in small biomedical datasets [23]. In each case, the mean or mode is used (in the event of a tie in the mode version, a random selection is used) to fill in the missing values, based on the particular attribute in question, conditioned on the particular decision class the attribute belongs to. There are many variations on this theme, and the interested reader is directed to [24] [25] for an extended discussion on this critical issue. Once missing values are handled, the next step is to discretise the dataset. Rarely is the data contained within a DT all of ordinal type - they generally are composed of a mixture of ordinal and interval data. Discretisation refers to partitioning attributes into intervals - tantamount to searching for "cuts" in a decision tree. All values that lie within a given range are mapped onto the same value, transforming interval into categorical data. As an example of a discretisation technique, one can apply equal frequency binning, where a number of bins n is selected and after examining the histogram of each attribute, n-1 cuts are generated so that there is approximately the same number of items in each bin. See the discussion in [26] for details on this and other methods of discretisation that have been successfully applied in rough sets. Now that the DT has been pre-processed, the rough sets algorithm can be applied to the DT for the purposes of supervised classification.

The basic philosophy of rough sets is to reduce the elements (attributes) in a DT based on the information content of each attribute or collection of attributes (objects) such that the there is a mapping between similar objects and a corresponding decision class. In general, not all of the information contained in a DT is required: many of the attributes may be redundant in the sense that they do not directly influence which decision class a particular object belongs to. This is the basis of the notion of equivalence classes. One of the primary goals of rough sets is to eliminate attributes that are redundant. Rough sets use the notion of the lower and upper approximation of sets in order to generate decision boundaries that are employed to classify objects. What we wish to do is to approximate X by the information contained in B by constructing the

B-lower (BL) and B-upper (BU) approximation of X. The objects in B-lower can be classified with certainty as members of X, while objects in B-upper are not guaranteed to be members of X. The difference between the two approximations: BU - BL, determines whether the set is rough or not: if it is empty, the set is crisp otherwise it is a rough set. What we wish to do then is to partition the objects in the DT such that objects that are similar to one another (by virtue of their attribute values) are treated as a single entity. One potential difficulty arises in this regard is if the DT contains inconsistent data. In this case, antecedents with the same values map to different decision out-comes (or the same decision class maps to two or more sets of antecedents). This is unfortunately the norm in the case of small biomedical datasets, such as the one used in this study. There are means of handling this and the interested reader should consult [16] for a detailed discussion of this interesting topic. The next step is to reduce the DT to a collection of attributes/values that maximises the information content of the decision table. This step is accomplished through the use of the indiscernibility relation IND(B) and can be defined for any subset of the DT. The elements of IND(B) correspond to the notion of an equivalence class. The advantage of this process is that any member of the equivalence class can be used to represent the entire class - thereby reducing the dimensionality of the objects in the DT. This leads directly into the concept of a reduct, which is the minimal set of attributes from a DT that preserves the equivalence relation between conditioned attributes and decision values. It is the minimal amount of information required to distinguish objects with in U. The collection of all reducts that together provide classification of all objects in the DT is called the CORE(A). The CORE specifies the minimal set of elements/values in the DT which are required to correctly classify objects in the DT. Removing any element from this set reduces the classification accuracy. It should be noted that searching for minimal reducts is an NP-hard problem, but fortunately there are good heuristics that can compute a sufficient amount of reducts in reasonable time to be usable. In the software system that we employ an order based genetic algorithm (o-GA) which is used to search through the decision table for approximate reducts [27]. The reducts are approximate because we do not perform an exhaustive search via the o-GA which may miss one or more attributes that should be included as a reduct. Once we have our set of reducts, we are ready to produce a set of rules that will form the basis for object classification. Rough sets generates a collection of 'if..then..' decision rules that are used to classify the objects in the DT. These rules are generated from the application of reducts to the decision table, looking for instances where the conditionals match those contained in the set of reducts and reading off the values from the DT. If the data is consistent, then all objects with the same conditional values as those found in a particular reduct will always map to the same decision value. In many cases though, the DT is not consistent, and instead we must contend with some amount of indeterminism. In this case, a decision has to be made regarding which decision class should be used when there are more than 1 matching conditioned attribute values. Simple voting may work in many cases, where votes are cast in proportion to the support of the particular class of objects. In addition to inconsistencies within the data, the primary challenge in inducing rules from decision tables is in the determination of which attributes should be included in the conditional part of the rule. If the rules are too detailed (i.e. they incorporate reducts that are maximal in length), they will tend to overfit the training set and classify weakly on test cases. What is generally sought in this regard are rules that possess low cardinality, as this makes the rules more generally applicable. This idea is analogous to the building block hypothesis used in genetic algorithms, where the search tends to select for chromosomes which are accurate and contain short, low defining length genes. There are many variations on rule generation, which are implemented through the formation of alternative types of reducts such as dynamic and approximate reducts. Discussion of these ideas is beyond the scope of this chapter and the interested reader is directed towards [28] [29] for a detailed discus-sion of these alternatives. In the next section, we describe the experiments that were performed on this dataset, along with the principal results of this study.

## 3 Methods

The structure of the dataset consisted of 14 attributes, including the two decision attributes which was displayed for convenience in Table 1. There were 4,410 entries in the table with no missing values. The attributes contained a mixture of categorical (e.g. Sex) and continuous (e.g. age) values, both of which can be used by rough sets without difficulty. The principal issue with rough sets is to discretise the attribute values - otherwise an inordinately large number of rules are generated. We employed an entropy preserving minimal description length (MDL) algorithm to discretise the data into ranges. This resulted in a compact description of the attribute values which preserved information while keeping the number of rules to a reasonable number (see the results section for details). We determined the Pearson's Correlation Coefficient of each attribute with respect to the each decision class. The correlation values can be used to determine if one or more attributes are strongly correlated with a decision class. In many cases, this feature can be used to reduce the dimensionality of the dataset prior to classification. As can be observed from Table 2, there are no attributes that were highly correlated (positively or negatively) with either decision attribute. In general, the correlations of the attributes with retinol were of the same order as that for beta-carotene - although there was a trend towards a lower value. In previous studies, we have found that if an attribute was highly correlated (in either direction), we could select those attributes with the largest correlation values without sacrificing classification accuracy significantly [30] [31]. In this study, two analyses were performed: one with all attributes and one where the absolute value of the correlation coefficient (was greater than some threshold (0.1 in this case), thus reducing the dataset to the items indicated with an asterisk in Table 2.

**Table 2.** Pearson correlation coefficient for all attributes (excluding the decision attribute) in the dataset with respect to the decision classes. Note the left hand side of column 2 corresponds to beta-carotene and the right hand side to retinol. Also note that correlations marked with n asterisk '*' were used in experiments labelled 'reduced at-tribute set.'

| Attribute name | Attribute type |
|---|---|
| AGE: | 0.089 0.102* |
| SEX: | -0.134*0.013 |
| SMOKSTAT: | -0.229* -0.135* |
| QUETELET: | -0.224* -0.217* |
| VITUSE: | -0.022 0.0321 |
| CALORIES: | -0.099 -0.035 |
| FAT: | 0.235* 0.193* |
| FIBER: | -0.022 0.002 |
| ALCOHOL: | -0.135* -0.054 |
| CHOLESTEROL: | 0.225* 0.242* |
| BETADIET: | -0.046 0.004 |
| RETDIET: | -0.012 -0.087 |
| BETAPLASMA:(BetaCarotene) | (decision attribute) |
| RETPLASMA: (Retinol) | (decision attribute) |

The next stage is the development of a decision table, where the last column is the decision value. Rough sets operates most effectively when the decision attributes are discrete - which necessitated discretisation in this particular dataset. We examined the decision attribute statistically and found that the mean value for the carotene decision class was 183, with a variance of +/-78 and for plasma retinol the values were 139 +/- 59. We initially selected the mean as the threshold value for the discretisation process - mapping all values below the mean to a decision of '0' and all those above the mean to a decision outcome of '1'. The rest of the attributes were discretised using the MDL algorithm within RSES. We then processed the dataset with this particular set of decision values to completion. We iterated this process, moving +/- 2% from the mean in an exhaustive search. We selected the value for the decision class that provided the largest classification accuracy (after taking the average of 10 instances of 5-fold validation for each threshold for the decision class). In this work, the beta-carotene decision attribute threshold was set as the mean - 8% and that for retinol was the mean + 4%. Reducts were generated using the dynamic reduct option, as experience with other rough sets based reduct generating algorithms has indicated this provides the most accurate result [30]. In brief, dynamic reducts partitions the decision table into multiple subtables, and for each subtable generates a set of reducts. The reducts that appear most often across all subtables are retained as the proper reducts. Lastly, decision rules were generated for the purpose of classification. The results of this process are presented in the next section.

## 4 Results

After separating the beta-carotene decision from the retinol decision attribute, the rough set algorithm was applied as described above from an implementation available from the internet (http://logic.mimuw.edu.pl/ rses). In brief, 5-fold cross validation strategy was employed in order to generate decision rules and classify the objects in the decision table (repeated 10 times and averaged, unless otherwise indicated). Since beta-carotene and retinol have different distributions within plasma and tissues, extracting each attribute as a separate decision attribute was a reasonable assumption.. The next question was to partition each of the decision classes into two bins - indicating low and high levels for the decision attributes. As previously mentioned, this was performed by empirical analysis. This approach is consistent with all known literature reports. These results were superior to providing equal frequency binning with two bins (data not shown).

In Table 3, samples of the resulting confusion matrices are displayed that were generated using the full datasets (all attributes). A confusion matrix provides data on the reliability of the results, indicating true positives/negatives and false positives/negatives. From these values, one can compute the accuracy, positive predictive value and the negative predictive value of the results. The results indicate an overall classification accuracy of approximately 90%. In Table 4, we present a sample of the resulting rules that were generated during the classification of the full beta-carotene dataset (full indicating that all conditional attributes were used. The support values are listed as well, indicating the number of instances that followed the particular rule. Note that the rules generated are in an easy to read format: if attribute X0 = A and attribute X1 = B then consequent = C. In Table 5, a subset of the rules for the full retinol dataset are presented, along with support values (indicated parenthetically next to each rule). With the reduced dataset, the classification accuracy for beta-carotene and retinol were 86.1% and 83.7% respectively. The number of rules were reduced somewhat (16,451 and 13,398 respectively for beta-carotene and retinol datasets).

**Table 3.** Randomly selected confusion matrices selected from a series of 10 classifications run on the full dataset. Note that the upper confusion matrix is for the beta-carotene dataset and the lower one is for the retinol dataset.

| Decision | Low | High | Result |
|---|---|---|---|
| Low | 32 | 6 | 0.84 |
| High | 3 | 38 | 0.93 |
| | 0.91 | 0.86 | *0.89* |
| Low | 21 | 8 | 0.72 |
| High | 0 | 50 | 1.00 |
| | 1.00 | 0.86 | *0.90* |

**Table 4.** A sample of the rules produced by the rough sets classifier on the beta-carotene full dataset. The rules combine attributes in conjunctive normal form and map each to a specific decision class. The '*' corresponds to an end point in the discretised range - the lowest value if it appears on the left hand side of a sub-range or the maximum value if it appears on the right hand side of a sub-range. Note the support values are indicated parenthetically for each rule.

| Antecedents | Decision |
|---|---|
| Age([*, 45)) AND SmokeStat(1) = 0 | (low levels)(support = 27) |
| Age([50,*)) AND SmokeStats(3) AND Cholesterol([100,*]= 1 | (high levels) (support = 39) |
| Age ([*,45)) AND SmokeStats(1) AND Cholesterol([100,*))= 0 | (low levels) (support = 18) |
| BMI ([*,25.1))) AND Cholesterol ([100,*)) = 0 | (low levels) (support = 25) |
| DailyFibre ([*,35.7)) AND Alcohol ([*, 1.3)) = 1 | (high levels) (support = 31) |

**Table 5.** A sample of the rules produced by the rough sets classifier on the retinol full dataset. The rules combine attributes in conjunctive normal form and map each to a specific decision class. The '*' corresponds to an end point in the discretised range - the lowest value if it appears on the left hand side of a sub-range or the maximum value if it appears on the right hand side of a sub-range. Note the support values are indicated parenthetically for each rule.

| Antecedents | Decision |
|---|---|
| Age([*, 45)) AND SmokeStat(1) = 0 | (low levels) (support = 22) |
| Age([50,*)) AND SmokeStats(3) AND Quetelet ([28.1,*]= 1 | (high levels) (support = 19) |
| Age ([*,45)) AND Vituse(1) AND Alcohol([6.3,*))= 0 | (low levels) (support = 18) |
| SmokeStats(3) = 0 | (low levels) (support = 15) |
| DailyFibre ([*,35.7)) AND Alcohol ([*, 1.3)) = 1 | (high levels) (support = 21) |

Since the cardinality of the rule was quite large, the decision rules were filtered based on right hand support. This process reduces the number of rules - and care must betaken to find the balance between the total number of rules and classification accuracy. The results from this experiment are presented in Table 6. As can be observed, removing all RHS support of 6 or less reduced the number of rule by a factor of 100, with only a minimal reduction in classification accuracy. Generally, filtering tends to eliminate rules that have a low frequency (i.e. a support of 1 or so), which add little information to the overall classifier except in extreme cases.

**Table 6.** Results of filtering based on right hand support (RHS) support, and the effect on the number of rules and resulting classification accuracy. The RHS support is an exclusion process, where rules with less then a specified amount of support are excluded in the classification process.

A) Beta-carotene dataset

| RHS Support | Number of Rules | Accuracy |
|---|---|---|
| 0 | 26,544 | 89% |
| 0-2 | 3,947 | 88% |
| 0-4 | 714 | 87% |
| 0-6 | 219 | 83% |

B) Retinol dataset

| RHS Support | Number of Rules | Accuracy |
|---|---|---|
| 0 | 30,018 | 89% |
| 0-2 | 11,219 | 86% |
| 0-4 | 2,317 | 84% |
| 0-6 | 689 | 81% |

The reduced dataset was generated based on the Pearson correlation coefficient - those attributes with a correlation coefficient below a given threshold (in this case 0.1) were eliminated from the decision table. The results indicate that the resulting classification accuracy was not greatly affected by this process. The correlation coefficient is a linear relationship between the attribute and the decision class. To determine if this filtering is sufficient to extract all correlated attributes, the attributes extracted from the complete dataset should be compared with those selected by the filtering process. The result of this analysis indicated that some of the attributes selected via decision rules were not selected based on the threshold criteria. The results are listed in Table 7.

Lastly, as an independent verification process, the dataset was examined using the built-in (RSES v 2.2.1) LTF-C classifier. The default parameters were employed, and the resultant classifier was used to evaluate both the full, the reduced dataset, and the dataset that consisted solely of the attributes from the decision attribute (see left column in table 7). The classification accuracies for both beta-carotene and retinol as the decision attributes are listed in Table 8. These values were consistent with the classification accuracy gener-ated using rough sets. These result may indicate that the full decision table contained some redundant and possibly conflicting attributes that reduced the classification accuracy with the LTF-C algorithm (and hence increased the number of rules for the rough sets based analysis). Although not displayed, the area under the ROC curves for the datasets (beta-carotene and retinol) was calculated - resulting in

**Table 7.** Summary of the resulting attributes generated from decision rules and those based on a threshold value for the Pearson correlation coefficient. Note that there are attributes that are not included in both categories.

| Decision Rules | Correlation Coefficient |
|---|---|
| AGE | SEX |
| SMOKESTAT | SMOKESTAT |
| ——— | QUETELET |
| ——— | FAT |
| ALCOHOL | ALCOHOL |
| CHOLESTEROL | CHOLESTEROL |
| BMI | ——— |
| DAILYFIBRE | ——— |

**Table 8.** Summary of the resulting attributes generated from decision rules and those based on a threshold value for the Pearson correlation coefficient. Note that there are attributes that are not included in both categories.

| Full Attribute Set Decision Rules | Reduced Attribute Set Conditional Attributes | Reduced Attribute Set Correlation Coefficient |
|---|---|---|
| 86.3% | 84.6% | 81.7% |
| 91.3% | 88.3% | 87.5% |

values of 0.88/0.92. The criteria for the ROC was based on the values for the midpoint of beta-carotene and retinol, repeated 5 times for each value (within the range of the mean +/- 10%).

## 5 Discussion

In this study, a clinical dataset containing information on factors that have been reported to influence plasma levels of the common anti-oxidants beta-carotene and retinol was examined using rough sets and an LTF-C neural network. The results show that many of the attributes, especially age, alcohol consumption, Quetelet ($weight/height^2$), dietary fat, and cholesterol intake correlated (either directly or inversely) with plasma anti-oxidant levels. These results are consistent with reports - though these results are not completely supported by any particular study. In particular, there are conflicting reports with regards to dietary consumption, cigarette smoking, and gender with respect to plasma levels of there anti-oxidants [2] [8]. The results from this study indicate that smoking and age are inversely correlated with anti-oxidant levels, consistent with literature reports [3]. A study by Nierenberg et al., has analysed a similar dataset, from a patient cohort with nonmelanoma skin cancer, without matched control subjects [9]. Using correlational analysis, the authors implicated dietary carotenoid intake and gender as positively correlated with carotenoid levels, while cigarette smoking and Quetelet were negatively corre-lated with carotenoid levels. There was a major confounding issue: the patients in that study were under a variety of controlled medication regimes, which may have significantly influenced the results. In addition, the Nierenberg study only examined dietary carotenoid levels. These factors have confounded the usefulness of the Nierenberg study - making direct comparison of their results with this study very difficult. In addition, this study was based solely on factors that influence plasma carotenoid levels - all patients were cancer free. The Nierenberg study contained patients with a variety of cancers. It would be of great interest if the plasma levels of these anti-oxidants could then be correlated with the presence or absence of cancer. This would provide a causal link from life-style habits, to plasma levels of anti-oxidants, through to the health status of the subjects.

The significant result of this study is the attributes that were selected for classification (see Table 7) and their values. The data from this study suggests that cigarette smoking is inversely correlated with plasma levels of carotenoids. Whether this result is clinically useful depends on the relationship between plasma carotenoid levels and disease status. This dataset did not address that issue - as all patients were cancer free. Alcohol consumption was inversely correlated with plasma levels of carotenoids - a finding supported by other studies. Besides providing information regarding the attributes involved in the decision class - the magnitude of the values is also generated using rough sets. In the final analysis, a principal finding of this study is that the attribute set can be reduced to six - and their correlations with the decision classes are clearly defined.

To the author's knowledge, no other machine learning based study of this dataset has produced these or related results for this specific dataset. Not only are the attributes highlighted that correlate with the decision attributes, but quantitative information re-garding the values of the attributes are automatically produced by the rough sets approach. Most clinical studies only give a correlation between attributes and decision classes based on some mean/median value for the attribute. The ability to manipulate the values of attributes can only be accomplished through a large subject population. This is not a feasible option in many cases - this dataset consisted of 315 subjects - which is considered a large study in clinical terms. The cost and effort required to work with such large populations of subjects means that studies of this size are rarely performed.

In addition to classification accuracy and quantitative estimates of attribute values, roughs sets is also able to remove attributes that are not sufficiently involved in the decision process. In this study, at most 6 of the 12 attributes were required to provide a classification accuracy that is near optimal. Reducing the dimensionality of the dataset is a very significant feature of rough sets - and is by definition one of its primary effects within the classification process. Generally, large clinical studies tend to extract as much information as they can - knowing that such studies are few and far between. This renders the resultant data difficult to analyse because of the inclusion of many possible superfluous attributes. In addition, a set of readily interpreted rules such as those listed in table 4 means the results can be interpreted more readily than those generated by neural networks. In addition, rough sets can be employed with missing data (although imputation is required) and when the attributes are of variable types (e.g. ordinal or continuous).

The primary concern when employing rough sets is the need to discretise the decision classes in order to reduce the number of rules. Filtering on support is a clear way of reducing the number of rules, and this can usually be accomplished without a significant reduction in the classification accuracy. Lastly, through standard validation techniques such as N-fold validation, our results produced results that are better than those published elsewhere in the literature. The area under the ROC was approximately 88% for beta-carotene and 92% for retinol. These promising results indicate that rough sets can be a useful machine learning tool in the automated discovery of knowledge, even from small and often sparse biomedical datasets.

The next stage in this analysis would be to apply the decision rules generated from this study to a clinical trial that contained patients with and without a particular type of cancer. This would allow mapping from life style/dietary habits to plasma levels, and from plasma levels to disease status. This requires the collaboration of both the machine learning and the medical communities. When this union of disciplines occurs, we can expect to extract the maximal amount of useful information from these types of studies.

## Acknowledgement

The author would like to acknowledge the source of this dataset:
http://lib.stat/cmu.edu/datasets/PlasmaRetinol.

## References

1. Olson, J.A.: Biological actions of carotenoids. J. Nutr. 119, 94–95 (1988)
2. Krinsky, N.I., Johnson, E.J.: Department of Biochemistry, School of Medicine, Tufts University, 136 Harrison Avenue, Boston, MA 02111-1837, USA; Jean Mayer USDA Human Nutrition Research Center on Aging at Tufts University, 136 Harrison Avenue, 711 Washington St, Boston, MA 02111-1837, USA
3. Palozzo, E.R., Byers, T., Coates, R.J., Vann, J.W., Sowell, A.L., Gunter, E.W., Glass, D.: Effect of smoking on serum nutrient concentrations in African-American women. Am J. Clin. Nutr. 59, 891–895 (1994)
4. Peto, R., Doll, R., Buckley, J.D., Sporn, M.B.: Can dietary beta-carotene materially reduce human cancer rates. Nature 290, 201–208 (1981)
5. Goodman, G.E., Alberts, D.S., Peng, Y.M., et al.: Plasma kinetics of oral retinol in cancer patients. Cancer Treat Rep. 68, 1125–1133 (1984)
6. Michaud, D.S., Feskanich, D., Rimm, E.B., Colditz, G.A., Speizer, F.E., Willett, W.C.: Intake of specific carotenoids and risk of lung cancer in 2 prospective US cohorts. Am. J. Clin. Nutr. 72, 990–997 (2000)
7. Ziegler, R.G.: A review of epidemiologic evidence that carotenoids reduce the risk of cancer. J. Nutr. 119(1), 116–122 (1989)

8. Moon, R.C.: Comparative aspects of carotenoids and retinoids as chemopreventive agents for cancer. J. Nutr. 119(1), 127–134 (1989)
9. Slattery, M.L., Benson, J., Curtin, K., Ma, K.-N., Schaeffer, D., Potter, J.D.: Carotenoids and colon cancer. Am. J. Clin. Nutr. 71, 575–582 (2000)
10. Terry, P., Jain, M., Miller, A.B., Howe, G.R., Rohan, T.E.: Dietary carotenoids and risk of breast cancer. Am. J. Clin. Nutr. 76, 883–888 (2002)
11. Nierenberg, D.W., Stukel, T.A., Baron, J.A., Dain, B.J., Greenberg, E.R.: Determinants of plasma levels of beta-carotene and retinol. American Journal of Epidemiology 130, 511–521 (1989)
12. Albanes, D.: $\beta$-Carotene and lung cancer: A case study. Am. J. Clin. Nutr. 69 (suppl.), 1345S–1350S (1999)
13. ATBCCPSG: The effect of vitamin E and beta carotene on the incidence of lung cancer and other cancers in male smokers. The Alpha-Tocopherol, Beta Carotene Cancer Prevention Study Group. New England Journal of Medicine 330, 1029–1035 (1994)
14. Bendich, A.: From 1989 to 2001: What have we learned about the Biological Actyions of Beta-Carotene? J. Nutr. 134, 225S–230S (2004)
15. Jandacek, R.J.: The cancary in the cell: A sentinel role for $\beta$-carotene. J. Nutr. 130, 648–651 (2000)
16. Zhang, S., Tang, G., Russell, R.M., Mayzel, K.A., Stamfer, M.J., Willett, W.C., Hunter, D.J.: Measurements of retinoids and carotenoids in breast adipose tissue and a comparison of concentrations in breast cancer cases and control subjects. Am. J. Clin. Nutr. 66, 626–632 (1997)
17. Wojnarski, M.: LTF-C: Architecture, training algorithm and applications of new neural classifier. Fundamenta Informaticae 54(1), 89–105 (2003)
18. Bazan, J., Szczuka, M.: The Rough Set Exploration System. In: Peters, J.F., Skowron, A. (eds.) Transactions on Rough Sets III. LNCS, vol. 3400, pp. 37–56. Springer, Heidelberg (2005), `http://logic.mimuw.edu.pl/rses`
19. Pawlak, Z.: Rough Sets. International Journal of Computer and Information Sciences 11, 341–356 (1982)
20. Pawlak, Z.: Rough sets - Theoretical aspects of reasoning about data. Kluwer, Dordrecht (1991)
21. Nguyen, H.S., Skowron, A.: Quantization of real-valued attributes. In: Proc. Second International Conference on Information Science, pp. 34–37 (1995)
22. Øhrn, A.: Discernibility and Rough Sets in Medicine Tools and Applications. Department of Computer and Information Science. Trondheim, Norway, Norwegian University of Science and Technology 239 (1999)
23. Revett, K.: Data-mining Small Biomedical Datasets Using Rough Sets. In: HCMC 2005. The First East European Conference on Health Care Modelling and Computation, Craiova, Romania, pp. 231–241 (2005)
24. Slezak, D.: Approximate Entropy Reducts. Fundamenta Informaticae (2002)
25. Bazan, J.G., Skowron, A., Synak, P.: Dynamic reducts as a Tool for Extracting Laws from Decision tables. In: Proceeding of the Third International Workshop on Rough Sets and Soft Computing, San Jose, California, pp. 526–533 (1994)
26. Slezak, D.: Approximate Entropy Reducts. Fundamenta Informaticae (2002)
27. Wroblewski, J.: Theoretical Foundations of Order-Based Genetic Algorithms. Fundamenta Informaticae 28(3-4), 423–430 (1996)
28. Nguyen, S.H., Polkowski, L., Skowron, A., Synak, P., Wróblewski, J.: Searching of Approximate Description of Decision Classes. In: RSFD 1996. Proc. of The Fourth International Workshop on Rough Sets, Fuzzy Sets and Machine Discovery, Tokyo, November 6-8, pp. 153–161 (1996)

29. Komorowski, J., Pawlak, Z., Polkowski, L., Skowron, A.: Rough sets: A tutorial. In: Pal, S.K., Skow-ron, A. (eds.) Rough Fuzzy Hybridization - A New Trend in Decision Making, pp. 3–98. Springer, Heidelberg (1999)
30. Revett, K., Gorunescu, F., Gorunescu, M.: A Rough Sets Based Investigation of a Beta-Carotene/Retinol Dataset, ISFUROS, ISBN 959-250-308-7
31. Revett, K.: A Rough Sets Based Classifier for Primary Biliary Cirrhosis Using RS to datamine a PCB dataset. In: IEEE Conference on Eurocon 2005, November 22-24, 2005, Belgrade, Serbia and Montenegro, pp. 1128–1131 (2005)

# Rough Text Assisting Text Mining: Focus on Document Clustering Validity

Leticia Arco[1], Rafael Bello[1], Yailé Caballero[2], and Rafael Falcón[1]

[1] Department of Computer Science, Central University of Las Villas
Carretera a Camajuaní, km 5 1/2, Santa Clara, Villa Clara, Cuba
{leticiaa,rbellop,rfalcon}@uclv.edu.cu

[2] Faculty of Informatics, University of Camagüey
Circunvalación Norte, km 5 1/2, Camagüey, Cuba
yailec@yahoo.com

**Summary.** In this chapter, the applications of rough set theory (RST) in text mining are discussed and a new concept named "Rough Text" is presented along with some RST-based measures for the evaluation of decision systems. We will focus on the application of such concept in clustering validity, specifically cluster labeling and multi-document summarization. The experimental studies show that the proposed measures outperform several internal measures existing on literature. Additionally, the application of Rough Text is illustrated.

## 1 Introduction

Rough set theory (RST) has many interesting applications. It is turning out to be methodologically significant to artificial intelligence and cognitive science, especially in the representation of and reasoning with vague and/or imprecise knowledge, machine learning, knowledge acquisition, decision analysis, knowledge discovery from databases, expert systems and pattern recognition [16] [18] [19] [20] [25]. It seems of particular importance to decision support systems and data mining. Contrary to other approaches, the main advantage of RST is that it does not need any preliminary or additional data about information systems.

Text mining or knowledge discovery from textual databases [8] is a technology for analyzing large collections of unstructured documents for the purposes of extracting interesting and non-trivial patterns or knowledge. The field is interdisciplinary, involving information retrieval, text analysis, information extraction, clustering, categorization, visualization, summarization, database technology, machine learning, and data mining.

Text mining is a challenging task as it involves dealing with text data that are inherently unstructured and fuzzy. Rough sets can be considered sets with fuzzy boundaries - sets that cannot be precisely characterized using the available set of attributes [20]. The basic concept of the RST is the notion of approximation space. Two advantages of RST can be used in Text Mining applications: (i) it

R. Bello et al. (Eds.): Granular Computing, STUDFUZZ 224, pp. 229–248, 2008.
springerlink.com 

does not need any preliminary or additional information about data, and (ii) it is a tool for use in computer applications in circumstances which are characterized by vagueness and uncertainty. Thus, the goal of this research is to define the new concept "Rough Text", in order to apply the advantages of RST to some Text Mining tasks. New RST-based measures for the evaluation of decision systems come along with the introduction of the "Rough Text" concept. The starting point of this concept is a classified corpus of documents (e.g., the results of the application of a document clustering algorithm to a corpus of texts). The "Rough Text" concept will allow us to obtain the upper and the lower approximation of each document cluster. Considering the complexity and the task-dependence of Text Mining processes, it can be stated that it is difficult to decompose these processes. In this chapter, we focus on the advantage of applying the presented concept in the clustering validity task. We propose the usage of "Rough Text" for assisting two other important text mining tasks, namely cluster labeling and multi-document summarization.

This chapter is organized as follows. Section 2 presents the general concepts about RST and the new measures for decision systems that rely on RST. We propose and describe the definition of "Rough Text" in Section 3. The application of "Rough Text" in clustering validity and the evaluation of the suggested measures are outlined in Section 4, whereas cluster labeling and multi-document summarization are briefly detailed in Section 5 as well as the proposal of using "Rough Text" to aid these tasks. Conclusions and further remarks finish the chapter.

## 2 Rough Set Theory

Rough set theory, introduced by Z. Pawlak [20], has often proved to be an excellent mathematical tool for the analysis of a vague description of objects. The adjective "vague", referring to the quality of information, means inconsistency or ambiguity which is caused by the granularity of information in a knowledge system. The rough sets philosophy is based on the assumption that with every object of the universe there is associated with certain amount of information (data, knowledge) expressed by means of some attributes used for object description. Objects having the same description are indiscernible with respect to the available information. The indiscernibility relation modeling the indiscernibility of objects thus constitutes a mathematical basis of RST; it induces a partition of the universe into blocks of indiscernible objects, called elementary sets that can be used to build knowledge about a real or abstract world. The use of the indiscernibility relation results in information granulation [16][19][20]. In this section we recall some basic notions related to rough sets and the extension of RST using similarity relations. Also, we mention some measures of closeness of concepts and measures of decision systems. Finally, we propose two new measures of decision systems using RST.

An information system is a pair $IS = (U, A)$, where $U$ is a non-empty, finite set called the universe and $A$ is a non-empty, finite set of attributes. Elements of $U$ are called objects. A decision system is a pair $DS = (U, A \cup \{d\})$, where $d \in A$ is the decision attribute.

The basic concepts of RST are the lower and upper approximations of a subset $X \subseteq U$. These were originally introduced with reference to an indiscernibility relation $R$. Let $R$ be a binary relation defined on $U$ which represents indiscernibility. By $R(x)$ we mean the set of objects which are indiscernible to $x$. Thus, $R(x) = \{y \in U : yRx\}$.

In classic RST, $R$ is defined as an equivalence relation (reflexive, symmetric and transitive). $R$ induces a partition of $U$ into equivalence classes corresponding to $R(x)$, $x \in U$. This classic approach to RST is extended by accepting that objects which are not indiscernible but sufficiently close or similar can be grouped in the same class [26]. The aim is to construct a similarity relation $R'$ from the indiscernibility relation $R$ by relaxing the original conditions for indiscernibility. This relaxation can be performed in many ways, thus giving many possible definitions for similarity. However, this similarity relation $R'$ must satisfy some minimal requirements. $R$ being an indiscernibility relation (equivalence relation) defined on $U$, $R'$ is a similarity relation extending $R$ iff $\forall x \in U$, $R(x) \subseteq R'(x)$ and $\forall x \in U$, $\forall y \in R'(x)$, $R(y) \subseteq R'(x)$, where $R'(x)$ is the similarity class of $x$, i.e. $R'(x) = \{y \in U : yR'x\}$. $R'$ is reflexive, any similarity class can be seen as a grouping of indiscernibility classes and $R'$ induces a covering of $U$ [23].

Notice that $R'$ is not imposed to be symmetric. Even if most definitions of similarity usually involve symmetry. Notice also that $R'$ is not imposed to be transitive. Unlike non-symmetry, non-transitivity has been often assumed for similarity. This clearly shows that an object may belong to different similarity classes simultaneously. It means that the covering induces by $R'$ on $U$ may not be a partition. The requirement of any similarity relation is reflexivity. $R'$ can always be seen as an extension of the trivial indiscernibility relation $R$ defined by $R(x) = \{x\}, \forall x \in U$.

The rough approximation of a set $X \subseteq U$, using an indiscernibility relation $R$, has been introduced as a pair of sets called $R$-lower and $R$-upper approximations of $X$. We consider here a more general definition of approximations which can handle any reflexive $R'$. The $R'$-lower and $R'$-upper approximations of $X$ are defined respectively by (1) and (2):

$$R'_*(X) = \{x \in X : R'(x) \subseteq X\} \tag{1}$$

$$R'^*(X) = \bigcup_{x \in X} R'(x) \tag{2}$$

When a similarity relation is used instead of the indiscernibility relation, other concepts and properties of RST (approximation measures, reduction and dependency) remain valid.

### 2.1 Measures for Decision Systems Using Rough Set Theory

RST offers measurement techniques for the analysis of information systems. Accuracy and quality of approximation and quality of classification measures are three representatives of these techniques.

**Accuracy of Approximation.** A rough set $X$ can be characterized numerically by the following coefficient called the accuracy of approximation, where $|X|$ denotes the cardinality of $X \neq \varnothing$.

$$\alpha(X) = \frac{|R'_*(X)|}{|R'^*(X)|} \tag{3}$$

Obviously $0 \leq a(X) \leq 1$. If $\alpha(X) = 1$, $X$ is crisp (exact) with respect to set of attributes, if $\alpha(X) < 1$, $X$ is rough (vague) with respect to set of attributes [23].

**Quality of Approximation.** The following coefficient

$$\gamma(X) = \frac{|R'_*(X)|}{|X|} \tag{4}$$

expresses the percentage of objects which can be correctly classified into class $X$. Moreover,$0 \leq \alpha(X) \leq \gamma(X) \leq 1$ and $\gamma(X) = 0$ if $\alpha(X) = 0$ while $\gamma(X) = 1$ if $\alpha(X) = 1$ [23].

**Quality of Classification.** If $C_1, \ldots, C_m$ are the decision classes of the decision system $DS$, the following coefficient describes the inexactness of approximation classifications

$$\Gamma(DS) = \frac{\sum_{i=1}^{m} |R'_*(C_i)|}{|U|} \tag{5}$$

The quality of classification expresses the percentage of objects which can be correctly classified in the decision system. If this coefficient is equal to 1, the decision system is consistent, otherwise is inconsistent [23].

The accuracy of approximation and quality of approximation are associated to the respective classes of a decision system; but in most cases, it is necessary to evaluate the entire decision system (e.g., quality of classification measure, see (5)). Thus, we propose two new functions in order to calculate the accuracy of entire decision system. The first one defines the accuracy of classification measure, which calculates the accuracy average per class. See formula (6). Because each class has a different influence in the quality of the decision system we propose the weighted accuracy of classification measure, in the meaning of weighted mean of the accuracy per class. See formula (7).

**Accuracy of Classification.** If $C_1, \ldots, C_k$ are the decision classes of the decision system $DS$, the following coefficient describes the accuracy of classifications.

$$\mathrm{A}(DS) = \frac{\sum\limits_{i=1}^{k} \alpha(C_i)}{k} \tag{6}$$

Obviously, $0 \leq A(DS) \leq 1$. If $A(X) = 1$, each decision system class is crisp (exact) with respect to the set of attributes; if $A(X) < 1$, at least one class of the decision system is rough (vague) with respect to the set of attributes.

**Weighted Accuracy of Classification.** If $C_1, \ldots, C_k$ are the decision classes of the decision system $DS$, the following coefficient describes the weighted accuracy of classifications. This weighing is carried out considering that bigger classes must exercise a bigger influence than classes having fewer elements when computing the accuracy of the approximation, therefore the weight is represented by the cardinality of each class.

$$\mathrm{A}_{Weighted}(DS) = \frac{\sum (\alpha(C_i) \cdot |C_i|)}{|U|} \tag{7}$$

If the decision system is a multiclassified one, we can replace $|U|$ with $\sum\limits_{i=1}^{k} |C_i|$.

The cardinality of the classes is not the only way of weighing the classes. Thus, we introduce the generalized variations of both the accuracy and quality of classification measures, because the experts can either ponder classes or use heuristics to define the importance of classes in a lot of applications.

**Generalized Accuracy of Classification.** If $C_1, \ldots, C_k$ are the decision classes of the decision system $DS$, the next expression computes the weighted mean of the accuracy per class. The experts can decide the weight of classes or they can use heuristics to define the importance of classes.

$$\mathrm{A}_{Generalized}(DS) = \frac{\sum\limits_{i=1}^{k} (\alpha(C_i) \cdot w(C_i))}{k} \tag{8}$$

**Generalized Quality of Classification.** If $C_1, \ldots, C_k$ are the decision classes of the decision system $DS$, the following expression computes the weighted mean of the quality of approximation per class.

$$\Gamma_{Generalized}(DS) = \frac{\sum\limits_{i=1}^{k} (\gamma(C_i) \cdot w(C_i))}{k} \tag{9}$$

Notice that in both expressions (see (8) and (9)), $w(C_i)$ is a value between 0 and 1 representing the weight of class $C_i$.

The rough membership function quantifies the overlapping degree between set $X$ and the $R'(x)$ class that $x$ belongs to. It can be interpreted as a frequency-based estimate of $\Pr(x \in X \mid x, R'(x))$, the conditional probability that object $x$ belongs to set $X$ [11]. It is defined as follows:

$$\mu_X(x) = \frac{|X \cap R'(x)|}{R'(x)} \tag{10}$$

But it is not only interesting to compute the ratio introduced in (10). There exists another gauge that can reflect the involvement of objects to classes. Thus, taking into consideration the characteristics of the rough membership function, the rough involvement function is introduced hereafter (see (11)).

**Rough Involvement Function.** The following ratio quantifies the percentage of objects correctly classified into the class $X$ which are related to the object $x$.

$$\nu_X(x) = \frac{|X \cap R'(x)|}{|X|} \tag{11}$$

In order to count on a measurement of the membership degree and the involvement degree of objects into classes, it is necessary to calculate the mean of the rough membership and the mean of the rough involvement per classes. Thus, the following novel measures have been designed.

**Mean of Rough Membership.** If $C_1, \ldots, C_k$ are the decision classes of the decision system $DS$, the following expression computes the mean of the rough membership per class. Notice that the rough membership for a class $X$ is the mean of the rough membership for each object $x$ belonging to $X$.

$$\mathrm{M}(DS) = \frac{\sum_{i=1}^{k} \mu_{class}(C_i)}{k} \tag{12}$$

$$\mu_{class}(X) = \frac{\sum_{x \in X} \mu_X(x)}{|X|}$$

**Mean of Rough Involvement.** If $C_1, \ldots, C_k$ are the decision classes of the decision system $DS$, the following expression computes the mean of the rough involvement per class. Notice that the rough involvement for a class $X$ is the mean of the rough involvement for each object $x$ belonging to $X$.

$$\mathrm{Y}(DS) = \frac{\sum_{i=1}^{k} \nu_{class}(C_i)}{k} \tag{13}$$

$$\nu_{class}(X) = \frac{\sum_{x \in X} \nu_X(x)}{|X|}$$

An emphasis has been put in the fact that the influence of all the classes is not to be considered equally when evaluating decision systems. Expressions (14) and (15) portray weighted variants of the rough membership and rough involvement measures, respectively.

**Weighted Mean of Rough Membership.** If $C_1, \ldots, C_k$ are the decision classes of the decision system $DS$, the following expression computes the weighted mean of the rough membership per class.

$$\mathrm{M}_{General}(DS) = \frac{\sum_{i=1}^{k} (\mu_{class}(C_i) \cdot |C_i|)}{|U|} \tag{14}$$

**Weighted Mean of Rough Involvement.** If $C_1, \ldots, C_k$ are the decision classes of the decision system $DS$, the following expression computes the weighted mean of the rough involvement per class.

$$\mathrm{Y}_{General}(DS) = \frac{\sum_{i=1}^{k} (\nu_{class}(C_i) \cdot |C_i|)}{|U|} \tag{15}$$

We take into account the cardinality in order to weight both expressions. Notice that if the decision system is a multi-classified one, we can replace $|U|$ with $\sum_{i=1}^{k} |C_i|$.

This section has made us become acquainted with novel RST-based measures by means of (6)-(7) and (12)-(15) which allow to perform a better characterization of decision systems and, hence, of the previously clustered textual corpora, as it is shown below.

## 3 "Rough Text" Definition

The vector space information retrieval system represents documents as vectors in a Vector Space Model (VSM) [21]. The document set comprises an $n \times m$ document-term matrix $M$, in which each column represents a term, each row represents a document, and each entry $M(i, j)$ represents the weighted frequency of term $j$ in document $i$. If we apply a clustering algorithm to this VSM, we can consider the combination of VSM and the clustering results as a decision system $DS = (U, A \cup \{d\})$, where $U = \{D_1, D_2, \ldots, D_n\}$ is the document collection, $A$ is a finite set of keywords or key phrases that describe this document collection and $d \notin A$ represents the clustering results (decision attribute). See Table 1.

We use a similarity relation $R'$ in our "Rough Text" concept, because two documents of $U$ can be similar but not equal. There is a variety of distance and similarity measures for comparing document vectors. Dice, Jaccard and Cosine coefficients are the most used in document clustering, because they have the attraction of simplicity and normalization [9]. Let $s : U \times U \longrightarrow \mathbf{R}$ a function that measures the similarity between objects of $U$, we consider the following definition of document similarity relation $R'$. See the following formula where $R'(x)$ is the similarity class of document $x$

$$R'(x) = \{y \in U : yR'x, \text{ i.e. } y \text{ is related with } x \text{ iff } s(x, y) > \xi\} \tag{16}$$

**Table 1.** A decision system consisting of a corpus and its clustering results. Each cell represents the weighted frequency of a term $j$ in a document $i$.

| | Term 1 | Term 2 | ... | Term $m$ | Cluster |
|---|---|---|---|---|---|
| Document 1 | $tf_{d_1}(t_1)$ | $tf_{d_1}(t_2)$ | | $tf_{d_1}(t_m)$ | Cluster$_{t_1}$ |
| Document 2 | $tf_{d_2}(t_1)$ | $tf_{d_2}(t_2)$ | | $tf_{d_2}(t_m)$ | Cluster$_{t_2}$ |
| ... | | | ... | | ... |
| Document $n$ | $tf_{d_n}(t_1)$ | $tf_{d_n}(t_2)$ | | $tf_{d_n}(t_m)$ | Cluster$_{t_k}$ |

where $\xi$ is a similarity threshold. We have to calculate $R'(x)$ for each document in $U$.

It is necessary to define $R'$-lower and $R'$-upper approximations for each similarity class (i.e., for each cluster) by taking into account formulas 1 and 2. Thus, the lower and upper approximations of a cluster of documents are defined in formulas 17 and 18 respectively.

$$R'_*(C_j) = \{D_i \in C_j : R'(D_i) \subseteq C_j\} \tag{17}$$

$$R'^*(C_j) = \bigcup_{D_i \in C_j} R'(D_i) \tag{18}$$

$R'_*(C_i)$ includes all documents that belong to $cluster_i$ and that are only similarity related to documents contained in the $cluster_i$. $R'_*(C_i)$ includes all documents that are similarity related of documents members of the $cluster_i$. The documents in the $R'_*(C_i)$ can be classified with certainty as member of $cluster_i$, while the documents in $R'_*(C_i)$ can be classified as possible members of $cluster_i$. The set $R'_*(C_i) - R'_*(C_i)$ is called the boundary region of $cluster_i$ and consists of those documents that on the basis of the knowledge in the terms those describe the document collection, cannot be unambiguously classified into $cluster_i$. The set $U - R'_*(C_i)$ is called the outside region of $cluster_i$ and consists of those objects which can be with certainty classified as not belonging to $cluster_i$. Thus, in RST each vague concept is replaced by pair of precise concepts called its lower and upper approximations; the lower approximation of a $cluster_i$ in "Rough Text" consists of all documents which surely belong to $cluster_i$, whereas the upper approximation of the $cluster_i$ consists of all documents which possibly belong to the $cluster_i$.

Thereby, it is possible to use the "Rough Text" concept in order to extract the lower and upper approximations of each cluster and to apply the measurement techniques to determine of closeness of concepts and the quality of decision systems. This approach can improve cluster labeling, summarization and document clustering validity.

## 4 "Rough Text" and Clustering Validity Measures

Clustering is a class of techniques that fall under the category of machine learning. The aim of a cluster analysis is to partition a given set of data or objects into clusters (subsets, groups, classes). Clusters are collections of similar data items, and they can be created without prior training on labeled examples (unsupervised learning). This partition should have the following properties: homogeneity within the clusters and heterogeneity between clusters [13] [27].

Many clustering techniques were developed especially for the recognition of structures in data in higher dimensional spaces (e.g. clustering of document collections) [13]. Applying clustering procedures to document collections is useful in a lot of applications. However, it is very difficult to evaluate a clustering of documents. The cluster validity problem is the general question whether the underlying assumptions (e.g., clusters shapes, number of clusters, etc.) of a clustering algorithm are satisfied at all for the considered data set. In order to solve this problem, several clustering quality (validity) measures have been proposed [13].

A clustering validity measure maps a clustering on a real number. The number indicates to what degree certain structural properties are developed in the clustering. Each validity measure can not capture all good clustering properties. Some measures are used for evaluating the number of clusters in the data set; others measure compactness, isolation and density of clusters. The performance of a clustering algorithm may be judged differently depending on which measure are used. Any new measure of, or view on, clustering quality might add to the understanding of clustering [7] [24]. To be more confident in results one should use several measures [12] [13] [15]. There are external, internal and relative measures [5] [12] [13] [15] [22] [28]. The external measures use a human reference classification to evaluate the clustering. In contrast, internal measures base their calculations solely on the clustering that has to be evaluated. Relative measures can be derived from internal measures by evaluating different clustering and comparing their scores [13].

Document clustering is characterized by vagueness and uncertainty; and the most of the document collections are not previously labeled. Thus, it is necessary to apply internal measures for validating document clustering. Internal measures (e.g., the similarity measure) are based on the representation. The basic idea behind internal measures stems from the definition of clusters. A meaningful clustering solution should group objects into various clusters, so that objects within each cluster are more similar to each other than objects from different clusters. In particular, intra-cluster similarity, is defined as the average similarity between objects of each cluster, and inter-cluster similarity, is defined as the average similarity between objects within each cluster and the remainder of the objects in the data set [5] [28]. In the absence of any external information, such as class labels, the cohesiveness of clusters can be used as a measure of cluster similarity. Overall Similarity is an internal measure based on the pairwise similarity of documents in a cluster [29]. This measure has some disadvantages, because it does not consider the relation between clusters, neither the size of the clusters.

Others internals are Dunn Indices measures. The Original Dunn and Dunn-Bezdek are two particular cases of Dunn Indices measures depending on the criteria of calculating distance measure between clusters and the cluster diameter measure

[3]. Original Dunn measure yields high values for clustering with compact and very well separated clusters. Bezdek recognized that the Original Dunn measure is very noise sensitive; thus, he proposed a new kind of calculating the Original Dunn index. The features of the original Dunn measure do not allow us to evaluate crisp and overlapped clustering results.

Davies-Bouldin measure is a function of the ratio of the sum of within-cluster scatter to between-cluster separation [6]. The Dunn index and the Davies-Bouldin index are related in that they have a geometric (typically centroidic) view on the clustering. All of these measures work well if the underlying data contains clusters of spherical form, but they are susceptible to data where this condition does not hold.

The $\Lambda$-measure and the measure $\bar{\rho}$ of expected density were proposed in [28]. These internal measures interpret a data set as a weighted similarity graph: they analyze the graph's edge density distribution to judge the quality of a clustering. The calculation of these measures has an expensive computational complexity. Another dark side of the weighted partial connectivity $\Lambda$-measure is that it is not a normalized measure.

The aforementioned measures are not able to capture all of the desirable properties when evaluating a clustering result; besides, in so many cases they assume a definite shape of the clusters to be assessed and also consider the existence of cluster centroids, as it is the case for Dunn-Bezdek and Davies-Bouldin measures. They have a different nature; hence, in order to evaluate a clustering employing all of these measures it is necessary to perform quite a few expensive and various computations, being not possible to reuse part of the computations done for calculating a measure as a subtask of computing another measure.

### 4.1 The Application of "Rough Text" in Clustering Validity

Considering the disadvantage of the above internal measures, we propose new technique for clustering validity using our "Rough Text" concept. These measures base their calculations solely on the clustering that has to be evaluated. Our approach arises from the following facts: (i) any new measure of, or view on, clustering quality might influence in the understanding of clustering, and (ii) external measures are not applicable in real world situations since reference classifications are usually not available. We propose to use the measures expressed in (5)-(9), (8)-(9) and (12)-(15) to validate the clustering results.

Below we present a method for measuring the quality of document clustering using the "Rough Text" thought.

**Algorithm 1.** *"Rough Text" for measuring the quality of document clustering.*
***Input:*** *Document collection, clustering results, similarity threshold and similarity function between documents.*
***Output:*** *Values of quality, accuracy and weighted accuracy of classification measures.*

1. *Create the decision system corresponding to the input.*
2. *Obtain the similarity class for each document in the decision system. according to (16).*
3. *Calculate lower and upper approximation for each cluster. Use (17)-(18).*
4. *Calculate the accuracy, quality, means of rough membership, and means of rough involvement measures for each cluster. Compute them by using (3),(4),(10),(11) respectively.*
5. *Reckon the measures defined in (5)-(9) and (12)-(15) for the decision system.*

If the boundary region is small, we will obtain better results of the proposed measures. Thus, we can measure the quality of the clustering using the "Rough Text" definition, because we can measure the vagueness of each cluster. Higher values of the measures indicate a better clustering.

### 4.2 Illustrating the Use of "Rough Text" in Clustering Validity

The following example is used to explain how using "Rough Text" in document clustering validity. The example has been conducted with 31 news of Reuters-21578 text categorization test collection . For our experiment, we select texts from only four topics: money-supply, trade, cocoa and acq. Table 2 depicts the description of the document collection that we want to use in our example. Its second column shows the subset of documents that belong to each topic, considering the referenced classification of the Reuters collection[1].

**Table 2.** Description of the document collection

| Topics | Relation of documents in each topic |
|---|---|
| Money-supply | 7, 8, 20, 24, 27 and 28 |
| Trade | 2, 3, 5, 9, 10, 14, 18, 19 and 29 |
| Cocoa | 1, 4, 6, 22, 23, 25 and 26 |
| Acq | 11, 12, 13, 15, 16, 17, 21, 30 and 31 |

The preprocessing of the documents was done including the following operations. (i) The transformation of documents includes: from British to American spelling, lemmatization, and substitution of abbreviations and contraction for their full forms. (ii) The generation of a normalized and weighted VSM representation for each document collection using frequency normalizer and a variant of TF-IDF formula. (iii) The dimensionality reduction involves the stop-words elimination and selection of the best ranking values using term quality II measure [2].

[1] The Reuters-21578 test collection is available at David D. Lewis' professional home page http://www.research.att.com/~lewis

**Table 3.** Description of clustering results. The lower and upper approximations as well as the corresponding measures for each cluster.

| Clusters | Relation of documents in each cluster | Measures |
|---|---|---|
| Money-Supply | $C_1$={7,8,**10**,20,24,28}<br>$R'_*(C_1)$={7,8,20,24}<br>$R'^*(C_1)$={3,7,8,**9**,10,**14**,**19**,20,24,**27**,28} | $\alpha(C_1) = 0.3636$<br>$\gamma(C_1) = 0.6667$ |
| Trade | $C_2$={2,3,5,9,14,18,19,**27**,29}<br>$R'_*(C_2)$={2,5,18,19}<br>$R'^*(C_2)$={2,3,5,9,**10**,14,18,19,27,**28**,29}} | $\alpha(C_2) = 0.3636$<br>$\gamma(C_2) = 0.4444$ |
| Cocoa | $C_3=R'_*(C_3)=R'^*(C_3)$={1,4,6,22,23,25,26} | $\alpha(C_3)=\gamma(C_3)=1$ |
| Acq | $C_4=R'_*(C_4)=R'^*(C_4)$={11,12,15,16,17,21,30,31} | $\alpha(C_4)=\gamma(C_4)=1$ |

**Table 4.** An excerpt of the second clustering results. Description of clusters 1, 4 and 5. The lower and the upper approximations of these clusters.

| Cluster | Relation of documents in each cluster |
|---|---|
| Cluster 1 | $C_1$={10,27} $R'_*(C_1)=\varnothing$ $R'^*(C_1)$={**2**,**3**,**5**,**7**,**8**,**9**,10,**14**,**19**,27,**28**} |
| Cluster 4 | $C_4$={7,8,20,24,28} $R'_*(C_4)$={20,24}$R'^*(C_4)$={7,8,**12**,**13**,20,24,27,28} |
| Cluster 5 | $C_5$={2,3,5,9,14,18,19,29} $R'_*(C_5)$={29}<br>$R'^*(C_5)$={2,3,4,5,10,11,14,18,19,29} |

We employed Simultaneous Keyword Identification and Clustering of Text Documents (SKWIC) algorithm [2]. It uses a deterministic crisp cluster analysis technique. It is an extension of classic k-means using a modification of Cosine coefficient. For that reason, the document similarity relation $R'$ was created using Cosine coefficient [29]. Thus, $R'(d_i) = \{d_k \in U : d_k R' d_i$, i.e. $d_j$ is related with $d_i$ iff $s_{Cosine}(d_i, d_j) > \xi\}$, where $d_i$ and $d_j$ are document vectors and $\xi$ is the means of the distances between all pairs of documents.

When we apply SKWIC algorithm to our example, we obtain four clusters, one for each topic; nevertheless, the algorithm made some mistakes in clusters 1 and 2. The piece of news 10 is about trade, but SKWIC algorithm assigned this piece of news to cluster 1; thus, this document does not belong to lower approximation of cluster 1, because it is related with documents in cluster 2. We can see the same situation with the piece of news 27. For that reason, the accuracy and quality of the approximations of clusters 1 and 2 have low values. Clusters 3 and 4 are correct; thus, these are not rough concepts. The combination of these results produces the values 0.7742 and 0.6818 for the quality and the accuracy of classification, respectively. See Table 3.

If we run the SKWIC algorithm with the same document collection but initializing it with another number of clusters, for example, six clusters, we need to merge clusters and the rough set measures can help us do it. Cluster 1 is a bad cluster because it has only two documents: the former one should belong to cluster 5 whereas the latter should belong to cluster 4. The upper approximations indicate how to merge the clusters as depicted in Table 4.

It is remarkable that only the accuracy and quality of classification measures were used in our example because we preferred to illustrate it with a few measures in order to highlight the application of the “Rough Text” concept. The evaluation and the bright sides of the proposed measures are explained in the following section.

### 4.3 Evaluating the New Method Using “Rough Text” for Clustering Validity

Evaluation is an exhausting task within the text mining field. To evaluate the proposed method for clustering validity a study case was designed. The study case includes 50 textual corpora from Reuters Agency news collection. The following steps for the experimental study were applied to each of the 50 created corpora.

1. **Textual representation of each corpus.** The corpora were transformed and the VSM representation was carried out on each transformed corpus, with a weighting based on a variation of the formula TF-IDF [2]. The dimensionality reduction was performed by the stop-words elimination and the selection of those 800 better terms, that is to say, terms that have the 800 higher qualities according to the Term Quality II measurement [2].
2. **Clustering of each corpus using three kinds of algorithms.** The selected clustering algorithms were Extended Star [10] and SKWIC [2]. The first one uses a crisp and overlapped clustering technique while the last one can be treated as hard and deterministic methods. We apply both algorithms to the 50 corpora.
3. **Clustering validity process.** We apply the Overall Similarity, Dunn Indices, Davies-Bouldin, Expected Density and Weighted Partial Connectivity measures to each clustering result for each corpus (see Sect. 4). Also, we apply the “Rough Text” concept by considering the algorithm proposed in Sect. 4.1, which includes all proposed and cited measures (see Sect. 2). The weight used to calculate the Generalized Accuracy of Classification measure (see formula 8) is the means of rough membership measure per class (see formula 10). The weight considered to calculate the Generalized Quality of Classification (see formula 9) is the means of the rough involvement measure per class (see formula 11).
4. **Statistical correlations.** Through the Pearson’s correlation method we found that there is a correlation coefficient between the quoted internal measures (see Sect. 4) and the results of the “Rough Text” concept application through the RST-based measures. Look at tables 5 and 6 wherein the results of the statistical correlations between the quoted internal measures and the novel RST-based measures for each different clustering result are outlined. Each row corresponds to six internal measures listed in the following order: (1) Overall Similarity, (2) Original Dunn index, (3) Dunn-Bezdek index, (4) Davies-Bouldin measure, (5) measure of Expected Density and (6) Weighted Partial Connectivity $\Lambda$-measure. The first subrow stands for the correlation coefficient and second one displays the correlation significance between each pair of internal and proposed measures.

**Table 5.** The Pearson correlation values between the internal measures and the introduced validity measures for the Extended Star algorithm results

| | A | $\Gamma$ | $A_{weighted}$ | $A_{general}$ | $\Gamma_{general}$ | M | $M_{general}$ | Y | $Y_{general}$ |
|---|---|---|---|---|---|---|---|---|---|
| 1 | .627 | .637 | .643 | .626 | .612 | .589 | .597 | .554 | .522 |
| | .000 | .000 | .000 | .000 | .000 | .000 | .000 | .000 | .000 |
| 3 | .264 | .338 | .343 | .291 | .289 | .220 | .272 | .204 | .283 |
| | .064 | .016 | .015 | .040 | .042 | .124 | .056 | .156 | .046 |
| 4 | -.890 | -.865 | -.880 | -.943 | -.954 | -.864 | -.853 | -.915 | -.916 |
| | .000 | .000 | .000 | .000 | .000 | .000 | .000 | .000 | .000 |
| 5 | -.408 | -.480 | -.453 | -.404 | -.409 | -.390 | -.416 | -.359 | -.408 |
| | .003 | .000 | .001 | .004 | .003 | .005 | .003 | .010 | .003 |
| 6 | -.102 | -.182 | -.160 | -.088 | -.089 | -.081 | -.114 | -.053 | -.117 |
| | .480 | .205 | .266 | .544 | .537 | .576 | .430 | .716 | .420 |

Make no mistake in interpreting the meaning of the correlation coefficient. While for Overall Similarity, Dunn Index, Weighted Partial Connectivity and Expected Density measures it is desirable to get a value as high as possible, in Davies-Bouldin the opposite meaning is sought. As to the proposed measures, since they deal with accuracy and quality of the classification as well as membership and involvement of objects to classes, the higher value the measure yields, the better the clustering is.

Be aware that we can find positive correlations between the suggested measures and the Overall Similarity measure for the two utilized clustering algorithms. The mean of rough involvement and weighted mean of rough involvement measures bear a positive correlation with respect to the original Dunn measure. Further note that it is tough to find correlations between our measures and the Dunn-Bezdek one when the clustering has been carried out with the SKWIC algorithm. Such gauge takes into consideration the cluster centroids to compute the distance between one another. This might remarkably falsify the results because it depends on the fact that the cluster centroids might have been correctly computed and the clusters shape as well, leading to the logical conclusion that the correlations between the new measures and the previously existing ones are very poor. Nevertheless, it is noticeable that the accuracy of and quality of classification measures and their weighted and generalized variants do correlate with Dunn-Bezdek for the Extended Star algorithm results, where the chosen stars were regarded as the cluster centroids. However, the SKWIC algorithm computes ideal cluster centroids and that may be the cause of the differences found when looking at correlations. Additionally, the weighted mean of rough involvement measure achieves a good correlation with Dunn-Bezdek measure both for Extended Star and SKWIC algorithms results.

Negative correlations are beheld between our measures and the Davies-Bouldin measure. You can expect such an outcome because of the desirable minimization behavior in regard to Davies-Bouldin's meaning, whereas a maximization tendency is looked forward concerning our measures. The best

**Table 6.** The Pearson correlation values between the internal measures and the introduced validity measures for the SKWIC algorithm results

| | A | $\Gamma$ | $A_{weighted}$ | $A_{general}$ | $\Gamma_{general}$ | M | $M_{general}$ | Y | $Y_{general}$ |
|---|---|---|---|---|---|---|---|---|---|
| 1 | .317 | .313 | .385 | .553 | .742 | .201 | .322 | .790 | .535 |
| | .008 | .027 | .006 | .000 | .000 | .161 | .023 | .000 | .000 |
| 2 | -.039 | -.028 | -.015 | .019 | .135 | -.054 | .004 | .374 | .410 |
| | .789 | .849 | .920 | .896 | .348 | .710 | .977 | .007 | .003 |
| 3 | .114 | .045 | .053 | .149 | .206 | .125 | .056 | .167 | .370 |
| | .432 | .757 | .714 | .303 | .151 | .387 | .697 | .246 | .008 |
| 4 | -.257 | -.181 | -.259 | -.443 | -.662 | -.098 | -.186 | -.861 | -.801 |
| | .072 | .208 | .070 | .001 | .000 | .498 | .196 | .000 | .000 |
| 5 | .262 | .332 | .298 | .164 | -.048 | .321 | .327 | -.510 | -.665 |
| | .066 | .018 | .036 | .254 | .743 | .023 | .020 | .000 | .000 |
| 6 | .371 | .313 | .385 | .553 | .742 | .201 | .322 | .790 | .535 |
| | .008 | .027 | .006 | .000 | .000 | .161 | .023 | .000 | .000 |

correlation coefficients are observed linked to both the generalized versions of the accuracy and quality of classification measures, where the weighing per class represented by the mean of rough membership and mean of rough involvement measures plays a vital role. The weighted mean of rough involvement also accomplishes good correlation coefficients with regard to Davies-Bouldin gauge.

The expected density measure considers in the computations the size of the clusters to be evaluated, that's why the best correlation coefficients are gotten with the quality of classification y weighted accuracy of classification measures for the assessment of the SKWIC algorithm's clustering results. However, when correlating this measure with the novel ones for the results of the Extended Star algorithm, negative values were yielded. Recall that this method utilizes a crisp and overlapping technique, undoubtedly exercising influence over the results achieved during the evaluation. Besides, this algorithm might produce many single-element clusters, therefore contributing to the misunderstanding of the evaluation outcomes. A strong point in favor of the RST-based measures is that they are insensitive to the effect provoked by single-element clusters during the evaluation.

The $\Lambda$-measure of weighted partial connectivity correlates to a high extent with the generalized versions of both accuracy and quality of classification measures for clustering results having a crisp and deterministic technique, consolidating the belief that the weighing form in the generalized variant is correct based on the evidence provided by the validation results. On the other hand, no correlation was attained when evaluating the clustering having a crisp and overlapping technique.

In a nutshell, every measure contained in the suggested method for applying the "Rough Text" concept is able to seize the features of the previously clustered textual collection. Good results were accomplished with the accuracy and quality measures for clustering characterization (see formulas 5 and 6) but, in spite of this, a better overall characterization was obtained by pondering classes according to their cardinality. The use of expressions 12 and 13 for weighing classes yielded even more praiseworthy results in the generalized versions of accuracy and quality of classifications (see formulas 8 and 9). Gauges denoted by expressions 14 and 15 reach to reflect the rough membership and involvement degrees of documents to each cluster, providing an accurate description of clusters from two different standpoints, thus performing the evaluation from another sense, also necessary in the assessment.

## 5 Other Applications of "Rough Text" in Text Mining

In this section we describe briefly two important text mining tasks: cluster labeling and multi-document summarization. We propose the usage of the "Rough Text" concept in order to assist these tasks.

One of the main problems with the two-phase framework is the gap between the clustering result of the representative dataset and the requirement of retrieving cluster labels for the entire large dataset. Traditionally, the post-clustering stage is named labeling process [4]. However, labeling is often ignored by the clustering researchers. Part of the reason is the clustering problem itself is still not well solved. A foreseeable problem in labeling large amount of data is that the cluster boundary will be extended more or less by incorporating the labeled data points [4]. Boundary extension might result in the connection of different clusters and thus we may need to merge them. Since the boundary is extending, the outliers around the boundary should also be treated carefully. Thereby, it is necessary to represent the clusters. Existing cluster representations can be classified into four categories: centroid-based, boundry-point-based (representative point-based), classification-tree-based and rule-based representations [14]. Representative point-based approach works better than centroids since it describes the clusters in more detail. But how to define the representative points precisely for arbitrarily shaped clusters is as difficult as the clustering problem.

Summarization is the process of condensing a source text into a shorter version preserving its information content. It can serve several goals – from survey analysis of a scientific field to quick indicative notes on the general topic of a text. There are single and multi document summarizations techniques [17]. An automatic multi-document summarization system generally works by extracting relevant sentences from the documents and arranging them in a coherent order [1]. Note that the document clustering and the text summarization algorithms can be used together. For instance, a user can first perform a clustering of some documents, to get an initial understanding of the document base. Then, supposing this user finds a particular cluster interesting, this user can perform a summarization of the documents in that cluster. If a user wants to extract relevant sentences from document clusters, it is possible to define the most representative documents in each cluster.

Remark that both in clustering labeling and multi-document summarization it might be useful to draw the most representative documents of the textual clusters. We suggest to use the "Rough Text" concept in order to extract

**Table 7.** An excerpt of the most representative document extraction in clusters 7 and 8 by changing the similarity threshold for the lower approximations construction

| Topic | Cluster 7 (Diabetes Mellitus) = {32, 35, 34, 37, 39, 40, 42, 43, 44, 47, 49, 51, 54, 56, 57, 58, 59, 60, 61, 64} | Cluster 8 (Cystic Fibrosis) = {1, 2, 3, 4, 5, 6, 7, 8, 9, 10, 11, 12, 13, 14, 15, 16, 28} |
|---|---|---|
| 0.22 | {35, 57, 58, 61, 64} | {1, 3, 4, 6, 8, 9, 10, 11, 12, 13, 14, 15, 16, 28} |
| 0.20 | {35, 57, 58, 61, 64} | {1, 3, 4, 6, 8, 9, 11, 14, 15, 28} |
| 0.18 | {35, 57, 58, 64} | {1, 3, 4, 6, 8, 9, 11, 15} |
| 0.16 | {57, 58, 64} | {3, 4, 6, 9, 15} |
| 0.14 | {57, 58} | {4, 6, 9, 15} |
| 0.12 | {57, 58} | {4, 9} |

the most representative documents for each cluster. The vague concept about representativity is replaced by the concept called "lower approximation", because in the "Rough Text" concept, the lower approximation of a cluster consists of all documents which surely belong to this cluster, thus these documents are the most representative ones of this cluster. Because of the one hand, the lower approximation of a cluster includes all documents that belong to this cluster and on the other hand their set of similarity related documents is contained in the set of documents of this cluster; thus, the documents in the lower approximation can be definitely classified as member of this cluster. We also can control the boundary region of each cluster, because we consider a threshold in our similarity relation R' in the "Rough Text" concept. Thus, we can regulate the granularity of the representative set of documents of each cluster.

The main advantage of using "Rough Text" in both tasks (cluster labeling and multi-document summarization) is that users can specify the size of the set of the most representative documents for each cluster. Thus, we can use the specified size in order to calculate the needed threshold in the similarity relation. The size of the lower approximation can influence in the size of the summary. The following example illustrates how to change the set of representative documents by changing the similarity threshold.

We create a corpus from BioMed Central's open access full-text corpus[2] for illustrating the above approach. Tables 7 and 8 present an excerpt of the clustering results. Look at the effect produced by the modification of the threshold for building the similarity relations when applying "Rough Text" in the clustering results, which is precisely the obtaining of more or less specific sets made up from the most representative documents for each cluster. When raising the similarity thresholds, the lower approximations of each cluster become smaller and therefore, only those documents that make up the kernel of each studied cluster are found.

[2] BioMed Central has published 22003 articles so far. http://www.biomedcentral.com/info/about/datamining/

**Table 8.** An excerpt of the most representative document extraction in clusters 9 and 12 by changing the similarity threshold for the lower approximations construction

| Topic | Cluster 9 (Lung Cancer) = {10, 98, 29, 100, 102, 103, 105, 106, 109, 110, 112, 113, 114, 119, 120} | Cluster 12 (AIDS) = {32, 71, 72, 73, 75, 79, 81, 83, 82, 84, 85} |
|---|---|---|
| 0.22 | {98, 29, 100, 102, 105, 106, 109, 110, 112, 113} | {79, 83, 82, 84, 85} |
| 0.20 | {98, 29, 100, 102, 105, 109, 110, 112, 113} | {79, 83, 84, 85} |
| 0.18 | {98, 100, 102, 105, 109, 110, 112, 113} | {79, 83, 84, 85} |
| 0.16 | {100, 102, 105, 110} | {79, 84, 85} |
| 0.14 | {100, 102, 105, 110} | {85} |
| 0.12 | {105} | {85} |

In multi-document summarization, the principal sentences may be extracted from the lower approximation of each cluster; thus, it is needless to process all documents of each cluster. If users like to obtain an extended summary, we can draw the sentences from the upper approximation of each cluster.

## 6 Conclusions

We have proposed a formal definition of "Rough Text"; which allows us the characterization of the previously clustered corpus of texts. The "Rough Text" concept consists in calculating the lower and upper approximations of each document cluster, and some evaluation measures can be calculated depending on the application of the concept.

We elaborated on the application of the novel definition in clustering validity. It has been demonstrated that our RST-based measures are able to seize at least the same good clustering properties than the quoted internal evaluation measures. A great profit of the introduced measures is that they all start computing the same initial concept ("Rough Text", similarity relations, lower and upper approximations) whereas the benchmark measures come up from different sources and so as to seize every clustering property it is mandatory to perform computationally complex and diverse calculations (e.g., compute centroids, represent the corpus as a graph, compute minimal expansion trees...). Also, we proposed the usage of the "Rough Text" concept in order to assist both cluster labeling and multi-document summarization tasks by extracting the most representative documents of each cluster.

## Acknowledgement

The support provided by the collaboration project between the VLIR (Flemish Interuniversity Council, Belgium) and the Central University of Las Villas (UCLV, Cuba) is gratefully recognized.

## References


1. Barzilay, R., Elhadad, M.: Using lexical chains for text summarization. In: Advances in Automatic Text Summarization, pp. 111–121. MIT Press, Cambridge (1999)
2. Berry, M.: Survey of text mining. In: Clustering, classification, and retrieval, Springer, Heidelberg (2004)
3. Bezdek, J.K., Li, W.Q., Attikiouzel, Y., Windham, M.: A geometric approach to cluster validity for normal mixtures. Soft Computing 1 (1997)
4. Chen, K., Liu, L.: ClusterMap: Labeling clusters in large datasets via visualization. In: CIKM 2004. Proceedings of the ACM Conference on Information and Knowledge Management, pp. 285–293 (2004)
5. Conrad, J., Alkofahi, K., Zhao, Y., Karypis, G.: Effective document clustering for large heterogeneous law firm collections. In: Proceedings of the 10th international Conference on Artificial Intelligence and Law, pp. 177–187 (2005)
6. Davies, D.L., Bouldin, D.W.: IEEE Transactions on Pattern Analysis and Machine Learning 1(4), 224–227 (1979)
7. Eugenio, B., Glass, M.: Computational Linguistics 30(11), 95–101 (2004)
8. Feldman, R., Dagan, I.: Knowledge discovery in textual databases (KDT). In: KDD 1995. Proceedings of the First International Conference on Knowledge Discovery and Data Mining, pp. 112–117. AAAI-Press, Stanford, California, USA (1995)
9. Frakes, W., Baeza-Yates, R.: Information retrieval. In: Data structures & algorithms, Prentice Hall, New Jersey (1992)
10. Gil-García, R., Badía-Contelles, J.M., Pons-Porrata, A.: In: Sanfeliu, A., Ruiz-Shulcloper, J. (eds.) CIARP 2003. LNCS, vol. 2905, pp. 480–487. Springer, Heidelberg (2003)
11. Grabowski, A.: Basic properties of rough sets and rough membership function. Journal of Formalized Mathematics 15 (2003)
12. Halkidi, M., Batistakis, Y., Vazirgiannis, M.: Journal of Intelligent Information Systems 17, 107–145 (2001)
13. Höppner, F., Klawonn, F., Kruse, R., Runkler, T.: Fuzzy clustering analysis. In: Methods for classification, data analysis and image recognition, John Wiley & Sons, Chichester (1999)
14. Jain, A., Murty, M.N., Flynn, P.J.: ACM Computing Surveys 31(3), 264–323 (1999)
15. Kaufman, K., Rousseuw, P.J.: Finding groups in data: An introduction to cluster analysis. Wiley, United Kingdom (2005)
16. Kryszkiewicz, M.: Information Sciences 112, 39–49 (1998)
17. Leuski, A., Lin, C.Y., Hovy, E.: iNeATS: Interactive multi-document summarization. In: Proceedings of the 41st Annual Meeting of the Association for Computational Linguistics, pp. 125–128 (2003)
18. Liang, J.Y., Xu, Z.B.: International Journal of Uncertainty, Fuzziness and Knowledge-Based Systems 24, 95–103 (2002)
19. Lingras, P.J., Yao, Y.: Journal of American Society for Information Science 49, 415–422 (1998)
20. Pawlak, Z., Grzymala-Busse, J.W., Slowinski, R., Ziarko, W.: Communications ACM 38(11), 89–95 (1995)
21. Polkowski, L., Skowron, A.: Rough sets in knowledge discovery. In: Methodology and applications, Physica-Verlag, Heidelberg (1998)
22. Rosell, M., Kann, V., Litton, J.E.: Comparing comparisons: Document clustering evaluation using two manual classifications. In: Proceedings of the International Conference on Natural Language processing (ICON) (2004)

23. Salton, G.: The SMART retrieval system. Prentice-Hall, Englewood Cliffs (1971)
24. Schenker, A., Last, M., Bunke, H., Kandel, A.: A comparison of two novel algorithms for clustering web documents. In: Proceedings of the 2nd International Workshop on Web Document Analysis WDA (2003)
25. Skowron, A., Stepaniuk, J.: Intelligent systems based on rough set approach. In: Proceedings of the International Workshop Rough Sets: State of the Art and Perspectives. Extended Abstracts. pp. 62–64 (1992)
26. Slowinski, R., Vanderpooten, D.: Advances in Machine Intelligence and Soft-Computing 4, 17–33 (1997)
27. Stein, B., Meyer, S.: In: Günter, A., Kruse, R., Neumann, B. (eds.) KI 2003. LNCS (LNAI), vol. 2821, pp. 254–266. Springer, Heidelberg (2003)
28. Stein, B., Meyer, S., Wißbrock, F.: On cluster validity and the information need of users. In: 3rd IASTED Conference on Artificial Intelligence and Applications, pp. 216–221. ACTA Press (2003)
29. Steinbach, M., Karypis, G., Kumar, V.: Neural Computation 14, 217–239 (2000)

# Construction of Rough Set-Based Classifiers for Predicting HIV Resistance to Nucleoside Reverse Transcriptase Inhibitors

Marcin Kierczak[1], Witold R. Rudnicki[2], and Jan Komorowski[1,2,3]

[1] The Linnaeus Centre for Bioinformatics, Uppsala University BMC, Box 598 Husargatan 3, SE-751 24 Uppsala, Sweden
[2] Interdisciplinary Centre for Mathematical and Computational Modelling Warsaw University Pawinskiego 5a, 02-106, Warsaw, Poland
[3] To whom correspondence should be addressed
jan.komorowski@lcb.uu.se

**Summary.** For more than two decades AIDS remains a terminal disease and no efficient therapy exists. The high mutability of HIV leads to serious problems in designing efficient anti-viral drugs. Soon after introducing a new drug, there appear HIV strains that are resistant to the applied agent. In order to help overcoming resistance, we constructed a classificatory model of genotype-resistance relationship. To derive our model, we use rough sets theory. Furthermore, by incorporating existing biochemical knowledge into our model, it gains biological meaning and becomes helpful in understanding drug resistance phenomenon. Our highly accurate classifiers are based on a number of explicit, easy-to-interpret IF-THEN rules. For every position in amino acid sequence of viral enzyme reverse transcriptase (one of two main targets for anti-viral drugs), the rules describe the way the biochemical properties of amino acid have to change in order to acquire drug resistance. Preliminary biomolecular analysis suggests the applicability of the model.

**Keywords:** HIV resistance, rough sets, NRTI.

## 1 Introduction

More than twenty years ago Barre-Sinoussi et al. [1] identified Human Immunodeficiency Virus (HIV-1) as an agent responsible for Acquired Immuno Deficiency Syndrome (AIDS). Since then, extremely rapid emergence of drug resistant mutants remains one of the major obstacles in setting-up efficient therapies [20, 21, 18, 7]. Currently, four classes of anti-HIV agents are available for clinical use: Nucleoside Reverse Transcriptase Inhibitors (NRTI), Non-Nucleoside Reverse Transcriptase Inhibitors (NNRTI), Protease Inhibitors (PI) and fusion inhibitors (FI). Both NRTI and NNRTI drugs are targeted against viral enzyme Reverse Transcriptase (RT) [19, 7].

Unfortunately, complete eradication of the virus from an infected individual still remains impossible and AIDS is considered to be a chronic, terminal disease [7]. This is mainly due to the extraordinarily high mutation rate in the

R. Bello et al. (Eds.): Granular Computing, STUDFUZZ 224, pp. 249–258, 2008.
springerlink.com 

HIV genome. Mutations in the RT coding sequence inevitably lead to the rapid emergence of RT-inhibitors resistant strains and is responsible for almost all treatment failures. It is a well established fact that HIV replication rate is sufficient to produce a population containing viruses capable of overcoming any form of currently existing therapy as soon as in the third day post-infection. Drug-resistant mutants are usually characterized by lower replication rate and by administering combination of three or four different drugs simultaneously, a significant slow down in the disease progress can be achieved [14, 13].

In order to maximize treatment efficacy it is necessary to perform drug resistance tests for every new isolated HIV strain. Despite the existence of different classifiers, it is still impossible to fully explain resistance observed in many newly emerging strains. A deeper understanding of the mechanisms underlying the drug resistance phenomenon is necessary.

To this end, we constructed a model of genotype-resistance relationship for the NRTI class of anti-HIV drugs. We wanted to incorporate existing *a priori* biological and biochemical knowledge into our model. The constructed model should be based on easy-to-understand formalism such as, for instance, rules that can be interpreted by a molecular biologist. Our long-term goal is a construction of classifiers useful in every-day clinical practice and possibly in the design of new drugs.

We assume that the reader is acquainted with the basic concepts of rough set theory [12]. In the first part of the article, we discuss selected aspects of HIV biology that are important for understanding our work. In the following section, we introduce the datasets used for construction of the models. After introducing the methods used in this work, we present the obtained results. This is followed by a discussion, where we make conclusions and compare our results with related work. The very last section of this article contains references to all the cited works.

## 2 HIV Biology and Lifecycle

Every single HIV particle contains two copies of viral RNA attached to the viral proteins. During the course of infection, this RNA has to be transformed into viral DNA that can be easily incorporated into the genetic material of the host cell. The process of transforming RNA to DNA is mediated by the viral RT enzyme. After incorporating its DNA into the host genome, the virus reprograms the cell to produce viral proteins (including RT) and RNA. These are assembled into new HIV particles that are subsequently released from the cell and can infect a new host [8].

As a result of the process called replication, approximately $10^9$ new viral particles are produced in an infected individual every day [14, 6]. Like other retroviral reverse transcriptases, the HIV-1 enzyme does not correct errors by exonucleolytic proofreading. When compared to other transcriptases, RT is exceptionally inaccurate: on average 1 error occurs per every 1700 nucleotides incorporated while reverse transcriptase from murine leukemia virus commits 1 error per 30,000 nucleotides [16]. This leads to the constant emergence of new mutations. Some of these mutations lead to impairment of the viral replication and are fatal to the virus, whereas the other promote the emergence of drug resistant strains. These drug resistant strains can appear even in untreated individuals and, under selection pressure of the antiviral drugs, they are selected as the dominant population [14].

Currently, two methods of testing RT resistance to anti-viral drugs are in use. One of them, phenotyping, is based on the direct measurement of the enzyme activity in presence of the drug in question. While accurate, it is a relatively slow and costly procedure [7].

The other one, genotyping, is based on the analysis of the RT gene sequence. In this method a sampled HIV genome is amplified by Polymerase Chain Reaction (PCR) and its sequence is determined either by direct sequencing or by specific hybridization with oligonucleotides. The sequences are subsequently translated into amino acids and analyzed to predict the drug resistance value [7].

Some attempts have been made to predict HIV drug-resistance using genotyping results, i.e. on the basis of the RT sequence. Draghici and Potter [5] developed a classifier based on neural networks, Beerenwinkel et al. [2, 3] based their classifier on decision trees. In contrast to our method, these aproaches do not attempt to incorporate available biochemical knowledge into the underlying models. Although accurate, they work like a 'black box' that does not give any deeper insight into the nature of the resistance phenomenon.

## 3 Data Material and Methods

Here we present a method for constructing classifiers that predict HIV-1 drug resistance from the RT amino acid sequence. We follow the usual [22] steps where the data is preprocessed and discretized. Every amino acid in the dataset is than described with selected (in our case 5) biochemical properties. Clearly the problem is ill-defined (782 sequences, 2800 attributes each) and direct use of all the attributes will result in poor-quality classifiers. The classifiers are based on features selected by human experts. We construct rough set-based classifiers and evaluate their quality in 10-fold cross validation and compute AUC (Area Under Receiver Operated Curve) values. Discretization step is done inside the cross validation loop. Next, we train the classifiers on randomized data and apply the t-Student test in order to assess the probability of obtaining these or better results by chance. Normality of the AUC means distributions obtained in the randomization test is confirmed by the Shapiro-Wilk test.

We used 782 aligned HIV-1 RT amino acid sequences from the publicly available Stanford HIV Resistance Database (http://hivdb.stanford.edu, rel. 38). Following Rhee et al. [15], we used only the PhenoSense assay-derived data. All the sequences were annotated with the resistance fold value (resistance value related to the one of the wild-type). An example of the database entry is given in Table 1.

For each single drug we created a separate training set. Sequences with missing resistance fold annotations were excluded from construction of the training sets.

**Table 1.** Sample of the database

| Sequence | Pos1 | Pos2 | Pos3 | Pos... | Pos559 | Pos560 | Resistance fold |
|---|---|---|---|---|---|---|---|
| 1 | P | I | R | ... | V | V | 3.8 |
| 2 | L | Q | G | ... | M | L | 2.9 |
| 3 | P | I | S | ... | V | I | 38 |
| ... | ... | ... | ... | ... | ... | ... | ... |
| 781 | P | R | S | ... | Q | L | 5.6 |
| 782 | L | I | T | ... | V | A | 0.8 |

**Table 2.** Cut-off values for the particular drugs

| Drug | Cut-off value |
|---|---|
| Lamivudine | 8.5 |
| Zidovudine | 8.5 |
| Abacavir | 2.5 |
| Tenofovir | 2.5 |
| Stavudine | 2.5 |
| Didanosine | 2.5 |

**Table 3.** Summary data on the susceptible and resistant sequences

| Drug | Susceptible | Resistant | Total |
|---|---|---|---|
| Lamivudine | 147 | 209 | 356 |
| Zidovudine | 235 | 118 | 353 |
| Abacavir | 128 | 226 | 354 |
| Tenofovir | 78 | 18 | 96 |
| Stavudine | 290 | 66 | 356 |
| Didanosine | 322 | 32 | 354 |

Discretization of the decision attribute was based on a set of cutoff-values that are well-established among clinicians [7] (Table 2).

Sequences where resistance value was greater than cut-off were labelled "resistant". All the remaining ones were considered to be "susceptible". The exact numbers of sequences in each training set are given in Table 3.

Subsequently, each amino acid in the training set was described with the appropriate 5-tuple: $[D_1, D_2, D_3, D_4, D_5]$ representing its important biochemical properties. The descriptors were selected following Rudnicki and Komorowski [17] and are presented in Table 4.

We assigned special descriptors to the missing values and insertions. Every sequence was compared to the wild-type virus in the following manner:

```
dataset = Load(sequences from the database)
wt_sequence = Load(consensus B strain wild type sequence)
described_wt_sequence = Describe_sequence(wt_sequence)
foreach(sequence in dataset){
  described_sequence = Describe_sequence(sequence)
  compared = ""
  foreach(position in described_sequence){
    compared[] = described_wt_sequence[position] -
                 described_sequence[position]
  }
final_dataset[] = compared
}
return final_dataset
```

**Table 4.** The set of descriptors

| Descriptor | Name |
|---|---|
| $D_1$ | normalized frequency of alpha-helix |
| $D_2$ | average reduced distance for side chain |
| $D_3$ | normalized composition of membrane proteins |
| $D_4$ | transfer free energy from vap to chx |
| $D_5$ | normalized frequency of extended structure |

After this step, the dataset contained values relative to the so-called Consensus B sequence. Consensus B sequence is used as a reference in drug-resistance testing [7] and is considered to be the sequence of the wild-type virus.

Following an international panel of experts (www.hivfrenchresistance.org, Table of Rules 2006), we took all the 19 positions (95 attributes) that are known to contribute to the resistance to NRTI drugs.

Such a preprocessed data were discretized using Equal Frequency Binning algorithm as implemented in the ROSETTA [9] system:

$$]-\infty, a_1] \rightarrow A_1, ]a_1, a_2] \rightarrow A_2, ..., ]a_{g-1}, \infty[\rightarrow A_g,$$

where there are $g$ classes.

To construct the model, we computed reducts using Genetic Algorithm as implemented in the ROSETTA system. Subsequently we used RuleGroupGeneralizer algorithm as described by Makosa [11].

In parallel, we constructed decision trees-based classifiers using J48 algorithm, k-nearest neighbor clustering-based (k-NN) classifiers using IBk algorithm and multi-layer perceptron-based (MLP) classifiers using their WEKA [22] implementations.

In order to assess the quality of the classifiers, we applied 10-fold cross validation and computed AUC values. In case of each rough set-based classifier, we performed an additional randomization test by generating 1000 training sets with randomly rearranged decision attributes and, for each such a set, performing 10-fold cross-validation.

## 4 Results

The obtained results are presented in Table 5.

**Table 5.** Results of the classification (after rule generalization). Bold type shows the highest AUC value. For all the rough set classifiers: $p < 0.005$ and $W > 0.97$.

| Drug | Decision trees | | k-NN | | MLP | | Rough set | |
|---|---|---|---|---|---|---|---|---|
| | Accuracy | AUC | Accuracy | AUC | Accuracy | AUC | Accuracy | AUC |
| Lamivudine | 0.92 | 0.95 | 0.82 | 0.80 | 0.57 | 0.62 | 0.95 | **0.96** |
| Zidovudine | 0.84 | 0.85 | 0.78 | 0.74 | 0.75 | 0.80 | 0.86 | **0.88** |
| Abacavir | 0.87 | 0.85 | 0.80 | 0.77 | 0.73 | 0.75 | 0.84 | **0.88** |
| Tenofovir | 0.82 | **0.83** | 0.84 | 0.66 | 0.80 | 0.63 | 0.80 | 0.74 |
| Stavudine | 0.91 | 0.86 | 0.86 | 0.78 | 0.81 | 0.78 | 0.90 | **0.93** |
| Didanosine | 0.92 | 0.67 | 0.94 | 0.77 | 0.92 | 0.70 | 0.93 | **0.80** |

Our classifiers were based on a number of IF-THEN rules. Below we present examples of the rules used by HIV resistance to Lamivudine classifier. Before the application of the rule generalization with the RuleGroupGeneralizer algorithm, the classifier consisted of 629 rules and AUC value was 0.98. After rule generalization (alpha = 2.0, coverage = 4.0), the classifier consisted of 144 rules and AUC value decreased to 0.96 (ref. Table 5). Below we present two examples of the generalized rules:

1. `P41D5([0.19000, 44.19000)) AND P62D5([44.14500, *)) =>`
   `Fold(resistant)`
   ```
   LHS Support = 6 objects
   RHS Support = 6 objects
   Accuracy = 1
   LHS Coverage = 0.018518
   RHS Coverage = 0.032432
   ```

2. `P41D3([0.23500, 44.23500)) AND P44D1([*, 0.27000)) AND`
   `P44D5(0.00000) AND P62D5([*, 0.14500)) => Fold(susceptible)`
   ```
   LHS Support = 2 objects
   RHS Support = 2 objects
   Accuracy = 1.0
   LHS Coverage = 0.0061728
   RHS Coverage = 0.0143885
   ```

where P41D5 means 5-th property at the position 41. LHS and RHS stand for Left and Right Hand Side respectively.

## 5 Discussion

Our goal was to construct an easy-to-interpret model of the genotype-resistance relationship. We also wanted to incorporate existing biochemical knowledge into the model. The classifier constructed using rough-set theory fulfills all the initial requirements for the model: it allows for incorporation of the existing knowledge and it is constructed on a number of explicit, legible rules for classification. To our knowledge, all other published work gives classifiers and only discusses their quality. In contrast, our work leads to a better understanding of the biochemichal properties of RT that control drug resistance.

Since, NRTIs were the first anti-HIV drugs, this group is best studied and there is a number of data sets available describing resistance to NRTIs phenomenon.

We constructed accurate classifiers for predicting HIV-1 resistance to NRTI drugs from RT sequence and incorporated existing biological knowledge (biochemical properties of amino acids) into the model. This approach may give a new insight into the resistance mechanisms. Classifiers are built from minimal sets of rules. Each rule defines a resistance/susceptibility pattern. Explicit rules underlying each classifier are easy to read and to interpret. They can reveal some unknown mechanisms of the resistance. Cases where classifier cannot apply appropriate rules are classified to the 'unknown' category. These cases can be subject of further analysis in the molecular biology facility.

Rough sets-based classifiers performed at the level comparable with their decision trees-based counterparts. Although the decision trees can be easily interpreted by humans, the algorithms used to construct them do not guarantee minimal models.

To construct our descriptors we selected 5 amino acid properties. Each pair of the descriptors shows correlation coefficient $r \leq 0.2$. Therefore they span almost orthogonal coordinates frame in a 5-dimensional space. The triad of biologically important properties: polarity, hydrophobicity and size is considered to be the good representative for amino acid and is widely used among biological society [10]. The descriptors used in our work are related to the members of this important triad: 1) "normalized frequency of alpha-helix" reflects the local propensity to form alpha-helix, 2) "average reduced distance for side chain" is a good indicator of the size of amino acid and is highly ($r = 0.8$) correlated with the polarity of amino acid, 3) "normalized composition of membrane proteins" can be a good indicator of the hydrophobicity, 4) "transfer free energy from vap (vapour phase) to chx (cyclohexane)" describes electrostatic properties of the amino acid, 5) "normalized frequency of extended structure" correlates with the local propensity to form beta-sheet. Rule generalization, as exemplified on resistance-to-Lamivudine classifier, reduces number of rules, at the same time only slightly decreasing predictive power of the model. Generalized rules are legible and can be examined by the molecular biologists or clinicians. Analysis of the two example rules presented in the Results section can shed light on how our model can improve the understanding of the resistance phenomenon.

The first rule and its interpretation is presented below:

P41D5([0.19000, 44.19000)) AND P62D5([44.14500, *)) $\Rightarrow$ Fold(resistant)

*if change in normalized composition of extended structure at position 41 takes value between* [0.19000, 44.1900) *and change in normalized frequency of extended structure at position 62 takes value between* [44.14500, $\infty$) *then the virus is resistant to Lamivudine.*

Since normalized frequency of extended structure corresponds to the secondary structure of the protein, we can suspect that certain mutations at positions 41 and 62 may induce structural changes that lead to drug resistance development. By examining which amino acids fulfill the constraints described by the rule, we can determine the space of possible mutations leading to the resistance to the drug in question.

The second presented rule:

P41D3([0.23500, 44.23500)) AND P44D1([*, 0.27000)) AND
P44D5(0.00000) AND P62D5([*, 0.14500)) $\Rightarrow$ Fold(susceptible),

interestingly enough says that when mutations occur at positions 41, 44 and 62, propensity to form beta-strand at position 44 has to remain unchanged. "Normalized frequency of membrane proteins" at position 41 takes value from interval [0.23500, 44.23500]. This descriptor reflects hydrophobic properties of amino acid and position 41 is located in close vicinity to the newly synthesized viral DNA. We can suspect that a change in hydrophobicity at this position will influence protein-DNA interaction that may subsequently lead to the development of drug resistance.

Unfortunately, since various different datasets were used by the authors applying other AI methods to the HIV-RT drug resistance problem, it was impossible to make straight comparisons of these techniques with our method. However, we believe that AUC values are a reliable measurement of the performance of our classifiers and, to a great extent, are independent on the datasets used.

Bonet et al. [4] applied an interesting approach to construct classifiers capable of predicting HIV protease resistance to anti-viral drugs. They describe every amino acid in the protease sequence using amino acid contact energy since, to some extent, it corresponds to the protein 3D structure. In particular it determines folding/unfolding of the protein. Our descriptors (except normalized composition of membrane proteins) also reflect some properties of the 3D structure of the protein. While some descriptors may be more informative for chemists attempting to improve existing anti-viral drugs, the other alternative set will shed more light on the molecular basis of the resistance mechanism.

Subsequently, they perform features extraction and train classifiers using newly extracted features. We use an alternative approach, features selection, in order to find which particular biochemical properties contribute to the resistance. Since every amino acid was described with five different properties, the chance of loosing biologically relevant information in the process of features selection is in our case small.

Our future work will be focused on analyzing the rules and trying to reveal general patterns in RT sequences of the drug resistant HIV strains.

## Acknowledgements

We would like to thank Dr. Arnaud Le Rouzic and Aleksejs Kontijevskis from The Linnaeus Centre for Bioinformatics for fruitful discussions and critical reading of the manuscript.

## References

1. Barre-Sinoussi, F., Chermann, J., Nugeyre, F.R.M., Chamaret, S., Gruest, J., Dauguet, C., Axler-Blin, C., Vezinet-run, F., Rouzioux, C., Rozenbaum, W., Montagnier, L.: Isolation of a T-lymphotropic retrovirus from a patient at risk for acquired immune deficiency syndrome (AIDS). Science 220, 868–871 (1983)
2. Beerenwinkel, N., Daumer, M., Oette, M., Korn, K., Hoffmann, D., Kaiser, R., Lengauer, T., Selbig, J.: Walter Geno2pheno: Estimating phenotypic drug resistance from HIV-1 genotypes. Nucl. Acids. Res. 31, 3850–3855 (2003)
3. Beerenwinkel, N., Schmidt, B., Walter, H., Kaiser, R., Lengauer, T., Hoffmann, D., Korn, K., Selbig, J.: Diversity and complexity of HIV-1 drug resistance: a bioinformatics approach to predicting phenotype from genotype. Proc. Natl. Acad. Sci. 99, 8271–8276 (2002)
4. Bonet, I., Saeys, Y., Grau-Ábalo, R., García, M., Sanchez, R., Van de Peer, Y.: Feature Extraction Using Clustering of Protein. In: Progress in Pattern Recognition, Image Analysis and Applications, Springer, Heidelberg (2006)
5. Draghici, S., Potter, R.B.: Predicting HIV drug resistance with neural networks. Bioinformatics 19, 98–107 (2003)
6. Drake, J.W.: Rates of spontaneous mutation among RNA viruses. Proc. Natl. Acad. Sci. 90, 4171–4175 (1990)
7. Hoffman, Ch., Kemps, B.S.: HIV Medicine 2005. Flying Publisher, Paris, Cagliari, Wuppertal (2005)
8. Jonckheere, H., Anne, J., De Clercq, E.: The HIV-1 reverse transcription (RT) process as target for RT inhibitors. Med. Res. Rev. 20, 129–154 (2000)
9. Komorowski, J., Øhrn, A., Skowron, A.: The ROSETTA Rough Set Software System. In: Klsgen, W., Zytkow, J. (eds.) Handbook of Data Mining and Knowledge Discovery, Oxford Univ. Press, Oxford (2002)
10. Lesk, A.: Introduction to bioinformatics. Oxford Univ. Press, Oxford (2002)
11. Makosa, E.: Rule tuning. MA thesis. The Linnaeus Centre for Bioinformatics, Uppsala University (2005)
12. Pawlak, Z.: Rough sets: Theoretical aspects of reasoning about data. Kluwer Acad. Publ., Dordrecht, Boston (1992)
13. Perelson, A.S., Neumann, A.U., Markowitz, M.: HIV-1 dynamics in vivo: Virion clearance rate, infected cell life-span, and viral generation time. Science 271, 1582–1586 (1996)

14. Rezende, L.F., Prasad, V.R.: Nucleoside-analog resistance mutations in HIV-1 reverse transcriptase and their influence on polymerase fidelity and viral mutation rates. Int. J. Biochem. Cell Biol. 36, 1716–1734 (2004)
15. Rhee, S.Y., Taylor, J., Wadhera, G., Ben-Hur, A., Brutlag, D.L., Shafer, R.W.: Genotypic predictors of human immunodeficiency virus type 1 drug resistance. Proc. Natl. Acad. Sci. 103, 17355–17360 (2006)
16. Roberts, J.D., Bebenek, K., Kunkel, T.A.: The accuracy of reverse transcriptase from HIV-1. Science 4882, 1171–1173 (1988)
17. Rudnicki, W.R., Komorowski, J.: Feature Synthesis and Extraction for the Construction of Generalized Properties of Amino Acids. In: Tsumoto, S., Słowiński, R., Komorowski, J. (eds.) Rough Sets and Current Trends in Computing, Springer, Heidelberg (2004)
18. Seelamgari, A., Maddukuri, A., Berro, R., de la Fuente, C., Kehn, K., Deng, L., Dadgar, S., Bottazzi, M.E., Ghedin, E., Pumfery, A., Kashanchi, F.: Role of viral regulatory and accessory proteins in HIV-1 replication. Front Biosci. 9, 2388–2413 (2004)
19. Shaefer, R.W., Shapiro, J.M.: Drug resistance and antiretroviral drug development. J. Antimicrob. Chemother. 55, 817–820 (2005)
20. Sobieszczyk, M.E., Jones, J., Wilkin, T., Hammer, S.M.: Advances in antiretroviral therapy. Top HIV Med. 14, 36–62 (2006)
21. Sobieszczyk, M.E., Talley, A.K., Wilkin, T., Hammer, S.M.: Advances in antiretroviral therapy. Top HIV Med. 13, 24–44 (2005)
22. Witten, I.H., Frank, E.: Data Mining: Practical machine learning tools and techniques. Morgan Kaufmann, San Francisco (2005)

# Part III: Fuzzy and Rough Sets in Decision-Making

# Rough Set Approach to Information Systems with Interval Decision Values in Evaluation Problems

Kazutomi Sugihara[1] and Hideo Tanaka[2]

[1] Department of Management Information Science, Fukui University of Technology, Japan
sugihara@fukui-ut.ac.jp

[2] Department of Kansei Design, Faculty of Psychological Science, Hiroshima International University, Japan
h-tanaka@he.hirokoku-u.ac.jp

**Summary.** In this chapter, a new rough set approach to decision making problems is proposed. It is assumed that the evaluations given by a decision maker are interval values. That is, we deal with the information system containing ambiguous decision expressed as interval values. By the approximations of the lower and upper bounds with respect to decision values, the approximations with interval decision values are illustrated in this chapter. The concept of the proposed approach resembles the one of Interval Regression Analysis. Furthermore, we discuss the unnecessary divisions between the decision values based on these bounds. The aim is to simplify IF-Then rules extracted from the information system. The method for removing the divisions is introduced using a numerical example.

## 1 Introduction

Rough set theory proposed by Pawlak [1] can deal with uncertain and vague data expressed by various descriptions. In rough set theory, it is possible that we can extract some relations from the information system including these data [1][2]. These relations are obtained as IF-Then rules. Rough set approach is applied to many decision problems [3][4]. In the conventional rough set approach to decision making problems [4], the evaluation by a decision maker is assumed to be certain and clear, that is, the value of the decision attribute is given as only crisp values. However, in fact, there exist some cases in which the decisions are ambiguous and imprecise. Then, we consider the information system with ambiguous decisions consisting of more than one decision attribute value and it is assumed that a decision maker may give more than one decision attribute value as intervals.

Firstly, we show the method for obtaining the lower and upper approximations of the given decisions by means of the equivalence classes based on indiscernibility relations. Through the proposed approximations, IF-Then rules can be

R. Bello et al. (Eds.): Granular Computing, STUDFUZZ 224, pp. 261–267, 2008.
springerlink.com 

obtained by reflecting the lower and upper approximations of decisions. Next, unnecessary gaps between neighboring decision values can be deleted from the generated IF-Then rules, namely we can find the redundant divisions and remove them. Finally, using a simple numerical example, we verify the feasibility of our approach.

## 2 Conventional Rough Sets

Let $S = \{U, C \cup \{d\}, V, f\}$ be an *information system* (*decision system*), where $U$ is a finite set of *objects*, $C$ is a set of *conditional attributes*, $\{d\}$ is a *decision attribute* ($d \notin C$), $V_q$ is the domain of the attribute $q \in C \cup \{d\}$, and $f : U \times (C \cup \{d\}) \to V$ is a total function such that $f(x,q) \in V_q$ for each $q \in Q$, $x \in U$, called $information\ function$.

For any subset $P$ of $C$, *indiscernibility relations* $R_P$ are defined as follows:

$$R_P = \{(x,y) \in U \times U | f(x,q) = f(y,q),\ \forall q \in P\} \tag{1}$$

$R_P$ is an equivalence relation (reflexive, symmetric and transitive). $(x,y) \in R_P$ means that $x$ is indiscernible from $y$. The following *equivalence class* $R_P(x)$ can be generated from the indiscerniblity relation $R_P$.

$$R_P(x) = \{y \in U | (x,y) \in R_P,\ \forall q \in P\} \tag{2}$$

Given $X \subseteq U$ and $P \subseteq C$ in an information system, the lower and upper approximations of $X$ are computed as shown below

$$\underline{P}(X) = \{x \in U | R_P(x) \subseteq X\} \tag{3}$$

$$\overline{P}(X) = \bigcup_{x \in X} R_P(x). \tag{4}$$

These approximation sets satisfy the inclusion relation $\underline{P}(X) \subseteq X \subseteq \overline{P}(X)$. The boundary of $X$ is denoted and defined as $Bn_P(X)$. If $Bn_P(X) = \phi$, then the set $X$ is exact with respect to $P$, If $Bn_P(X) \neq \phi$, then the set $X$ is rough with respect to $P$.

## 3 Information System with Ambiguous Decisions in Evaluation Problems

In the conventional rough sets approach, it is assumed that the given values with respect to a decision attribute are univocally determined. That is, each object $x$ has only one decision value in the set of decision values. However, there exist some cases in which this assumption is not appropriate to real decision making problems.

In this chapter, the situations where decision values $d(x)$ are given to each object $x$ as interval values are considered. Let $Cl_n (n = 1, \cdots, N)$, be the $n$-th class with respect to a decision attribute. It is supposed that for all $s, t$, such that $t > s$, each element of $Cl_t$ is preferred to each element of $Cl_s$. The interval decision classes (values) $Cl_{[s,t]}$ are defined as

$$Cl_{[s,t]} = \bigcup_{s \le r \le t} Cl_r. \tag{5}$$

We assume that the decision value of each $x \in U$ belongs to one or more classes, that is, $d(x) = Cl_{[s,t]}$. By means of $Cl_{[s,t]}$, a decision maker expresses ambiguous judgments on each object $x$.

Based on the above equations, the decisions $d(x)$ with respect to the attribute set $P$ can be obtained by the lower and upper approximations as follows.

**Definition 1.** The lower bound $\underline{P}\{d(x)\}$ and the upper bound $\overline{P}\{d(x)\}$ of $d(x)$ are defined as

$$\underline{P}\{d(x)\} = \bigcap_{y \in R_P(x)} d(y) \tag{6}$$

$$\overline{P}\{d(x)\} = \bigcup_{\{d(y) | y \in R_P(x)\}} d(y) \tag{7}$$

$\underline{P}\{d(x)\}$ means that $x$ certainly belongs to the common classes which are assigned to all the elements of the equivalence class $R_P(x)$. $\overline{P}\{d(x)\}$ means that $x$ may belong to the classes which are assigned to each element of the equivalence classes $R_P(x)$, respectively. It is obvious that the following inclusion relation holds $\underline{P}\{d(x)\} \subseteq d(x) \subseteq \overline{P}\{d(x)\}$. Expressions (6) and (7) are based on the concept of *greatest lower* and *least upper*, respectively. This concept is similar to the one of Interval Regression Analysis proposed by Tanaka and Guo [5].

## 4 Numerical Example

Let us consider an example of *an evaluation problem in a sales department.* The director rates the staffs into four ordered classes: Excellent, Good, Fair and Bad. To clarify the evaluation rules, the director evaluated 10 staffs as shown in Table 1 where there are two attributes that are the ability in communication and ability in promotion, respectively, and there is an overall evaluation as the decision attribute. In this case, it is assumed that the director was allowed to give them ambiguous evaluations. For example, [Fair, Excel.] means that the decision value $d(x)$ has the evaluation “Fair” or “Good” or “Excellent”.

The indiscernibility relation with respect to $P = \{$Communication, Promotion$\}$ is

$$R_P = \{(3,4), (6,7), (8,9), (i,i), (i = 1, \cdots, 10)\}.$$

**Table 1.** An information system

| Staffs | Communication | Promotion | Evaluation |
|---|---|---|---|
| 1 | Bad | Bad | Bad |
| 2 | Bad | Fair | Bad |
| 3 | Fair | Bad | Bad |
| 4 | Fair | Bad | [Bad, Good] |
| 5 | Fair | Fair | [Fair, Good] |
| 6 | Fair | Good | Fair |
| 7 | Fair | Good | [Good, Excel] |
| 8 | Good | Fair | [Fair, Good] |
| 9 | Good | Fair | [Fair, Excel] |
| 10 | Good | Good | Excel |

Where:
$U = \{1, 2, 3, 4, 5, 6, 7, 8, 9, 10\}$
$C = \{$Communication, Promotion$\}$
$V_{\text{Comm.}} = \{$Good, Fair, Bad$\}$
$V_{\text{Prom.}} = \{$Good, Fair, Bad$\}$
$\{d\} = \{$Excellent, Good, Fair, Bad$\}$.

From the indiscernibility relations, the lower bounds of the decisions $d(x)$ for each object $x$ are obtained as follows:

$\underline{P}\{d(1)\} =$ "Bad"
$\underline{P}\{d(2)\} =$ "Bad"
$\underline{P}\{d(3)\} =$ "Bad"
$\underline{P}\{d(4)\} =$ "Bad"
$\underline{P}\{d(5)\} =$ ["Fair", "Good"]
$\underline{P}\{d(6)\} = \phi$
$\underline{P}\{d(7)\} = \phi$
$\underline{P}\{d(8)\} =$ ["Fair", "Good"]
$\underline{P}\{d(9)\} =$ ["Fair", "Good"]
$\underline{P}\{d(10)\} =$ "Excel".

Similarly, the upper bounds of the decisions $d(x)$ for each object $x$ are obtained as follows:

$\overline{P}\{d(1)\} =$ "Bad"
$\overline{P}\{d(2)\} =$ "Bad"
$\overline{P}\{d(3)\} =$ ["Bad", "Good"]
$\overline{P}\{d(4)\} =$ ["Bad", "Good"]
$\overline{P}\{d(5)\} =$ ["Fair", "Good"]
$\overline{P}\{d(6)\} =$ ["Fair", "Excel"]
$\overline{P}\{d(7)\} =$ ["Fair", "Excel"]
$\overline{P}\{d(8)\} =$ ["Fair", "Excel"]
$\overline{P}\{d(9)\} =$ ["Fair", "Excel"]
$\overline{P}\{d(10)\} =$ "Excel".

With respect to each object $x$, the following decision rules are induced from the lower bounds $\underline{P}\{d(x)\}$ of $d(x)$.

- If $f(x, q_{\text{Comm.}}) =$ "Bad" and $f(x, q_{\text{Prom.}}) =$ "Bad", then exactly $d(x) =$ "Bad". (supported by 1)
- If $f(x, q_{\text{Comm.}}) =$ "Bad" and $f(x, q_{\text{Prom.}}) =$ "Fair", then exactly $d(x) =$ "Bad". (supported by 2)
- If $f(x, q_{\text{Comm.}}) =$ "Fair" and $f(x, q_{\text{Prom.}}) =$ "Bad", then exactly $d(x) =$ "Bad". (supported by 3,4)
- If $f(x, q_{\text{Comm.}}) =$ "Fair" and $f(x, q_{\text{Prom.}}) =$ "Fair", then exactly $d(x) =$ ["Fair","Good"]. (supported by 5)
- If $f(x, q_{\text{Comm.}}) =$ "Good" and $f(x, q_{\text{Prom.}}) =$ "Fair", then exactly $d(x) =$ ["Fair","Good"]. (supported by 8,9)
- If $f(x, q_{\text{Comm.}}) =$ "Good" and $f(x, q_{\text{Prom.}}) =$ "Good", then exactly $d(x) =$ "Excel.". (supported by 10)

Similarly, with respect to each object $x$, the following decision rules are induced from the upper bounds $\overline{P}\{d(x)\}$ of $d(x)$.

- If $f(x, q_{\text{Comm.}}) =$ "Bad" and $f(x, q_{\text{Prom.}}) =$ "Bad", then possibly $d(x) =$ "Bad". (supported by 1)
- If $f(x, q_{\text{Comm.}}) =$ "Bad" and $f(x, q_{\text{Prom.}}) =$ "Fair", then possibly $d(x) =$ "Bad". (supported by 2)
- If $f(x, q_{\text{Comm.}}) =$ "Fair" and $f(x, q_{\text{Prom.}}) =$ "Bad", then possibly $d(x) =$ ["Bad","Good"]. (supported by 3,4)
- If $f(x, q_{\text{Comm.}}) =$ "Fair" and $f(x, q_{\text{Prom.}}) =$ "Fair", then possibly $d(x) =$ ["Fair","Good"]. (supported by 5)
- If $f(x, q_{\text{Comm.}}) =$ "Fair" and $f(x, q_{\text{Prom.}}) =$ "Good", then possibly $d(x) =$ ["Fair","Excel."]. (supported by 6,7)
- If $f(x, q_{\text{Comm.}}) =$ "Good" and $f(x, q_{\text{Prom.}}) =$ "Fair", then possibly $d(x) =$ ["Fair","Excel."]. (supported by 8,9)
- If $f(x, q_{\text{Comm.}}) =$ "Good" and $f(x, q_{\text{Prom.}}) =$ "Good", then possibly $d(x) =$ "Excel.". (supported by 10)

## 5 Removal of Unnecessary Divisions Between Decision Values

Now we consider the divisions between decision values. The discussion on the removal of redundant divisions stems from Definition 1. In the previous numerical example, ambiguous decisions are approximated by the lower bounds and the upper bounds. We remark that in the induced rules, there exist no decision rules coming to the conclusion that the object $x$ belongs to the crisp decision value "Fair" / "Good", the interval decision value "Fair or worse" / "Good or better". In this case, the division between "Fair" and "Good" with respect to the decision attribute is considered as unnecessary, because it makes no difference to a decision maker whether or not there exists the division. Therefore this fact means that the division between "Fair" and "Good" may be removed from the divisions between decision values. This situation comes when a decision maker sets redundant classes to strictly evaluate objects. The proposed approach is helpful in that the given redundant classes are reduced to the minimal ones. Based on the proposed approximations, the definition of removing the unnecessary divisions in decision values is illustrated as follows:

**Definition 2.** The division between $Cl_r$ and $Cl_{r+1}$ can be removed if the obtained approximations of each $x$ with the division correspond to the ones without it.

## 6 Concluding Remarks

In this chapter, a new rough set approach for decision making problems is proposed. In decision making problems, there are many cases where the decision maker's judgments are uncertain. In our method, the information system with interval decisions given by a decision maker is dealt with. By defining the lower bounds and the upper bounds of the given decisions, the method for approximating the interval decision is proposed in Section 3. The introduced method, which steps aside from the conventional rough set approach, is illustrated under the assumption that the decision values are comparable. From the definition of these bounds, we discuss the removal of unnecessary divisions between adjacent decision values. The discussion on *removal of unnecessary divisions* is not quite taken as an object of study. However, we are confident that it is natural for us to consider the discussion. We are interested in decision making problems based on rough sets by various binary relations. The proposed approach is one attempt to apply the rough sets to decision making problems.

## Acknowledgment

This research was supported by the Grant-In Aid for Young Scientist(B) No.18700278.

## References

1. Pawlak, Z.: Rough Classification. International Journal of Man-Machine Studies 20, 469–483 (1984)
2. Nguyen, H.S., Slezak, D.: Approximate reducts and association rules correspondence and complexity results. In: Zhong, N., Skowron, A., Ohsuga, S. (eds.) New Directions in Rough Sets, Data Mining, and Granular-Soft Computing. LNCS (LNAI), vol. 1711, pp. 137–145. Springer, Heidelberg (1999)

3. Sugihara, K., Ishii, H., Tanaka, H.: On conjoint analysis by rough approximations based on dominance relations. International Journal of Intelligent Systems 19, 671–679 (2004)
4. Greco, S., Matarazzo, B., Slowinski, R.: Rough sets theory for multicriteria decision analysis. European Journal of Operational Research 129, 1–47 (2001)
5. Tanaka, H., Guo, P.: Possibilistic Data Analysis for Operations Research. Physica-Verlag, Heidelberg (1999)

# Fuzzy Rule-Based Direction-Oriented Resampling Algorithm in High Definition Display

Gwanggil Jeon[1], Rafael Falcón[2], and Jechang Jeon[1]

[1] Dept. of Electronics and Computer Engineering, Hanyang University, Korea
windcap315@ece.hanyang.ac.kr

[2] Department of Computer Science, Central University of Las Villas, Cuba
rfalcon@uclv.edu.cu

**Summary.** This chapter is concerned with the introduction of a new resampling algorithm for high resolution display dependent upon upscaling methods. Our proposed algorithm performs dynamic image segmentation into regions with eight possible edge directions. The edge direction is determined by means of a fuzzy rule-based edge detector. The region classifier employs fuzzy rules during the edge detection process. The superior performance in terms of PSNR over the conventional methods is clearly demonstrated.

**Keywords:** upscaling, resampling, directional interpolation, fuzzy rules.

## 1 Introduction

Image resampling is utilized for image reconstruction [1], supersampling [2], improving appearance of image display for human viewers [3], zooming [4], reducing artifacts [5], etc. Because there is a large amount of different digital media formats, various applications require resampling approaches for resizing images. Therefore, it's worth studying how to evaluate accurate values for missing pixels in the large image using information from pixels in the small image. In these days, flat panel display (FPD) [6] such as thin film transistor (TFT), liquid crystal display (LCD), and plasma display panel (PDP) become more common display equipments than cathode-ray tube (CRT) in large display market. Although FPD has higher panel resolution, it is based on progressive scanning format, the interlaced signal cannot be displayed on FPD [7]. Thus, FPD requires more adaptive image signal processing for high resolution display and progressive signal. Many conventional image upscaling algorithms often yield blurred results. The quality of the upscaling, however, can be improved by applying some time-consuming algorithms.

Most of the resampling methods proposed in the literature can be classified into two categories: conventional and adaptive methods. In conventional methods, resampling functions are applied indiscriminately to the whole image. As a result, the modified image generally suffers from edge blurring, aliasing and other artifacts. On the other hand, adaptive methods are designed to avoid these

R. Bello et al. (Eds.): Granular Computing, STUDFUZZ 224, pp. 269–285, 2008.
springerlink.com 

problems by analyzing the local structure of the source image and applying different interpolation functions with dissimilar support areas. Our proposed method qualifies among the adaptive ones.

Recently, many different approaches that adopt fuzzy reasoning have been proposed within the engineering domain. Fuzzy reasoning methods have proved effective in image processing (e.g., filtering, interpolation, edge detection, and morphology), having countless practical applications [8] [9] [10]. In [11], a line interpolation method using an intra- and inter-field edge direction detector was proposed to obtain the correct edge information. This detector works by identifying small pixel variations in ten orientations and by using rules to infer the interpolation filter.

Nearest neighbor [12] and Bilinear interpolation [13] are two other typically consumed methods. They need less computational complexity in comparison to peers. However, the nearest neighbor method produces blocky edges, whereas the bilinear interpolation method yields blurry images. The whole image is applied the same interpolation procedure in both cases. In our approach, we classify the region to be filled into eight directions. The computationally intensive and edge-preserving technique is performed in $-30\circ$, $30\circ$, $-45\circ$, $45\circ$, $-60\circ$, and $60\circ$ regions, whereas more straightforward methods are employed in the remaining ($0\circ$ and $90\circ$) ones. The decision on a suitable resampling technique is made by analyzing the direction of the aforementioned region.

In this chapter, we present an adaptive resampling algorithm for images ranging from $176 \times 72$ to $352 \times 288$ in size. The algorithm's main idea is the image segmentation into eight kinds of regions and the adaptive interpolation of the missing regions as the second step. Its performance was properly compared to other methods reported in the literature.

The remainder of the chapter is structured as follows: in Sect. 2, the details of the region classifier will be described; the extended cubic curve fitting method is presented in Sect. 3 whereas we elaborate on the resamply strategy in Sect. 4. Empirical results and conclusions are finally outlined in Sects 5 and 6.

## 2 Region Classifier Based on Fuzzified Edge Detector

Fig. 1 shows the proposed direction-oriented resampling (DOR) algorithm, where $i$ and $j$ respectively represent the vertical and horizontal line number of the pixel. The pixel's intensity at location $(i, j)$ is denoted by $x(i, j)$. Pixels represented by ? are existing pixels having real values. Pixels $A, B, C, D, E, F$ and $G$ are pixels to be interpolated using the existing ones. The key to the success of DOR is an accurate estimation of the edge direction.

The edge pattern appears not only in the horizontal direction but also in the vertical and diagonal directions. Besides, the video sequences will be magnified by 2 in the horizontal direction and by 4 in the vertical direction. We assume that eight edge orientations can be selected in order to fill the region up. Differences

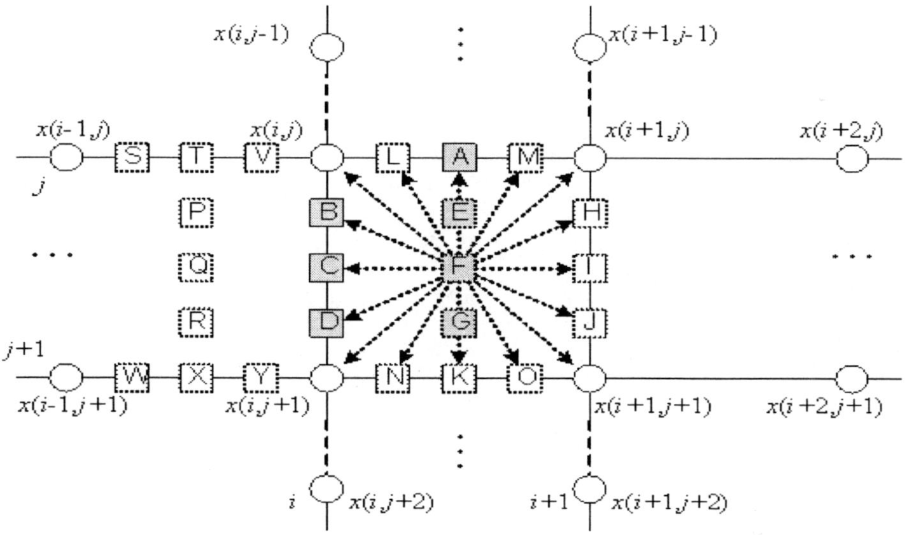


**Fig. 1.** Pixel window and illustration of the DOR algorithm

between two pixels through the pixel $x(i, j)$ according to the defined direction are computed as shown below:

$$\begin{aligned}
\Delta_{0^\circ}(i,j) &= |x(i,j) - x(i+1,j)| \\
\Delta_{90^\circ}(i,j) &= |x(i,j) - x(i,j+1)| \\
\Delta_{30^\circ}(i,j) &= |x(i,j+1) - x(i+2,j)| \\
\Delta_{-30^\circ}(i,j) &= |x(i,j) - x(i+2,j+1)| \\
\Delta_{45^\circ}(i,j) &= |x(i,j+1) - x(i+1,j)| \\
\Delta_{-45^\circ}(i,j) &= |x(i,j) - x(i+1,j+1)| \\
\Delta_{60^\circ}(i,j) &= |x(i,j+1) - x(i+1,j-1)| \\
\Delta_{-60^\circ}(i,j) &= |x(i,j) - x(i+1,j+2)|
\end{aligned} \tag{1}$$

where $\Delta_\Theta(i, j)$ is the pixel variation in the $\Theta$-th direction at the pixel $x(i, j)$. These inputs are turned into fuzzy variables represented by their associated fuzzy sets which are modeled as trapezoidal membership functions [11]. In order to compute the value that expresses the size of the fuzzy derivative in a certain direction, we define the use of the fuzzy set *SMALL* for each direction as shown in Fig. 2. Each one is defined differently by two parameters: $a_\Theta$ refers to the threshold value required for achieving maximum membership and $b_\Theta$ defines the upper bound of the function (that is, all differences greater than $b_\Theta$ do not belong to the function).

Notice that the membership value $\mu_\Theta$ increases as long as $\Delta_\Theta$ is less than $b_\Theta$ and gets closer to zero. If it is also smaller than $a_\Theta$, then it gets a membership degree of one. The membership degree $\mu_\Theta$ is obtained by (2).

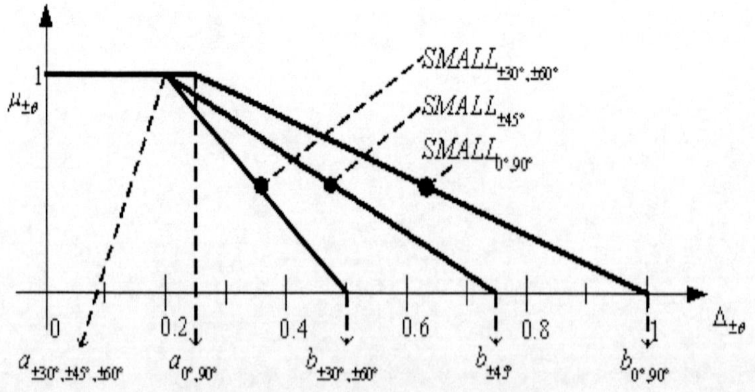


**Fig. 2.** *SMALL* membership functions

$$\mu_\Theta = \begin{cases} 1 & \text{if } \; 0 \le \Delta_\Theta \le a_\Theta \\ 1 - \dfrac{\Delta_\Theta - a_\Theta}{b_\Theta - a_\Theta} & \text{if } \; a_\Theta \le \Delta_\Theta \le b_\Theta \\ 0 & \text{if } \; b_\Theta \le \Delta_\Theta \end{cases} \tag{2}$$

A control rule is described as a conditional statement in which antecedents are the conditions and the consequence is a control decision. The conjunction of the antecedents' membership values provides the truth level of the rule's consequent. The fuzzified input is simultaneously broadcasted to all control rules to be compared with their antecedents. The fuzzy rule base characterizes the control policy needed to infer fuzzy control decisions, i.e., directions for our fuzzy detector. The fuzzy reasoning scheme adopted is the max-min composition. These rules are implemented using the minimum to represent the AND-operator, and the maximum for the OR-operator. All rules having any truth in their premises will fire and contribute to the output. Afterwards, the truth levels of the same consequences are unified using the fuzzy disjunction maximum. The fuzzy rule base used in this chapter is shown in Table 1.

The final process in the computation of the fuzzy filter is defuzzification. This process converts the output fuzzy value into a crisp value. To make the final decision about the edge direction at the pixel $x(i,j)$, our fuzzy detector chooses the direction with the maximum membership value, as described by (3).

$$Direction(i,j) = argmax_\Theta \left(\mu_{dir(i,j)=\Theta}\right) \tag{3}$$

**Table 1.** Rules base for fuzzy edge detection

| Input for predicting value $x(i,j)$ | Fuzzy sets | $Dir(i,j)$ |
|---|---|---|
| $\Delta$ ∘ | $SMALL$ ∘ | 0∘ |
| $\Delta$ ∘ | $SMALL$ ∘ | 90∘ |
| $\Delta_{3}$ ∘ | $SMALL_{3}$ ∘ | 30∘ |
| $\Delta_{-3}$ ∘ | $SMALL_{-3}$ ∘ | −30∘ |
| $\Delta_{45}$∘ | $SMALL_{45}$∘ | 45∘ |
| $\Delta_{-45}$∘ | $SMALL_{-45}$∘ | −45∘ |
| $\Delta_{6}$ ∘ | $SMALL_{6}$ ∘ | 60∘ |
| $\Delta_{-6}$ ∘ | $SMALL_{-6}$ ∘ | −60∘ |

where $Direction(i,j)$ is the edge direction. Each pixel $x(i,j)$ is classified into one of the eight possible regions according to the result of the edge direction.

## 3 Expanded Cubic Curve Fitting (ECCF) Method

Our proposed expanded cubic curve fitting (ECCF) resampling method is portrayed in Fig. 3. The resampling algorithm is based on the Fan's algorithm [14]. ECCF uses four pixels at the horizontal or vertical neighborhood of $x(i,j)$ to obtain better interpolation results. We assume that the luminance transition in the horizontal or vertical direction approximated as a third order function of $i$ (or $j$). The symbol $\diamond$ represents the existing pixels. In Fig. 3, $F(-5), F(-4), F(-3), F(-1), F(0)$ and $F(1)$ are pixels to be interpolated.

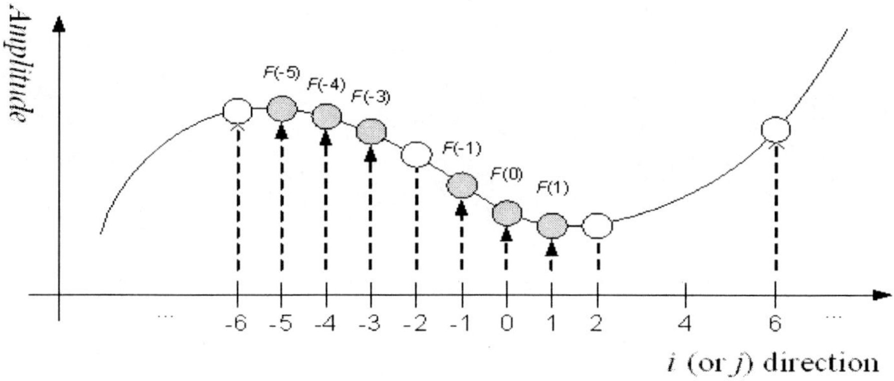


**Fig. 3.** Luminance transition in $i$ (or $j$) direction to obtain the six missing pixels

$$\begin{bmatrix} F(-5) \\ F(-4) \\ F(-3) \\ F(-1) \\ F(0) \\ F(1) \end{bmatrix} = \begin{bmatrix} 1 & j & j^2 & j^3 \\ 1 & j & j^2 & j^3 \\ 1 & j & j^2 & j^3 \\ 1 & j & j^2 & j^3 \\ 1 & j & j^2 & j^3 \\ 1 & j & j^2 & j^3 \end{bmatrix}_{j=\{-5,-4,-3,-1,0,1\}} \begin{bmatrix} \alpha \\ \beta \\ \gamma \\ \delta \end{bmatrix} = \begin{bmatrix} 1 & -5 & 25 & -125 \\ 1 & -4 & 16 & -64 \\ 1 & -3 & 9 & -27 \\ 1 & -1 & 1 & -1 \\ 1 & 0 & 0 & 0 \\ 1 & 1 & 1 & 1 \end{bmatrix} \begin{bmatrix} \alpha \\ \beta \\ \gamma \\ \delta \end{bmatrix} \tag{4}$$

$$= \begin{bmatrix} 1 & -5 & 25 & -125 \\ 1 & -4 & 16 & -64 \\ 1 & -3 & 9 & -27 \\ 1 & -1 & 1 & -1 \\ 1 & 0 & 0 & 0 \\ 1 & 1 & 1 & 1 \end{bmatrix} \begin{bmatrix} 1 & -6 & 36 & -216 \\ 1 & -2 & 4 & -8 \\ 1 & 2 & 4 & 8 \\ 1 & 6 & 36 & 216 \end{bmatrix}^{-1} \begin{bmatrix} F(-6) \\ F(-2) \\ F(2) \\ F(6) \end{bmatrix}$$

$$= \begin{bmatrix} 0.6016 & 0.6016 & -0.2578 & 0.0547 \\ 0.3125 & 0.9375 & -0.3125 & 0.0625 \\ 0.1172 & 1.0547 & -0.2109 & 0.0391 \\ -0.0547 & 0.8203 & 0.2734 & -0.0391 \\ -0.0625 & 0.5625 & 0.5625 & -0.0625 \\ -0.0391 & 0.2734 & 0.8203 & -0.0547 \end{bmatrix} \begin{bmatrix} F(-6) \\ F(-2) \\ F(2) \\ F(6) \end{bmatrix}$$

We regard $F(j) = \alpha + \beta j + \gamma j^2 + \delta j^3$ as a third order function of $j$. We suppose $F(-5), F(-4), F(-3), F(-1), F(0)$ and $F(1)$ are the pixels to be interpolated and $F(-6), F(-2), F(2)$ and $F(6)$ correspond to four sample pixels of the original field, respectively.

With these function values already known, the four equations, $F(-6) = \alpha - 6\beta + 36\gamma - 216\delta$, $F(-2) = \alpha - 2\beta + 4\gamma - 8\delta$, $F(2) = \alpha + 2\beta + 4\gamma + 8\delta$ and $F(6) = \alpha + 6\beta + 36\gamma + 216\delta$ can be obtained by simple substitutions of the value of j. Through the above equations, pixels $F(-5), F(-4), F(-3), F(-1), F(0)$ and $F(1)$ can be written as depicted in (4).

## 4 Resampling Strategy

According to the fuzzy edge detector, any edge direction can be assigned to the region. Later on, the most suitable resampling algorithm is employed adaptively. The region classifier decides whether the pixel is located within $R_1(90\circ)$, $R_2(0\circ)$, $R_3(45\circ)$, $R_4(-45\circ)$, $R_5(30\circ)$, $R_6(-30\circ)$, $R_7(60\circ)$ or $R_8(-60\circ)$ area.

### 4.1 Case 1 - $R_1$ Region: 90○ (Vertical) Direction

If the region is classified into $R_1$, we assume that there is a vertical direction edge through the region. Pixel $A$ is obtained using bilinear interpolation (BI) in the horizontal direction and pixels $B$, $C$ and $D$ are obtained using the ECCF method in the vertical direction. To interpolate pixels $E$, $F$ and $G$, we estimate the pixel

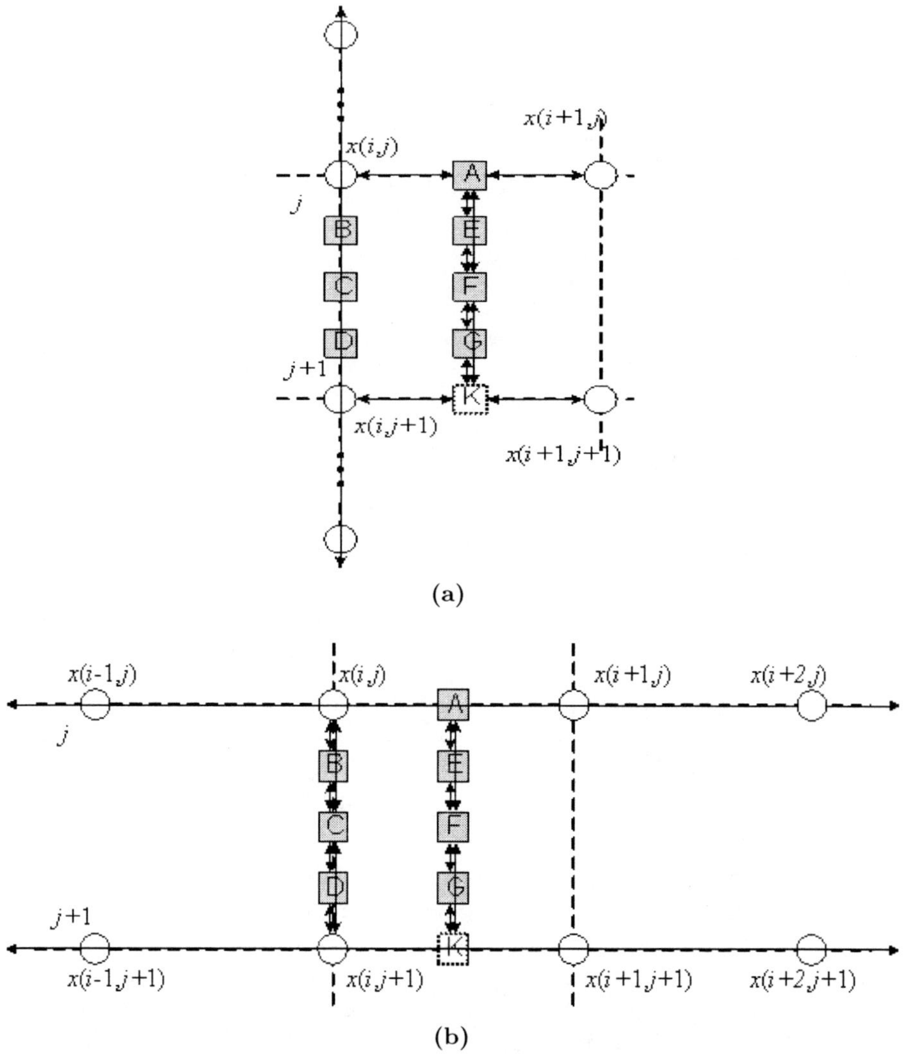


**Fig. 4.** Directional interpolation method in the region with (a) 90o direction (b) 0o direction

$K$ using BI in the horizontal direction. Thus, $F$ is obtained by the average value of $A$ and $K$. The midpoint of $A$ and $F$ becomes $E$ whereas the midpoint of $F$ and $K$ becomes $G$, respectively, as shown in Fig. 4(a) and expression (5).

$$\begin{bmatrix} B \\ C \\ D \end{bmatrix} = \begin{bmatrix} -0.0547 & 0.8203 & 0.2734 & -0.0391 \\ -0.0625 & 0.5625 & 0.5625 & -0.0625 \\ -0.0391 & 0.2734 & 0.8203 & -0.0547 \end{bmatrix} \begin{bmatrix} x(i,j-1) \\ x(i,j) \\ x(i,j+1) \\ x(i,j+2) \end{bmatrix}$$

$$A = \frac{x(i,j) + x(i+1,j)}{2}$$

$$K = \frac{x(i,j+1) + x(i+1,j+1)}{2}$$

$$F = \frac{A+K}{2} \qquad (5)$$

$$E = \frac{A+F}{2}$$

$$G = \frac{F+K}{2}$$

### 4.2 Case 2 - $R_2$ Region: 0○ (Horizontal) Direction

If the region is classified into $R_2$, we assume that there is a horizontal direction edge through the region. Pixel $A$ is obtained using the ECCF method in the horizontal direction whereas pixels $B$, $C$, and $D$ are obtained using BI in the vertical direction. To interpolate pixels $E$, $F$ and $G$, we estimate pixel $K$ using the ECCF method in the horizontal direction. $F$ is computed as the average value of $A$ and $K$. The midpoint of $A$ and $F$ becomes $E$ and the midpoint of $F$ and $K$ becomes $G$, respectively, as Fig. 4(b) and expression (6) both show.

$$A = \begin{bmatrix} -0.0625 \\ 0.5625 \\ 0.5625 \\ -0.0625 \end{bmatrix}^T \begin{bmatrix} x(i-1,j) \\ x(i,j) \\ x(i+1,j) \\ x(i+2,j) \end{bmatrix}, K = \begin{bmatrix} -0.0625 \\ 0.5625 \\ 0.5625 \\ -0.0625 \end{bmatrix}^T \begin{bmatrix} x(i-1,j+1) \\ x(i,j+1) \\ x(i+1,j+1) \\ x(i+2,j+1) \end{bmatrix}$$

$$C = \frac{x(i,j) + x(i,j+1)}{2} \qquad B = \frac{x(i,j) + C}{2}$$

$$D = \frac{x(i,j+1) + C}{2} \qquad F = \frac{A+K}{2} \qquad (6)$$

$$E = \frac{A+F}{2} \qquad G = \frac{F+K}{2}$$

### 4.3 Case 3 - $R_3$ Region: 45○ Direction

If the region is classified into $R_3$, we assume that there is a 45○ diagonal direction edge through the region. Pixels $L$, $A$, and $M$ are obtained using the ECCF method in the horizontal direction over the $j-$th row and pixels $W$, $X$, $Y$ and $N$ are obtained using the ECCF method in the horizontal direction over the $j+1$ row. Since the pixels $B$, $C$, $D$, $E$, $F$ and G are located in the region with diagonal 45○ direction edge as shown in Fig. 5(a), we compute pixels $B$, $C$, $D$, $E$, $F$ and $G$ as outlined in the expression (7).

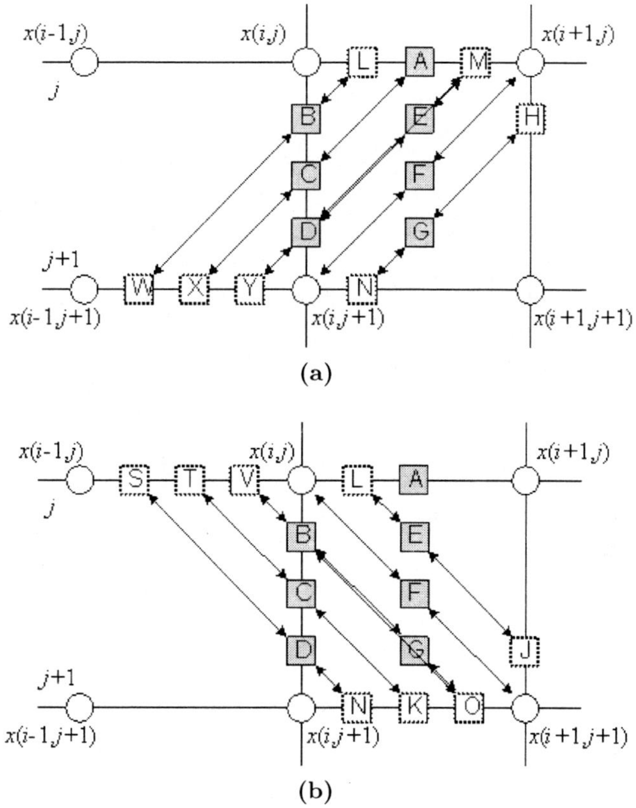


**Fig. 5.** Directional interpolation method in the region with (a) 45∘ direction (b) −45∘ direction

$$\begin{bmatrix} L \\ A \\ M \end{bmatrix} = \begin{bmatrix} -0.0547 & 0.8203 & 0.2734 & -0.0391 \\ -0.0625 & 0.5625 & 0.5625 & -0.0625 \\ -0.0391 & 0.2734 & 0.8203 & -0.0547 \end{bmatrix} \begin{bmatrix} x(i-1,j) \\ x(i,j) \\ x(i+1,j) \\ x(i+2,j) \end{bmatrix}$$

$$\begin{bmatrix} W \\ X \\ Y \\ N \end{bmatrix} = \begin{bmatrix} 0.6016 & 0.6016 & -0.2578 & 0.0547 \\ 0.3125 & 0.9375 & -0.3125 & 0.0625 \\ 0.1172 & 1.0547 & -0.2109 & 0.0391 \\ -0.0547 & 0.8203 & 0.2734 & -0.0391 \end{bmatrix} \begin{bmatrix} x(i-1,j+1) \\ x(i,j+1) \\ x(i+1,j+1) \\ x(i+2,j+1) \end{bmatrix} \tag{7}$$

$$H = \frac{3x(i+1,j) + x(i+1,j+1)}{4} \qquad B = \frac{3L + W}{4}$$

$$C = \frac{A + X}{2} \qquad D = \frac{3Y + M}{4} \qquad E = \frac{2M + D}{3}$$

$$F = \frac{x(i+1,j) + x(i,j+1)}{2} \qquad G = \frac{2N + H}{3}$$

### 4.4 Case 4 - $R_4$ Region: $-45\circ$ Direction

If the region is classified into $R_4$, we assume that there is a $-45\circ$ diagonal direction edge through the region. Pixels $S$, $T$, $V$, $L$ and $A$ are obtained using the ECCF method in the horizontal direction over the $j$−th row while pixels $N$, $K$ and $O$ are obtained using the ECCF method in the horizontal direction over the $j+1$ row. Since the pixels $B$, $C$, $D$, $E$, $F$ and $G$ are located in the region with $-45\circ$ diagonal direction edge as pictured in Fig. 5 (b), we estimate pixels $B$, $C$, $D$, $E$, $F$ and $G$ by means of (8):

$$\begin{bmatrix} S \\ T \\ V \\ L \\ A \end{bmatrix} = \begin{bmatrix} 0.6016 & 0.6016 & -0.2578 & 0.0547 \\ 0.3125 & 0.9375 & -0.3125 & 0.0625 \\ 0.1172 & 1.0547 & -0.2109 & 0.0391 \\ -0.0547 & 0.8203 & 0.2734 & -0.0391 \\ -0.0625 & 0.5625 & 0.5625 & -0.0625 \end{bmatrix} \begin{bmatrix} x(i-1,j) \\ x(i,j) \\ x(i+1,j) \\ x(i+2,j) \end{bmatrix}$$

$$\begin{bmatrix} N \\ K \\ O \end{bmatrix} = \begin{bmatrix} -0.0547 & 0.8203 & 0.2734 & -0.0391 \\ -0.0625 & 0.5625 & 0.5625 & -0.0625 \\ -0.0391 & 0.2734 & 0.8203 & -0.0547 \end{bmatrix} \begin{bmatrix} x(i-1,j+1) \\ x(i,j+1) \\ x(i+1,j+1) \\ x(i+2,j+1) \end{bmatrix} \tag{8}$$

$$J = \frac{3x(i+1,j+1) + x(i+1,j)}{4} \qquad B = \frac{3V + O}{4}$$

$$C = \frac{T+K}{2} \qquad D = \frac{3N+S}{4} \qquad E = \frac{2L+J}{3}$$

$$F = \frac{x(i,j) + x(i+1,j+1)}{2} \qquad G = \frac{2O+B}{3}$$

### 4.5 Case 5 - $R_5$ Region: $30\circ$ Direction

If the region is classified into $R_5$, we assume that there is a $30\circ$ diagonal direction edge through the region. Pixels $A$, $Q$, $X$, $C$ and $I$ are obtained using BI on the dotted lines. Pixel $H$ has the average value of $x(i+1,j)$ and $I$. Since the pixels $B$, $C$, $D$, $E$, $F$ and $G$ are located in the region with diagonal $30\circ$ direction edge as shown in Fig. 6(a), we compute them by (9).

$$\begin{aligned} A &= \frac{x(i,j) + x(i+1,j)}{2} & Q &= \frac{x(i,j) + x(i-1,j+1)}{2} \\ X &= \frac{x(i-1,j+1) + x(i,j+1)}{2} & C &= \frac{x(i-1,j+1) + x(i+1,j)}{2} \\ I &= \frac{x(i+1,j) + x(i+1,j+1)}{2} & H &= \frac{3x(i+1,j) + x(i+1,j+1)}{4} \\ B &= \frac{A+Q}{2} & D &= \frac{2X+H}{3} \\ E &= \frac{C + x(i+1,j)}{2} & F &= \frac{D+H}{2} \\ G &= \frac{x(i,j+1) + I}{2} \end{aligned} \tag{9}$$

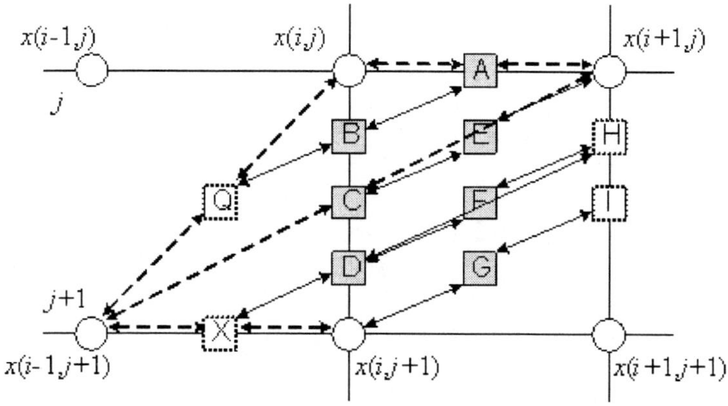


**(a)**

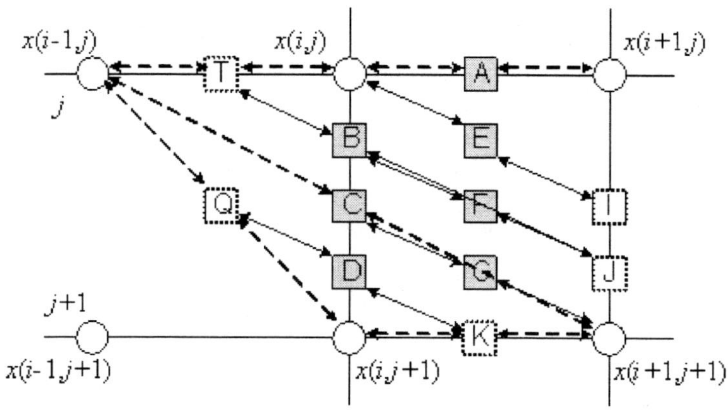


**(b)**

**Fig. 6.** Directional interpolation method in the region with (a) 30o direction (b) −30o direction

### 4.6 Case 6 - $R_6$ Region: −30o Direction

If the region is classified into $R_6$, we assume that there is a 30o diagonal direction edge through the region. Pixels $A$, $T$, $Q$, $K$, $C$ and $I$ are obtained using BI on the dotted lines. Pixel $J$ has the average value of $x(i+1, j+1)$ and $I$. Since the pixels $B$, $C$, $D$, $E$, $F$ and $G$ are located in the region with diagonal −30o direction edge as shown in Fig. 6(b), we calculate them by (10).

$$A = \frac{x(i,j) + x(i+1,j)}{2} \qquad T = \frac{x(i-1,j) + x(i,j)}{2}$$

$$Q = \frac{x(i-1,j) + x(i,j+1)}{2} \qquad K = \frac{x(i,j+1) + x(i+1,j+1)}{2}$$

$$
\begin{aligned}
C &= \frac{x(i-1,j)+x(i+1,j+1)}{2} & I &= \frac{x(i+1,j)+x(i+1,j+1)}{2} \\
J &= \frac{I+x(i+1,j+1)}{2} & B &= \frac{2T+J}{3} \\
D &= \frac{Q+K}{2} & E &= \frac{x(i,j)+I}{2} \\
F &= \frac{B+J}{2} & G &= \frac{C+x(i+1,j+1)}{2}
\end{aligned}
\tag{10}
$$

### 4.7 Case 7 - $R_7$ Region: 60○ Direction

If the region is classified into $R_7$, we assume that there is a 60○ diagonal direction edge through the region. Pixels $A$, $Q$, $X$ and $K$ are obtained using BI on the dotted lines. Pixels $C$ and $F$ are obtained using BI on the solid lines. We estimate pixels $B$, $D$, $E$ and $G$ by (11), as shown in Fig. 7(a).

$$
\begin{aligned}
A &= \frac{x(i,j)+x(i+1,j)}{2} & Q &= \frac{x(i,j)+x(i-1,j+1)}{2} \\
X &= \frac{x(i-1,j+1)+x(i,j+1)}{2} & K &= \frac{x(i,j+1)+x(i+1,j+1)}{2} \\
C &= \frac{A+X}{2} & F &= \frac{x(i+1,j)+x(i,j+1)}{2} \\
B &= \frac{x(i,j)+C}{2} & D &= \frac{C+x(i,j+1)}{2} \\
E &= \frac{A+F}{2} & G &= \frac{F+K}{2}
\end{aligned}
\tag{11}
$$

### 4.8 Case 8 - $R_8$ Region: −60○ Direction

If the region is classified into $R_8$, it is assumed that there is a −60○ diagonal direction edge through the region. Pixels $A$, $T$, $Q$ and $K$ are obtained using BI on the dotted lines. Pixels $C$ and $F$ are obtained using BI on the solid lines. We estimate pixels $B$, $D$, $E$, and $G$ in the way depicted below as shown in Fig. 7(b).

$$
\begin{aligned}
A &= \frac{x(i,j)+x(i+1,j)}{2} & T &= \frac{x(i-1,j)+x(i,j)}{2} \\
Q &= \frac{x(i-1,j)+x(i,j+1)}{2} & K &= \frac{x(i,j+1)+x(i+1,j+1)}{2} \\
C &= \frac{T+K}{2} & F &= \frac{x(i,j)+x(i+1,j+1)}{2}
\end{aligned}
\tag{12}
$$

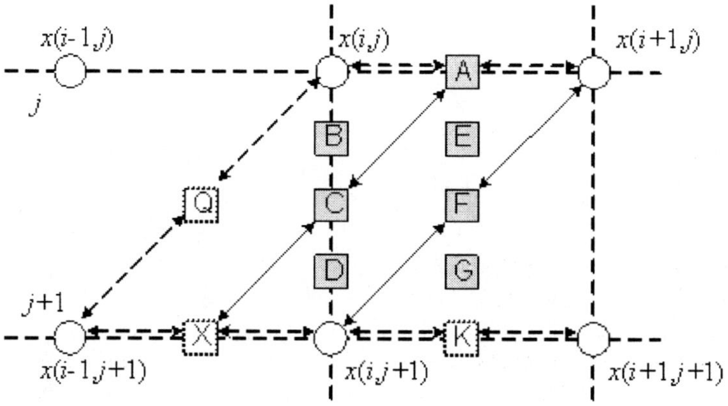


**(a)**

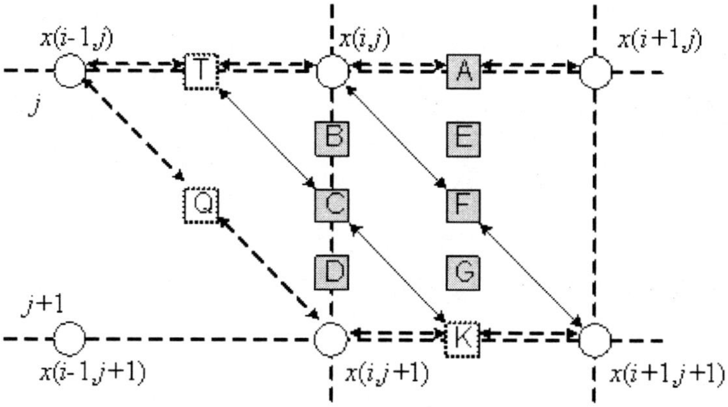


**(b)**

**Fig. 7.** Directional interpolation method in the region with (a) 60○ direction (b) −60○ direction

$$B = \frac{x(i,j) + C}{2} \qquad D = \frac{C + x(i,j+1)}{2}$$

$$E = \frac{A + F}{2} \qquad G = \frac{F + K}{2}$$

## 5 Experimental Results

As a measure of objective dissimilarity between a filtered image and the original one, we use the mean square error (MSE) and the peak signal to noise ratio (PSNR) in decibels:

$$MSE(Img, Org) = \frac{\sum_{i=1}^{N} \sum_{j=1}^{M} [Org(i,j) - Img(i,j)]^2}{NM} \tag{13}$$

$$PSNR(Img, Org) = 10 \log_{10} \frac{S^2}{MSE(Img, Org)} \tag{14}$$

where $Org$ is the original image, $Img$ is the deinterlaced image of size $N \times M$ and $S$ is the maximum possible pixel value (with 8-bit integer values, the maximum will be 255). We conducted an extensive simulation to test the performance of our algorithm using a Pentium IV processor (3.2 GHz). The algorithms were implemented in C++ and tested using five real-world sequences with a field size of 352 × 288. The test images were sub-sampled by a factor of four in the vertical direction and by a factor of two in the horizontal direction without anti-aliasing filtering. Then, we measured the performance of upsampling by using pictures that were converted from progressive pictures. These original progressive sequences were used as a benchmark to compare our algorithm with. Table 2 displays the test image characteristics.

For the objective performance evaluation, five CIF video sequences [15] were selected to challenge the four algorithms: nearest neighbor interpolation (NNI) [12], BI [13] and simple cubic curve fitting (SCCF) [14]. You may behold in Fig. 8 a subjective comparison of the five algorithms when resampling the original Flower image. Additionally, Tables 3 and 4 reflect the outcome of a comparison among the four methods in terms of PSNR and normalized average CPU time,

**Table 2.** Test image characteristics in resampling system [16]

| Test image | Characteristics |
|---|---|
| Akiyo | Low amount of spatial detail and low amount of motion |
| Foreman | Medium amount of spatial detail and medium amount of motion |
| Mobile | High amount of spatial detail and high amount of motion |
| News | Low amount of spatial detail and low amount of motion |
| T. Tennis | Medium amount of spatial detail and high amount of motion |

**Table 3.** Results of different upsampling methods for five CIF sequences in terms of the PSNR (in dB)

| Sample | NNI | BI | SCCF | DOR |
|---|---|---|---|---|
| Akiyo | 26.340 | 29.335 | 29.289 | 30.279 |
| Foreman | 23.126 | 26.884 | 26.904 | 27.838 |
| Mobile | 15.784 | 18.489 | 18.193 | 18.524 |
| News | 21.625 | 25.136 | 24.757 | 25.285 |
| T. Tennis | 22.145 | 24.491 | 24.085 | 24.650 |

**(a)** Original

**(b)** NNI

**(c)** BI

**(d)** SCCF

**(e)** DOR

**Fig. 8.** Subjective quality comparison of the Flower image

respectively. The computational CPU time of our proposed algorithm is almost the same or slightly greater than that of SCCF method. We observed that the DOR algorithm outperforms the above methods throughout all of the chosen sequences in terms of PSNR. For the "Akiyo" sequence, the suggested methods

**Table 4.** Results of different upsampling methods for five CIF sequences in terms of the normalized average CPU time (in seconds/frame)

| Sample | NNI | BI | SCCF | DOR |
|---|---|---|---|---|
| Akiyo | 0.158 | 0.353 | 0.822 | 1.000 |
| Foreman | 0.181 | 0.281 | 0.709 | 1.000 |
| Mobile | 0.158 | 0.276 | 0.819 | 1.000 |
| News | 0.230 | 0.272 | 0.891 | 1.000 |
| T. Tennis | 0.191 | 0.269 | 0.802 | 1.000 |

are superior to the SCCF method, in terms of PSNR, by up to 0.990 dB. From the experimental results, we observed that our proposed algorithm has good objective quality for different images and sequences (as illustrated in Fig. 8 with the Flower image), with a low computational CPU time required to achieve the real-time processing.

## 6 Conclusions

A new resampling method was introduced in this chapter. Our proposed algorithm performs a dynamic image segmentation into regions with eight possible edge directions (0○, −30○, 30○, −45○, 45○, −60○, 60○, and 90○). The edge direction is determined by means of the fuzzy rule-based edge detector. This method combines the advantage of bilinear interpolation, cubic curve fitting resampling method and direction-oriented interpolation algorithm. The proposed resampling algorithm yields a low proportion of staircase artifacts in comparison to peers. The algorithm's performance, measured in terms of PSNR and computational complexity, was compared to different methods and functions previously reported in the literature. The bright side of the algorithm is quite obvious at the edges in the actual resampling process.

## References


1. Nguyen, N., Milanfar, P., Golub, G.: A computationally efficient superresolution image reconstruction algorithm. IEEE Transactions on Image Processing 10, 573–583 (2001)
2. Klassen, R.V.: Increasing the apparent addressability of supersampling grids. IEEE Transactions on Computer Graphics and Applications 13, 74–77 (1993)
3. Iwamoto, K., Komoriya, K., Tanie, K.: Eye movement tracking type image display system for wide view image presentation with high resolution. In: International Conference on Intelligent Robots and Systems, pp. 1190–1195 (2002)
4. Smith, J.R.: VideoZoom spatio-temporal video browser. IEEE Transactions on Multimedia 1, 157–171 (1999)
5. Zou, J.J., Yan, H., Levy, D.C.: Reducing artifacts in block-coded images using adaptive constraints. SPIE Optical Engineering 42, 2900–2911 (2003)

6. Tannas, L.E.J.: Evolution of flat panel displays. In: Proceedings of the IEEE, vol. 82, pp. 499–509 (1994)
7. Keith, J.: Video Demystified A Handbook for the Digital Engineer. Elsevier, Amsterdam (2005)
8. Darwish, M., Bedair, M.S., Shaheen, S.I.: Adaptive resampling algorithm for image zooming. Proc. Inst. Electr. Eng. Vision, Image, Signal Processing 144, 207–212 (1997)
9. Russo, F., Ramponi, G.: Edge extraction by FIRE operators. In: Proc. 3rd IEEE International Conference on Fuzzy Systems, pp. 249–253 (1994)
10. Kimura, T., Taguchi, A.: Edge-preserving interpolation by using the fuzzy technique. SPIE Nonlinear Image Processing and Pattern Analysis 12, 98–105 (2001)
11. Jeon, G., Jeong, J.: A Fuzzy Interpolation Method using Intra and Inter Field Information. In: Proceedings of ICEIC 2006 (2006)
12. Veenman, C.J., Reinder, M.J.T.: The Nearest Subclass Classifier: A Compromise between the Nearest Mean and Nearest Neighbor Classifier. IEEE Transactions on Pattern Analysis and Machine Intelligence 27, 1417–1429 (2005)
13. Bellers, E.B., de Haan, G.: Advanced de-interlacing techniques. In: Proc. ProRisc/IEEE Workshop on Circuits, Systems and Signal Processing, pp. 7–16 (1996)
14. Fan, Y.C., Lin, H.S., Tsao, W., Kuo, C.C.: Intelligent intra-field interpolation for motion compensated deinterlacing. In: Proceedings of ITRE 2005, vol. 3, pp. 200–203 (2005)
15. `http://www.itu.int/rec/T-REC-H.261-199303-I/en`
16. `ftp://meru.cecs.missouri.edu/pub/sequences/`

# RSGUI with Reverse Prediction Algorithm

Julia Johnson[1] and Genevieve Johnson[2]

[1] Dept. of Math. and Computer Science
Laurentian University
Sudbury, ON, P3E 2C6, Canada
jjohnson@cs.laurentian.ca

[2] Department of Psychology
Grant MacEwan College
Edmonton, AB, T5J 4S2, Canada

**Summary.** Rough Set Graphical User Interface (RSGUI) is a software system appropriate for decision-making based on inconsistent data. It is unique in its capability to apply the rough set based reverse prediction algorithm. Traditionally, condition attribute values are used to predict decision attributes values. In reverse prediction, the decision attribute values are given and the condition attribute values that would lead to that decision are predicted. Reverse prediction was used in an electronic purchasing application to provide the characteristics of products that customers will purchase.

## 1 Introduction

In the traditional rough set prediction process, if-then rules are generated from inconsistent data. Given attribute values for a new case, the if-then rules are followed in making decisions. In an electronic purchasing application, there is need to do just the opposite. The vendor wishes to predict characteristics of products that would lead to customers making a purchase.

In reverse prediction, given decision attribute value $v$, condition attribute values that best imply $v$ are predicted. An introduction to reverse prediction within a rough set framework was provided in [1]. An algorithm was introduced that makes use of ordinary prediction to implement reverse prediction. The current contribution is to formulate reverse prediction within the broader context of an application and a system that performs data analysis. Particular attention is paid to ordinary prediction tasks (discretization, optimization of rules) and their counterparts in reverse prediction.

In this chapter, reverse prediction is explained in Section 2. Rough Set Reverse Prediction Algorithm (RSRPA) [1] is reviewed in Section 3. Use of reverse prediction in an electronic purchasing application is demonstrated in Section 4. Validation of the reverse prediction algorithm is discussed in Section 5. Implementation of ordinary prediction in RSGUI is demonstrated in Section 6. Two rough set based systems, RSES and Rosetta, are described and compared with RSGUI in Section 7. Conclusions are presented in Section 8.

R. Bello et al. (Eds.): Granular Computing, STUDFUZZ 224, pp. 287–306, 2008.
springerlink.com 

## 2 Reverse Prediction

Ordinary prediction is expressed as

$$C_1[\text{given}], C_2[\text{given}], \ldots, C_n[\text{given}] \rightarrow D[\text{predict}] \qquad (1)$$

where the $C_i$ are condition attributes and $D$, a decision attribute. For example, to predict a customer's response to a product (e.g., purchased, not purchased), features of the product are condition attributes. The customer's response is the decision attribute.

It is possible to interchange the roles played by the condition and decision attributes while still employing ordinary prediction:

$$D[\text{given}] \rightarrow C_1[\text{predict}], C_2[\text{predict}], \ldots, C_n[\text{predict}] \qquad (2)$$

For example, let us consider the customer's response as a property of the product, and to predict the color of products purchased by customers. The condition attributes are not the same in this problem as in the previous problem.

To derive syntactically a statement of reverse prediction from that of ordinary prediction (statement (1)), the task of predicting moves leftward across the symbol for material implication. The condition and decision attributes remain the same. The direction of the implication symbol remains the same.

$$C_1[\text{predict}], C_2[\text{predict}], \ldots, C_n[\text{predict}] \rightarrow D[\text{given}] \qquad (3)$$

Statement (3) reads: Given a value for the decision attribute, predict the condition attribute values that best imply the value of the decision attribute. Reverse prediction helps answer a question such as: To what extent does a product being purchased follow from it being blue? If the roles of condition and decision attributes were interchanged in (3), a question could be answered such as: To what extent does a product being blue follow from it being purchased?

$$D[\text{predict}] \rightarrow C_1[\text{given}], C_2[\text{given}], \ldots, C_n[\text{given}] \qquad (4)$$

Consider rule (4) obtained by changing the direction of the implication in (1). Rule (4) may be derived from (2) using reverse prediction, or from (3) by reversing the roles of attributes.

To summarize, two properties of attributes have been distinguished: condition (C) or decision (D) and given (G) or predicted (P). Two binary choices lead to the following four possibilities:

1. { C, G } $\longrightarrow$ { D, P }
2. { D, G } $\longrightarrow$ { C, P }
3. { C, P } $\longrightarrow$ { D, G }
4. { D, P } $\longrightarrow$ { C, G }

1 and 2 are ordinary prediction. 3 and 4 are reverse prediction.

It might be argued that reverse prediction cannot be used due to lack of information. Researchers are struggling to find enough information from data to do ordinary prediction. The following view is intended to illustrate that reverse prediction is no more difficult than ordinary prediction.

Assume that there are two conditions $a$ and $b$ each with two possible values (t)rue or (f)alse. Rules generated from ordinary prediction have the form

$$\textbf{if } a^{(t)} \textbf{ then } b^{(t \text{ or } f)}.$$

The superscripts show the amount of uncertainty in the given condition. Interchanging the conditions $a$ and $b$ does not change the form of the expression.

Rules generated by reverse prediction have the form

$$\textbf{if } a^{(t \text{ or } f)} \textbf{ then } b^{(t)}.$$

An amount of uncertainty resides in the antecedent of the statement for reverse prediction, but the same amount lies in the consequent of the expression for ordinary prediction. Uncertainty being on the left hand side of the implication symbol is not the reason that reverse prediction may seem intuitively to be more difficult. Such intuition may reflect the false sense that, whereas ordinary prediction gives if-then rules, reverse prediction gives only-if rules. However, both ordinary and reverse prediction are concerned with sufficiency, and neither with necessity.

The prediction problem is well defined because the data have already been interpreted, to some degree, by populating the information table with values placed in the appropriate columns. Difficulty arises when the attributes are not provided in advance.

A form of Bayes theorem is useful for inferring causes from their effects. A neuro-physiological problem involves vast amounts of brain imaging data [2] [3]. Multiple detectors receive signals from multiple sources. Each detector records a mixture of signals from the different sources. The goal is to recover estimates of the original signals.

In what is called the *forward problem*, the regions of activity or signal sources in the brain are known. The objective is to calculate the corresponding magnetic fields on the surface of the head. In the *inverse problem*, from the magnetic fields on the head, the locations and orientations of the sources in the brain must be calculated. A derivation within a Bayesian framework of a pre-existing algorithm was found for solving the forward problem. The inverse problem is more difficult to solve than the forward problem.

Pawlak used Bayes theorem to explain decisions reached by deductive inference [4]. Additionally, the Rough Bayesian Model was derived from the Rough Set model [5]. Reverse prediction is different from Pawlak's notion of explanation [6] in which the steps in decision making are presented in a decision tree. Subsequent research may demonstrate the utility of Pawlak's method to explain decisions reached by reverse prediction. The algorithm for reverse prediction is given in the next section.

## 3 Rough Set Reverse Prediction Algorithm

Preconditions and initializations follow:

INPUT:
  Decision table, attribute value $V$
SETUP:
  Let $U$ be the universe of objects, $C$ be the set of condition attributes,
  $X$ be a given concept with attribute value $V$
  Let $C' = C\breve{a}$
  **Let BCR = ø be a set of best condition rules**

A rule generated by reverse prediction will be distinguished from a regular deductive rule by using the term *condition rule* for reverse rules and *predictive rule* for regular deductive rules. BCR contains the highest quality (best) condition rules. Prior to the processing of a given concept, the set BCR is empty. $C'$ contains all remaining condition attributes that have not yet been synthesized into a reverse rule. Rules generated from a traditional rough set prediction method are evaluated using a measure for quality of the rules. An outer loop (not shown in the code fragment) allows processing of all possible concepts. There is one concept per possible combination of values for the decision attributes.

### 3.1 Reducts

Subsets of the columns of an information table may provide the same predictive power as all of the columns. A process of finding reducts (rule reductions, table reductions) is integral to the process of generating predictive rules.

Indiscernibility methodology is the process of viewing an information table as describing classes of individuals where members of a class cannot be distinguished from one another on the basis of values associated with their column names. Use of indiscernibility classes to eliminate redundant attributes from predictive rules is accomplished by means of the traditional prediction process. The process of reverse prediction begins with the reduced rules.

### 3.2 Coverage and Certainty

A rule is said to *cover* an example row of the information table if that row can be completely characterized by the rule. A row is completely characterized by a rule if the rule references in its left hand side at least those non-redundant attributes required to unambiguously place the example into its indiscernibility class.

Coverage, a commonly used quality measure for decision rules [7], is the number of rows covered by the rule divided by the number of rows in the decision table. This gives a ratio between 0 and 1, inclusive. Coverage expresses the level of generality of the rule with respect to the data set.

The certainty of a rule is calculated by dividing the number of rows covered by the rule that also appear in the concept, by the number of rows covered by the rule that do not appear in the concept. Deterministic rules have a certainty of 1 while non-deterministic rules have a certainty of less than 1 (but larger than 0).

In the context of reverse prediction, the verb *cover* refers to an attribute rather than an example. An attribute is covered by a condition rule if the attribute name appears on the left hand side of the rule.

### 3.3 Inductive Learning Algorithm

A method for traditional prediction is called from within the reverse prediction algorithm. The rough set algorithm used for traditional prediction is known as the RS1 inductive learning algorithm [8].

### 3.4 Reverse Prediction Algorithm

A condition rule has exactly one condition attribute on its left hand side.

RSRPA begins by executing the RS1 algorithm on the decision table for the concept under consideration. These rules are already optimized. If more than one rule is generated, the highest quality one is chosen using certainty and coverage. The attributes covered by the highest quality predictive rule are removed from the set $C'$, initially containing the entire set of condition attributes. RS1 is executed again and if a predictive rule is generated, its condition attributes are removed from the remaining condition attributes $C'$. RSRPA terminates when there are no condition attributes left to remove ($C1 =$ ). The RSRPA algorithm follows:

**/* Only attributes covered by the highest quality rules generated by traditional prediction are included in reverse rules*/**

RS1:
  Execute Traditional Prediction
  If more then one rule generated
    Pick the rule $R$ with the highest coverage and certainty
  BCR = BCR $\cup$ $R$
  For each condition attribute $C_i$ covered by rule $R$
  record the pair $(C_i, C_v)$;
  $C' = C'$ - $C_i$;
  If $C1' \neq$ Go to RS1
END.

The set BCR contains pairs <condition attribute, value> that are mutually exclusive with respect to the condition attributes covered. No two rules in BCR cover the same attribute. The overall algorithm is given in Fig. 1.

## 4 Electronic Purchasing Application

In the context of an electronic purchasing environment, customer satisfaction is critical. Therefore, vendors require information on the characteristics of products that would lead to customers being satisfied. It is often difficult to obtain precise

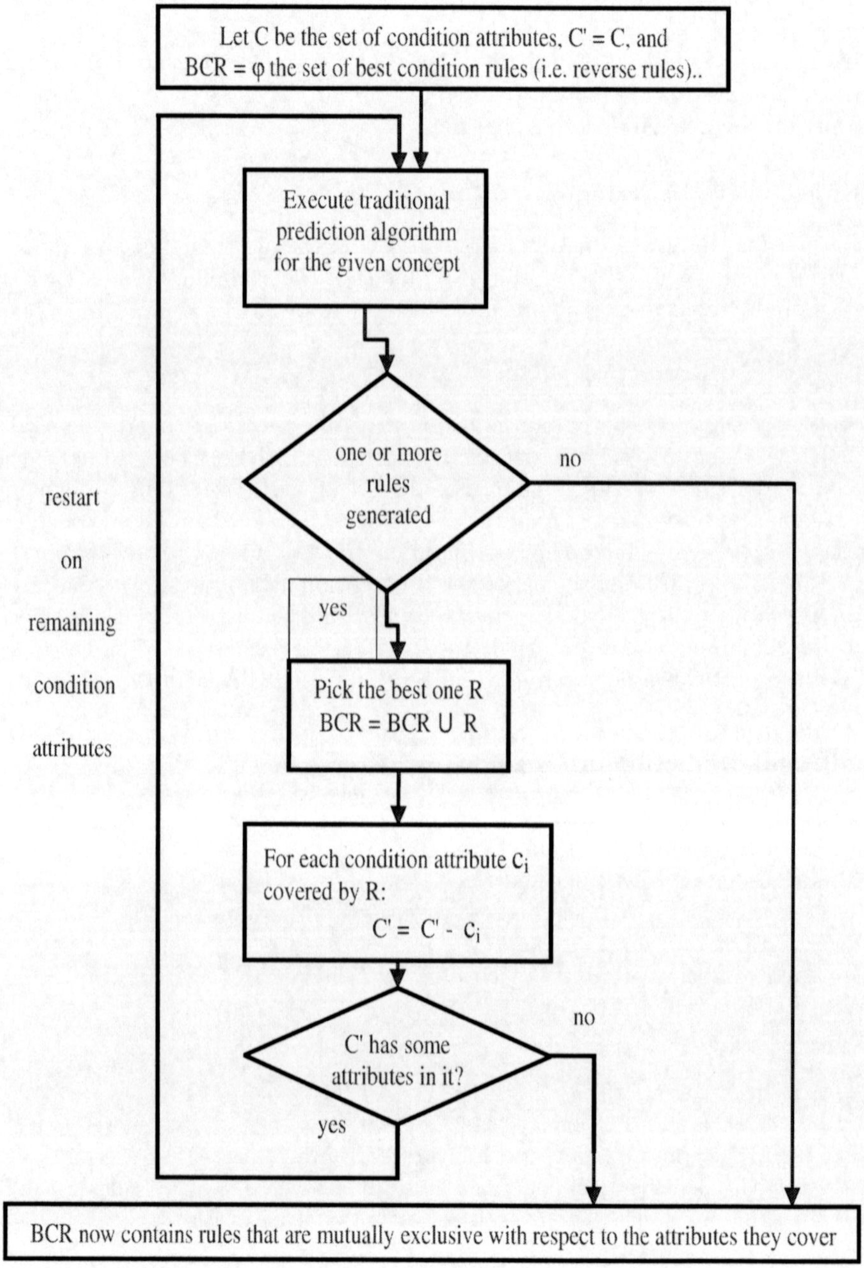


**Fig. 1.** Rough Set Reverse Prediction Algorithm: Rules generated from a traditional rough set prediction method are evaluated using a measure for quality of the rules. The highest quality rule is defined as the one that covers the largest number of rows of the table. Best condition rules (reverse rules) for a given concept are generated. An outer loop for processing all possible concepts has been omitted from the figure.

| Row | color | size | material | texture | purchased |
|---|---|---|---|---|---|
| 0 | blue | XL | cotton | sleek | yes |
| 1 | red | XL | cotton | smooth | no |
| 2 | yellow | XS | silk | smooth | yes |
| 3 | blue | S | cotton | smooth | yes |
| 4 | red | XXL | silk | sleek | no |
| 5 | blue | XL | cotton | sleek | yes |
| 6 | red | XL | cotton | smooth | no |
| 7 | yellow | L | silk | smooth | yes |
| 8 | blue | S | cotton | smooth | yes |
| 9 | red | XXL | silk | sleek | no |
| 10 | blue | XL | cotton | sleek | yes |
| 11 | red | XL | cotton | smooth | no |
| 12 | yellow | L | silk | smooth | yes |
| 13 | blue | S | cotton | smooth | yes |
| 14 | red | XXL | silk | sleek | no |
| 15 | blue | XL | cotton | sleek | yes |
| 16 | red | XL | cotton | smooth | no |
| 17 | yellow | L | silk | smooth | yes |
| 18 | blue | S | cotton | smooth | yes |
| 19 | red | XXL | silk | sleek | no |
| 20 | blue | XL | cotton | sleek | yes |
| 21 | red | XL | cotton | smooth | no |
| 22 | yellow | L | silk | smooth | yes |
| 23 | blue | S | cotton | smooth | yes |
| 24 | red | XXL | silk | sleek | no |

**Fig. 2.** Information table describing an electronic purchasing application. The products supplied by the vendor are sweaters. A vender wishes to know the properties of sweaters purchased by customers. The attributes of interest are color, size, material and texture. Each row of the information table describes a sweater. The table includes an attribute purchased to express whether or not a sweater with given properties was purchased by the customer.

information regarding customer needs and preferences. Buyers are not necessarily aware of the tangible properties of the products they prefer. However, vague and inconsistent data are typically available. A rough set reverse prediction approach was used to provide the best value for each product attribute that, taken together, would lead to the customers being satisfied.

RSGUI, the graphical user interface into which the reverse prediction algorithm is embedded, will be demonstrated next. RSGUI consists of tabs that permit the user to execute ordinary prediction as well as reverse prediction. The possible operations are listed horizontally, as illustrated in the screen shot of Fig. 2. The tab in use is labeled TABLE which results in output of a previously loaded table.

### 4.1 Condition Attributes

A process of discretization is frequently used in a preprocessing step to move from numeric to soft values (e.g., price: low, medium, high). Indiscernibility methodology requires a limited number of possible values for attributes for constructing an equivalence relation. Quantization and/or domain driven methods may be used to convert the data sets from continuous to discrete ones. The condition attribute values in our purchasing application were sufficiently generalized so as to avoid the need for discretization. Work is ongoing to permit the use of rough set methods on continuous data [9] [10].

### 4.2 Decision Attributes

Buyers are limited in their capacity to keep a large number of decision values in mind. Satisfaction rating scale options are subject to individual interpretation (e.g., extremely satisfied, very satisfied, somewhat satisfied, somewhat unsatisfied, very unsatisfied, extremely unsatisfied). Thus, as few options as possible is recommended (e.g., unsatisfied, satisfied and very satisfied). Bi-valued attributes *Texture* and *Purchased* used in the following illustrations lead to four possible concepts.

### 4.3 Reverse Prediction Algorithm

The rules printed in the status box of Fig. 3 were referred to as *best condition rules* in the previous section. Each decision (or concept) is printed once followed by a list of pairs <name, value>. The name identifies condition attributes whose associated values are computed to be the best predictors of the given concept.

## 5 Evaluation of Reverse Prediction Algorithm

A hockey game application requiring dynamic decision making was used to test RSRPA. This validation technique was presented in [1]. The problem was to determine the condition attribute values (behavior of individual team members) that lead to the desired decision attribute value (i.e., to win the game)? Behaviors were implemented as methods coded in Java. Sample behaviors follow:

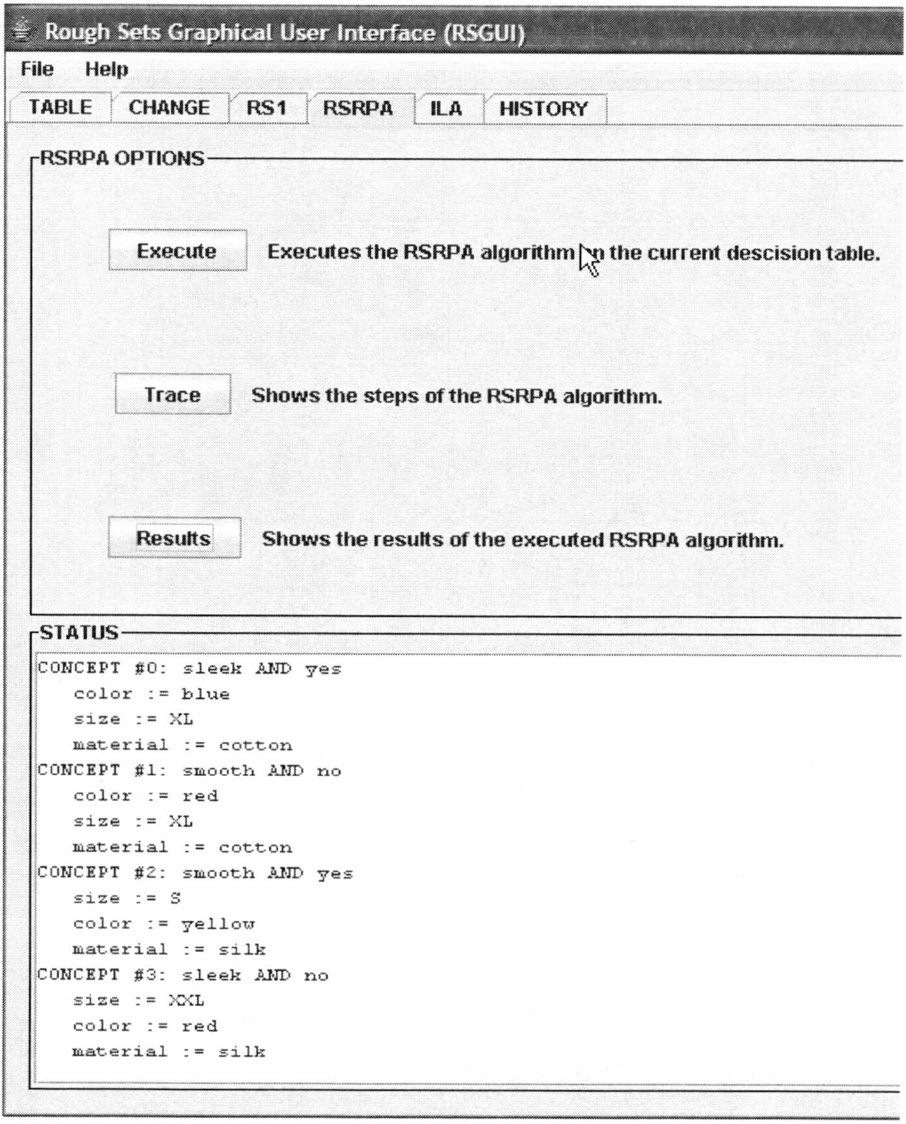


**Fig. 3.** RSRPA has been executed on a sweater database with decision attributes Texture and Purchased. Each of the decision attributes has two possible values. The condition attribute values for sweaters that imply each combination of decision values are predicted.

1. A1 - the player chases the puck (Chaser)
2. A4 - the player predicts how (s)he will get to the puck (Psychic Chaser)
3. B1 - the player shoots the puck directly at the net (Random Shooter)

The decision attribute measures the success or failure of a combination of behaviors. There were five players per team and a player may be in one of four possible states. There were twenty condition attributes in the information table. Player A may be in one of the following states at a given time:

1. the puck is in A's possession (mine)
2. the puck is in A's teammate's possession (mate's)
3. the puck is in A's opposing team's possession (foe's)
4. the puck is free (fate's)

The following table represents one row of the information table (minus the decision attribute) consisting of five groups of four behaviors. Within a group, the first field codes the behavior that a player uses when in state 1 (mine), the second when in state 2 (mate's), the third when in state (foe's), the fourth when in state (fate's).

| Results from RSRPA - This is NOT the Decision Table. This is the predicted combination of behaviors sufficient for a win. | | | | |
|---|---|---|---|---|
| player | mine | mates | foes | fates |
| 1 | $I1[B1]$ | $C1$ | $H1[A1, C7]$ | $A4$ |
| 2 | $B2$ | $C2$ | $H1[A1, C6]$ | $A4$ |
| 3 | $F1$ | $H1[A1, C7]$ | $G1$ | $A4$ |
| 4 | 12[$B3$] | $E6$ | $E6$ | $E6$ |
| 5 | $B3$ | $E9$ | $E9$ | $H2[A4, E9]$ |

The best combination of behaviors that will lead to a win were predicted using RSRPA to produce what became known as the rough set team. The benefit of using names of methods (with names of parameters also identifying behaviors) is that the predicted behaviors so named can be executed. The algorithm was evaluated by running the rough set team against hundreds of randomly generated teams.

Quantitative measures of RSRPA's success were obtained by computing the percentage of games won by the rough set team. Of the 1000 games played, the rough set team won 78.8%, lost 14.4% and tied 6.8%.

## 6 Algorithms for Ordinary Prediction

The RS1 algorithm [8] generates deterministic rules first, and if those are not sufficient to explain the concept, then non-deterministic rules are generated. It functions by incrementally selecting a series of attributes around which to pivot, generating rule sets of increasing complexity until all universe examples are covered.

At first, each attribute ($A_i$) is individually processed, and for each possible value ($V_{ij}$) of ($A_i$), a subset ($S_{ij}$) of the universe ($E$) is generated. These subsets can be part of the Upper Bound($\overline{Y}$), the Lower Bound ($\underline{Y}$) or neither.

$$\sum_{i=1}^{m}\sum_{j=1}^{n_i} S_{ij} = subset(E, A_i = V_{ij})$$

The set of all positive class examples is generated as a subset ($S_+$), and the attribute subset ($S_{ij}$) is part of the Lower Bound if it intersects with this class subset. Likewise, an attribute subset ($S_{ij}$) is part of the Upper Bound if it is included within this class subset.

$$S_{ij} \subseteq \overline{Y} \iff (S_{ij} \cup S_+)$$

$$S_{ij} \subseteq \underline{Y} \iff (S_{ij} \subseteq S_+)$$

A quality value represented by $\alpha$ is generated for each attribute. The attribute with the largest value of $\alpha$ becomes the pivot attribute for the next iteration. The universe of possible elements is cleared of rows that are covered by the rule set using the equation:

$$\alpha = 1 - \frac{|\overline{Y} - \underline{Y}|}{|E|}$$

$$E = E - [(E - \overline{Y}) \cup \underline{Y}]$$

Using the pivot attribute, the list of attributes is traversed again and new subsets are generated for each of the value combinations for pivot and attribute. The Lower and Upper bounds are again generated and the attribute with the best $\alpha$ is joined to the pivot, so that we now have a two attribute pivot. The process repeats again, adding attributes to the pivot, until we either run out of attributes or the universe becomes empty.

RS1 tends to produce rules that are over-specific resulting from optimizing rule sets each time new attributes are joined to the pivot. The application of local rather than global information at each iteration leads to unnecessary and irrelevant conditions included in the decision rules. Consequently, the rules lack ability to classify examples not previously seen in the training set and examples with missing attribute values. Solutions to the problem of missing attribute values can be found in [11][12]. A solution to the problem of unknown values comes at the cost of increasing the number of rules generated.

An alternate inductive learning algorithm (ILA [13][14]) for traditional, as opposed to reverse rule prediction, has been implemented in RSGUI. It produces if-then rules from training examples using global information. A rule is considered more general the fewer its number of conditions. The designers of ILA aimed at producing more general rules based on the premise that generality of rules increases their classification capability.

In fact, ILA produces a fewer number of rules than RS1. RSGUI was used to compare the rules generated by ILA and RS1 on the previously discussed sweater database. ILA generated five rules while RS1 generated six. RSGUI was used to compare RS1 and ILA on the data sets demonstrated in the ILA literature, and fewer rules from ILA were observed in all cases.

Local and global information for optimizing predictive rules based on rough sets is an active area of research [15]. An objective in the design of RSGUI was to implement a basic algorithm at both extremes so as to study the effect that different optimization approaches have on reverse prediction. Figure 5 illustrates

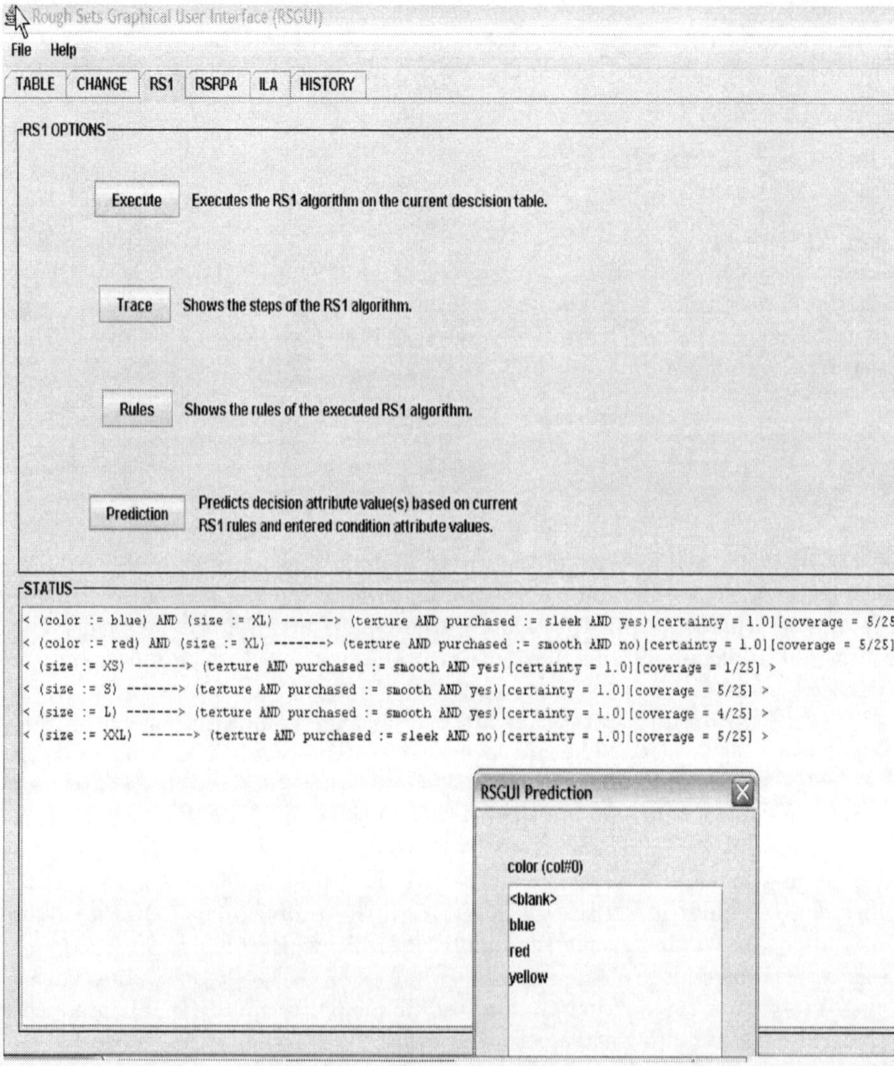


**Fig. 4.** RS1 algorithm was executed on the sweater database by clicking on the RS1 tab followed by the execute button. If-then rules were displayed using the rules button. The rules appear in the status window together with a measure of the quality of each rule given by certainty and coverage. Characteristics of a sweater may be entered using the menu that pops up by clicking the prediction button as shown. The rules are applied by matching the left hand side of the rule to the values entered.

the use of RSGUI to generate decision rules using ILA and Fig. 6 presents a trace of the execution of ILA during rule generation.

Briefly, the algorithm works as follows: The example set is divided into sub-tables, one for each decision attribute value. Condition attribute values that

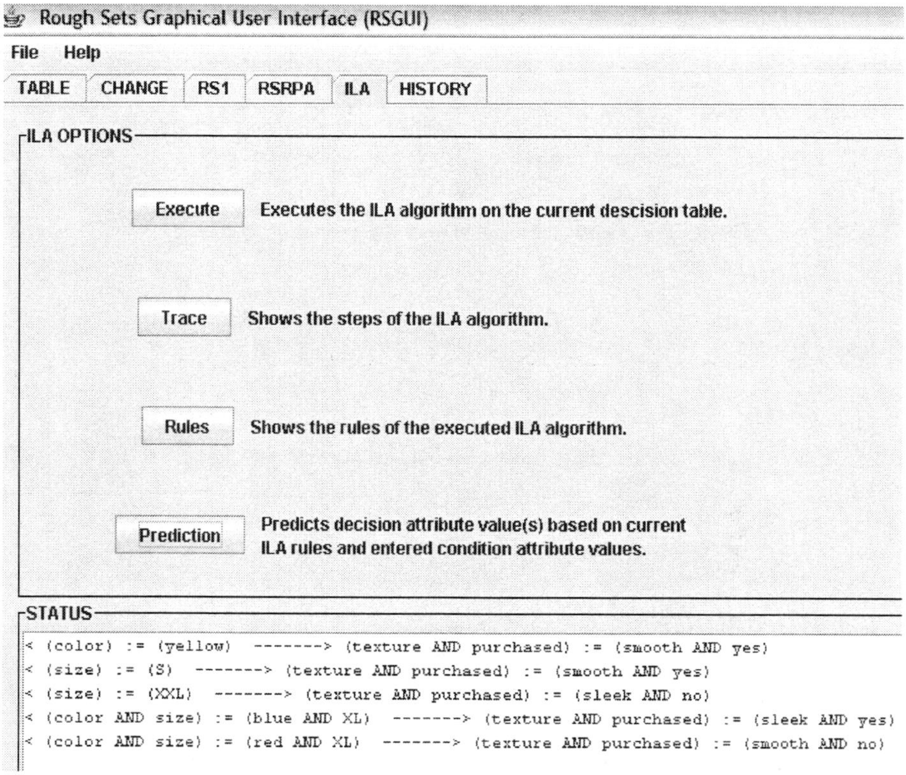


**Fig. 5.** Inductive learning algorithm (ILA) was executed on the sweater database using the ILA tab together with the execute button

occur in one subtable and not others are sought because such a rule is independent from rules derived from other classes. Combinations of the condition attributes of a subtable begin at combinations of length 1 and the length increments with each iteration. Having found a rule, the examples it covers are removed from the training set by marking them. The algorithm terminates when all examples have been marked. Refer to [13] [14] for details of the algorithm.

### 6.1 History Tab

The purpose of the history tab is to record a history of the operations that have been done in a current session. See Fig. 7.

## 7 RSES, Rosetta and RSGUI

Rough Set Exploration System (RSES) [16] and Rosetta [17] [18] are systems for reasoning under uncertain conditions within a rough set framework. The commercially available windows-based Rosetta system provides analysis of tabular

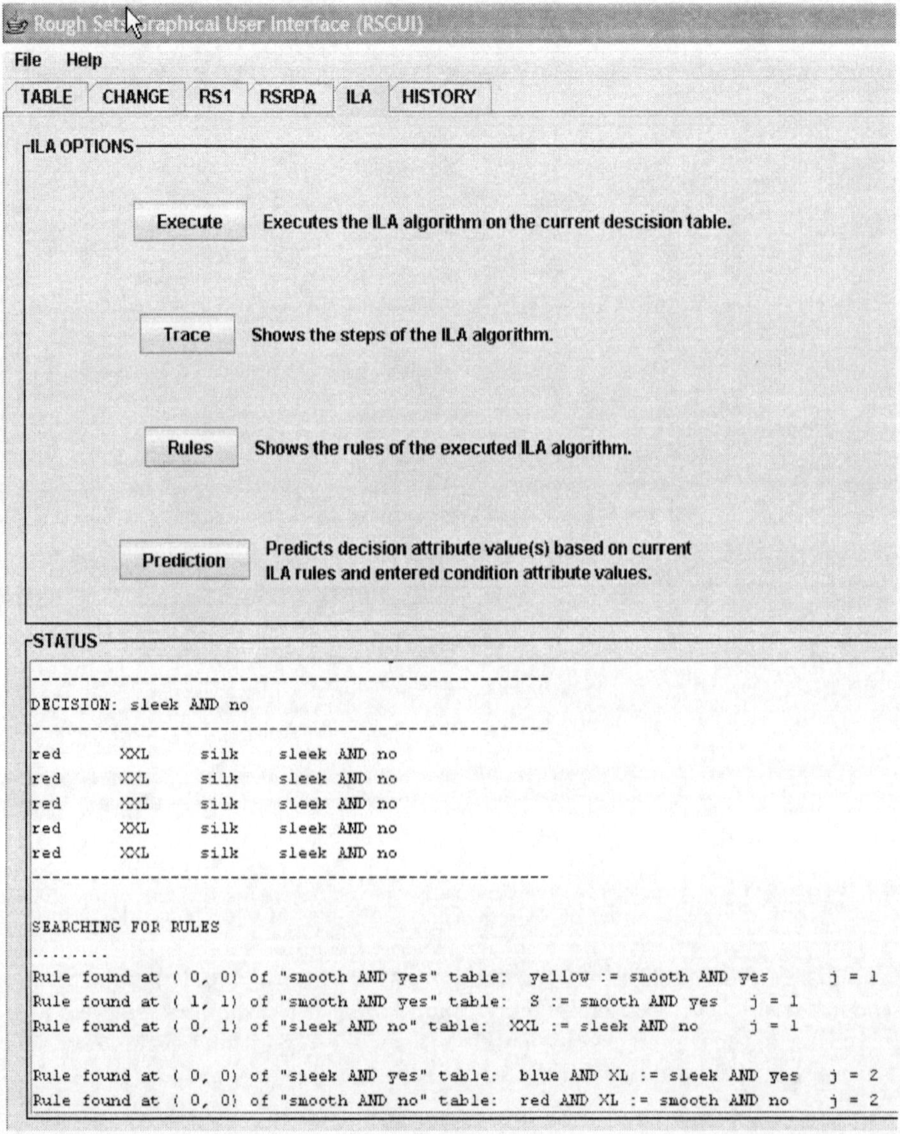


**Fig. 6.** A trace of the steps of the inductive learning algorithm for generating the rules of Fig 5 was displayed in the status window. The initial division results in four subtables. J initially equal to 1 records the number of attributes in the combinations currently under consideration. A row is marked when it is used to generate a rule. This process is repeated for all values of each attribute of each sub-table.

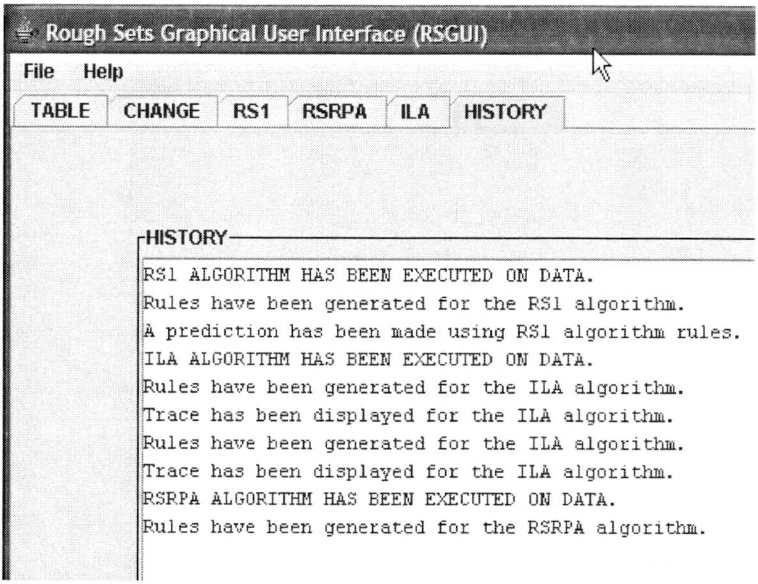


**Fig. 7.** Recording previous work. The RS1 algorithm was executed and the rules generated displayed. A prediction based on the rules was made by specifying a value for each condition attribute. The ILA algorithm was executed, a trace of the algorithm was displayed, and so on.

data based on indiscernibility modeling methodology which involves calculation of reducts. RSES is free for non-commercial data analysis and classification.

Like RSGUI, both RSES and Rosetta allow the user to specify which attributes are decision attributes and which are condition attributes. RSGUI is distinguished from RSES and Rosetta by the advantage of reverse prediction.

An overview of RSES and Rosetta in Subsections 7.1 and 7.2 allows for comparison with RSGUI in Subsection 7.3.

### 7.1 RSES

RSES is an extensive system introducing all aspects of data exploration. Table data may be decomposed into two disjoint parts where the parts are themselves table objects. A split factor between 0 and 1 specified by the user determines the size of each of the subtables. One subtable is a compliment of the other. In a train-and-test scenario, the data table is split into two parts based on the split factor. One part is used as the training set and the other as a test set. A table object resulting from a split operation is automatically assigned a name composed of the original table name and the value of the split factor.

The user interface was crafted to achieve uniformity of concepts by basing them on the notion of an object. In Fig. 8, an icon labeled with T denotes a table. The complete data set named Sweater contains information about all example sweaters. Sweater-0.6 and Sweater-0.4 radiating with shaded arrows from the

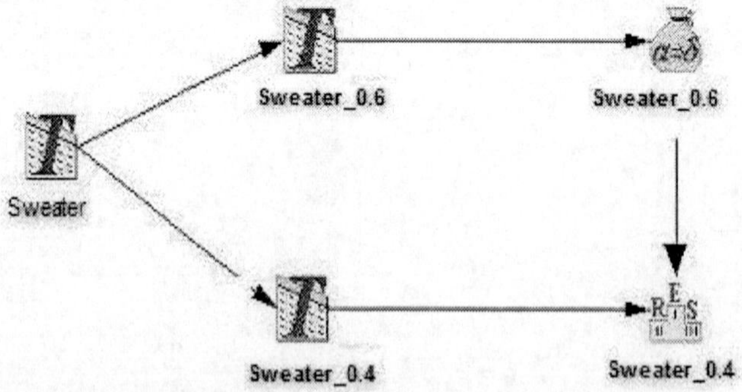


**Fig. 8.** RSES interface illustrating definition of a database containing information about customer preferences for sweaters

table icon are subtables of the complete Sweater table. Sweater-0.6 naming the icon labeled $\alpha \Rightarrow \delta$ denotes a classifier. The classifier obtained from the training set is automatically assigned the training set name, but can be distinguished from the training set by a special icon for classifiers. Sweater-0.4 together with the icon labeled above RES denotes the test results. Clicking on any of the icons results in the expansion of the icon to the object that it represents. A split factor 0.6 was used which means that 60% of the original table was used as training a set and the remainder as a test set.

RSES methods fall into two categories 1) train and test scenarios and 2) testing using the cross-validation method. The train and test scenarios can be broken down into different methods as follows:

(a) Rule based classifier
(b) Rule based classifier with discretization
(c) Decomposition tree
(d) k-NN (k Nearest Neighbor) classifier
(e) TF (Local Transfer Function) classifier

One of these methods is selected before data can be analyzed.(a) and (b) are effective for small data sets. Personal experience with large data sets (e.g., 500 records with approximately 40 attributes) suggests that methods (a) and (b) are inadequate for classifier construction despite memory increase. The complexity of calculating rules increased as the size of data sets increased limiting the use of the first two methods listed above. For large data sets, RSES researchers recommend methods (c), (d) and (e). We were successful in using cross-validation with fold factor 2 on 500 records. To date, our experience with the cross-validation method has not resulted in a single successful run with greater than 400 records and fold factor 10.

### 7.2 Rosetta

Rosetta has been used for selecting genes that discriminate between tumor subtypes [19]. Microarray technology generates vast amounts of data in the process of simultaneously measuring the behavior of thousands of genes. The genes types act as column headings of an information table.

The Rosetta authors distinguish information systems from decision systems. A Cartesian product of value sets in a given order defines an information system. Information about the association of attributes with their value sets resides in a data dictionary associated with the information system. Information about the attribute as condition or decision also appears in the data dictionary. A decision table is the information table together with its data dictionary.

The Rosetta user decides which attributes are conditional and which are decisional, as in RSES and RSGUI. Unique to Rosetta, attributes may be disabled which makes them invisible to algorithms that operate on the data set. Such a feature is required for real world databases, emphasizing the point that Rosetta is a production system.

If-then rules are generated and validated (i.e., the quality of rules is checked). The quality of rules for prediction is evaluated based on a choice of quality measures. Prediction rules may have an empty right hand side in which case they serve to find patterns in data. Such rules have the ability to classify objects. Two options are implemented for classifiers; 1) an algorithm for support based voting but with no tolerance for missing values and 2) an algorithm that allows the voting to incorporate user-defined distance values between decision classes.

### 7.3 Comparison with RSGUI

Most rough set systems including RSGUI have the ability to specify which attributes of the information table are conditions and which are decisions. Moreover, during operation of those systems, the role of an attribute as condition or decision may be changed. In addition to increasing the applicability of the decision system, this facility provides users with the ability to find the best rules to explain the data.

Whereas both RSES and Rosetta accept tables with no decision attribute, in RSGUI all tables are assumed to have at least one decision attribute. An error message occurs if the user attempts to generate rules from a table with zero decision attributes.

The other systems provide both classification and rule generation. RSGUI was intended as a system for experimenting with the notion of reverse prediction. For generalization to reverse classification, one must decide whether reverse prediction rules should have no right hand side or no left hand side.

RSGUI software includes the interface, two deterministic rough set algorithms for traditional prediction, and the RSRPA (reverse prediction algorithm). Similar to Rosetta, the user specifies the type of attribute (i.e., condition or decision). Unlike other rough sets software, however, RSGUI allows the user to specify which of the attributes are given and which are predicted. Once data are entered, the user is prompted for the number of decision attributes. The allocation of attributes as condition or decision can be changed while experimenting with a given information table.

In RSGUI, the user chooses one of two algorithms to generate predictive rules. A row or column may be removed permitting both horizontal and vertical projections of the data to be analyzed. The user can refine the rules generated by adjusting the characterization of one or more attributes as condition or decision, and in addition as predictor or predicted.

Rosetta carries out database management tasks such as data dictionary and data completion to resolve null values. RSES also allows for data completion. Both RSES and Rosetta require significant user training. RSGUI, in contrast, accomplishes less, but has a simple interface and requires minimal user training.

### 7.4 Incorporating Fuzzy Sets

RSES and Rosetta both provide a variety of hybrid techniques, for example, genetic algorithms. Future work on RSGUI involves adding a fuzzy set component to model the degree of customer satisfaction. The linguistic quantifier most permits an attribute such as purchased to be expanded to purchased by most:

$$\mu_{\text{most}}(X) = \begin{cases} 1 & x \geq 0.8 \\ \frac{x-0.3}{0.5} & 0.3 \leq x \leq 0.8 \\ 0 & x \leq 0.3 \end{cases}$$

The variable $x$ is the number of satisfied customers. The linguistic quantifier yields three fuzzy subsets which may be referred to as yes, no, maybe. The concept *most* is fully justified if at least 80% of the customers are satisfied, not justified if 30% or less are satisfied and, otherwise, partially justified to the degree of satisfaction given by the above expression. The cutoff points have been chosen arbitrarily for illustration.

## 8 Conclusion

RSRPA takes as input a decision table and the required decision (e.g., purchased) for which we aim. Output is a set of the predicted best condition attribute values of products that would lead to them being purchased.

The vendor does not necessarily know the features of a product that most customers prefer. The market scenario may vary dramatically from one region of the world to another, influenced, for example, by climate or terrain. Product appreciation values automatically generated from a sample of data for the region will be free of biases based on the vendor's own environment.

The rough set reverse prediction method provides the ability to automatically articulate desirable product attributes. The reverse prediction algorithm permits prediction of customer preferences in the form of rules.

Reverse prediction has been embedded in RSGUI. The best condition rules do not need to be optimized differently for reverse prediction. Their optimization derives from the execution of traditional prediction from within the RSRPA algorithm itself.

Comparison of RSGUI with two other rough set based systems, RSES and Rosetta, resulted mainly in showing what RSGUI is not. RSES is a hybrid system that permits users to run experiments on their data by choosing from a large array of rule generation algorithms and classification methods. Rosetta is a commercially available system that performs database management tasks, in addition to a full range of rough set operations.

However, RSGUI is unique in its ability to apply the reverse prediction algorithm. Traditionally, condition attribute values predict decision attribute values. In reverse prediction, the decision attribute values predict the condition attribute values. RSGUI is an interface for experimenting with the reverse prediction algorithm. The method of evaluating RSRPA needs improvement, although the implemented strategy gave positive results. Playing the rough set team against randomly generated teams may not give a fair evaluation of the success of RSRPA. Use of randomly generated teams assumes a uniform distribution. But randomly generated teams may exhibit a different distribution (e.g., normal). A more representative collection of opposing teams is needed to better test the effectiveness of reverse prediction.

Two properties of attributes have been discussed: condition (C) or decision (D) and given (G) or predicted (P) leading to four possibilities:

1. $\{ C, G \} \longrightarrow \{ D, P \}$
2. $\{ D, G \} \longrightarrow \{ C, P \}$
3. $\{ D, P \} \longrightarrow \{ C, G \}$
4. $\{ C, P \} \longrightarrow \{ D, G \}$

1 is ordinary prediction. 2 is derivable from 1, and 3 is derivable from 4, by interchanging the roles of condition and decision attributes. 4 is reverse prediction that has been the subject of this chapter.

## References

1. Johnson, J., Campeau, P.: In: Ślęzak, D., Wang, G., Szczuka, M., Düntsch, I., Yao, Y. (eds.) RSFDGrC 2005. LNCS (LNAI), vol. 3641, pp. 88–97. Springer, Heidelberg (2005)
2. Shah, A.S., Knuth, K.H., Lakatos, P., Schroeder, C.E.: Lessons from applying differentially variable component analysis (dVCA) to electroencephalographic activity. In: AIP Conference Proceedings, vol. 707, pp. 167–181 (2003)
3. Shah, A.S., Knuth, K.H., Truccolo, W.A., Ding, M., Bressler, S.L., Schroeder, C.E.: A Bayesian approach to estimate coupling between neural components: Evaluation of the multiple component event related potential (mcERP) algorithm. In: AIP Conference Proceedings, vol. 659, pp. 23–38 (2002)
4. Pawlak, Z.: International Journal of Intelligent Systems 18, 487–498 (2003)
5. Slezak, D.: Rough sets and Bayes' factor. In: Peters, J.F., Skowron, A. (eds.) Transactions on Rough Sets III. LNCS, vol. 3400, pp. 202–229. Springer, Heidelberg (2005)
6. Pawlak, Z.: Flow graphs and data mining. In: Peters, J.F., Skowron, A. (eds.) Transactions on Rough Sets III. LNCS, vol. 3400, pp. 1–36. Springer, Heidelberg (2005)

7. Warren, R.H., Johnson, J.A., Huang, G.H.: Application of rough sets to environmental engineering modeling. In: Transactions on Rough Sets I, vol. 3400, pp. 202–229 (2004)
8. Wong, S.K., Ziarko, W.: A machine learning approach to information retrieval. In: Proceedings of the 9th Int. Conf. on R & D in Information Retrieval, pp. 228–233 (1986)
9. Grzymala-Busse, J.: A comparison of three strategies to rule induction from data with numerical attributes. In: Proc. Int. Workshop on Rough Sets in Knowledge Discovery, vol. 82, pp. 132–140 (2003)
10. Grzymala-Busse, J.: MLEM2– discretization during rule induction. In: Proc. IIPWM 2003. Int. Conf. on Intelligent Information Processing and WEB Mining Systems, pp. 499–508 (2003)
11. Grzymala-Busse, J.: A rough set approach to data with missing attribute values. In: Wang, G.-Y., Peters, J.F., Skowron, A., Yao, Y. (eds.) RSKT 2006. LNCS (LNAI), vol. 4062, pp. 58–67. Springer, Heidelberg (2006)
12. Grzymala-Busse, J.: Incomplete data and generalization of indiscernibility relation, definability and approximations. In: Ślęzak, D., Wang, G., Szczuka, M., Düntsch, I., Yao, Y. (eds.) RSFDGrC 2005. LNCS (LNAI), vol. 3641, pp. 244–253. Springer, Heidelberg (2005)
13. Sever, H., Gorur, A., Tolun, M.R.: Text categorization with ILA. In: Yazıcı, A., Şener, C. (eds.) ISCIS 2003. LNCS, vol. 2869, pp. 300–307. Springer, Heidelberg (2003)
14. Tolun, M.R., Sever, H., Uludag, M., Abu-Soud, S.M.: Cybernetics and Systems. An International Journal 30(7), 609–628 (1999)
15. Grzymala-Busse, J., Rzasa, W.: Local and global approximations for incomplete data. In: Proc. of Rough Sets and Current Trends in Computing, pp. 244–225 (2006)
16. Bazan, J., Szczuka, M.: In: Peters, J.F., Skowron, A. (eds.) Transactions on Rough Sets III. LNCS, vol. 3400, pp. 37–56. Springer, Heidelberg (2005)
17. Weng, L., Hongyue, D., Zhan, Y., He, Y., Stepaniants, S.B., Bassett, D.E.: Bioinformatics 22(9), 1111–1121 (2006)
18. Menon, C., Lakos, C.: Towards a semantic basis for Rosetta. In: Estivill-Castro, V. (ed.) ACSC 2004. Computer Science 2004, Twenty-Seventh Australasian Computer Science Conference, vol. 26, pp. 175–184 (2004)
19. Midelfart, H., Komorowski, J., Nørsett, K., Yadetie, F., Sandvik, A.K., Laegreid, A.: Fundamenta Informaticae 53(2), 155–183 (2002)

# An Algorithm for the Shortest Path Problem on a Network with Fuzzy Parameters Applied to a Tourist Problem

Fábio Hernandes[1], Maria Teresa Lamata[2], José Luis Verdegay[2], and Akebo Yamakami[3]

[1] Dpto. de Ciência da Computação, Universidade Estadual do Centro-Oeste C.P. 3010, 85015-430, Guarapuava-PR, Brazil
hernandes@unicentro.br

[2] Dpto. de Ciencias de la Computación e I. A., E.T.S. de Ingeniería Informática, Universidad de Granada E-18071, Granada, Spain
{mtl,verdegay}@decsai.ugr.es

[3] Dpto. de Telemática, Faculdade de Engenharia Elétrica e de Computação Universidade Estadual de Campinas C.P. 6101, 13083-970, Campinas-SP, Brazil
akebo@dt.fee.unicamp.br

**Summary.** In problems of graphs involving uncertainties, the shortest path problem is one of the most studied topics as it has a wide range of applications in different areas (e.g. telecommunications, transportation, manufacturing, etc.) and therefore warrants special attention. However, due to its high computational complexity, previously published algorithms present peculiarities and problems that need to be addressed (e.g. they find costs without an existing path, they determine a fuzzy solution set but do not give any guidelines to help the decision-maker choose the best path, they can only be applied in graphs with fuzzy non-negative parameters, etc.). Therefore, in this chapter is presented an iterative algorithm with a generic order relation that solves the cited disadvantages. This algorithm is applied in a tourist problem. It has been implemented using certain order relations, where some can find a set of fuzzy path solutions while others find only the shortest path.

## 1 Introduction

The problem of finding the shortest path from a specified source node to the other nodes is a fundamental matter that appears in many applications, for example: transportation, routing, communications and recently in supply chain management.

Let $G = (V, E)$ be a graph, where $V$ is the set of vertices and $E$ is the set of edges. A *path* between two nodes is an alternating sequence of vertices and edges starting and ending with the vertices. The *length* (cost) of a path is the sum of the weights of the edges on the path. However, since there can be more than one path between two vertices, there is then the problem of finding a path with the minimum cost between these two specified vertices. In classical graph theory, the weight of each edge is a crisp number. However, most applications for this

R. Bello et al. (Eds.): Granular Computing, STUDFUZZ 224, pp. 307–320, 2008.
springerlink.com 

problem have parameters that are not naturally precise, i.e. costs, capacities, demands, etc, and in such cases, fuzzy numbers based on fuzzy set theory (see [1]) can be applied. This problem is called *fuzzy shortest path problem.*

In the fuzzy shortest path problem, the final costs (time) are fuzzy numbers, which is difficult to find a smaller path than all the other existing paths. It is therefore often hard to find a fuzzy cost, which is strictly smaller than the other costs.

In this chapter we will apply the fuzzy shortest path problem in a tourist problem, where the uncertainties are in the time (parameters of the arcs of the network). Then, the main objective is to find the shortest path between some tourist points of the San Salvador city, in Brazil.

In the literature there are various papers on this subject. The paper by Dubois and Prade [2] is one of the first on this topic; it considers extensions of the classic Floyd and Ford-Moore-Bellman algorithms. Nevertheless, it has been verified that both algorithms can return solutions with lengths without an associated path (see [3]), and these problems were outlined by Klein [4] with the fuzzy dominance set. Another algorithm for finding the shortest path was presented by Okada and Gen [5, 6], where there is a generalization of Dijkstra's algorithm in which the weights of the arcs are given as intervals. Okada and Soper [7] characterized the solution as a fuzzy set, where each element is a nondominated path or Pareto Optimal path with fuzzy edge weights. Blue et al. [8] presented an algorithm that finds a cut value to limit the number of analyzed paths, and then applied a modified version of the k-shortest path (crisp) algorithm proposed by Eppstein [9]. Okada [10] follows the idea of finding a fuzzy set solution; he introduced the concept of the degree of possibility of an arc being on the shortest path. Nayeem and Pal [11] presented an algorithm which gives a single fuzzy shortest path or a guideline for choosing the best fuzzy shortest path according to the decision-maker's viewpoint.

Analyzing these articles, it is clear that they present peculiarities and/or problems that warrant attention (i.e. they can find costs without an existing path; they determine a fuzzy solution set but do not provide decision-makers with any guidelines for choosing the best path; they can only be applied in graphs with fuzzy non-negative parameters, but there are certain real problems where negative parameters appear that need to be analyzed (see [12])). Consequently, Hernandes et al [13, 14] proposed an iterative algorithm for the shortest path problem in graphs with fuzzy parameters. This algorithm is based on the Ford-Moore-Bellman algorithm [15], and it is presented with a generic order relation, i.e. decision-makers can choose, or propose, the order relation that best suits their problem. It has some advantages, such as: it can be applied in graphs with negative parameters and can detect whether there are negative circuits; it can be implemented by using a variety of order relations in such a way that, first, when decision-maker only looks for a path, the algorithm can find it; but, second, if the decision-maker is looking for a diversity of such paths, also the algorithm, depending on the order relation selected, could find it.

In this chapter we present the generic algorithm, proposed by Hernandes et al [13, 14], and an application for it, based on a tourist problem.

This chapter is organized as follows: Section 2 introduces some basic concepts. Section 3 presents the proposed algorithm. Section 4 outlines an illustrative example in the tourist context whose results are commented and analyzed. Finally, Section 5 is devoted to outline the main conclusions.

## 2 Concepts and Terminology

In this Section some well-known concepts needed in the rest of the chapter are introduced.

### 2.1 Fuzzy Numbers

**Definition 1.** A *triangular fuzzy number* is represented by $\tilde{a} = (m, \alpha, \beta)$, with the membership function, $\mu_{\tilde{a}}(x)$, defined by the expression:

$$\mu_{\tilde{a}}(x) = \begin{cases} 0, & \text{if } x \leq m - \alpha \\ \frac{x-(m-\alpha)}{\alpha}, & \text{if } m - \alpha < x < m \\ 1, & \text{if } x = m \\ \frac{(m+\beta)-x}{\beta}, & \text{if } m < x < m + \beta \\ 0, & \text{if } x \geq m + \beta \end{cases} \quad (1)$$

where $m$ is the centre; $\alpha$ is the left spread and $\beta$ is the right spread.

***Remark:*** The membership degrees between the values $m - \alpha$ and $m$ will be represented by $f^L_{\tilde{a}}(x)$ and those between $m$ and $m + \beta$ by $f^R_{\tilde{a}}(x)$.

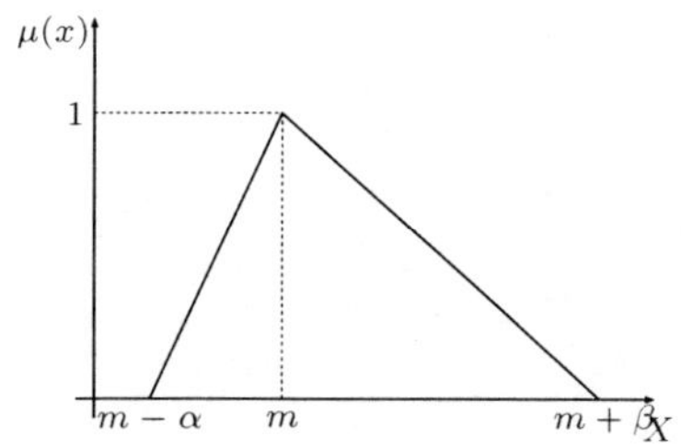


**Fig. 1.** Example of triangular fuzzy number

**Definition 2.** *Modal value* is the value $x \in [m - \alpha, m + \beta]$ when the membership function has the maximum value.

**Definition 3.** Let $A$ be a fuzzy subset of $X$, the $\alpha - cut$ of $A$, denoted by $A_\alpha$, is a set consisting of those elements of $X$ whose membership values exceed the level $\alpha$,

$$A_\alpha = \{x|A(x) \geq \alpha\} \quad (2)$$

where $\alpha \in [0, 1]$.

**Definition 4.** Let $\tilde{a}$ and $\tilde{b}$ be two fuzzy numbers, $\tilde{a} = (m_1, \alpha_1, \beta_1)$ and $\tilde{b} = (m_2, \alpha_2, \beta_2)$. Then the *fuzzy sum* of these two numbers is given by:

$$\tilde{a} \oplus \tilde{b} = (m_1, \alpha_1, \beta_1) \oplus (m_2, \alpha_2, \beta_2) = (m_1 + m_2, \alpha_1 + \alpha_2, \beta_1 + \beta_2). \quad (3)$$

### 2.2 Generic Order Relation

Various methods for ordering and ranking fuzzy numbers have been proposed. However, as mentioned above, in this work is considered an algorithm with a generic order relation, i.e. decision-maker will be able either to choose an order relation from the existing ones or to propose another for the problem. The order relation of the algorithm is defined as:

**Definition 5.** Let $\tilde{a}$ and $\tilde{b}$ be two triangular fuzzy numbers, then $\tilde{a}$ *is preferred to* $\tilde{b}$ $(\tilde{a} \prec \tilde{b})$ *iff* $\tilde{a} < \tilde{b}$.

The function $f(*)$ can be applied using any comparison criterion for fuzzy numbers.

In particular, in this chapter will be implemented the following comparison criterion: Yager's center of gravity method [16, 17, 18], García and Lamata order relation [19], Liou and Wang's index [20] and Okada and Soper's order relation [7]. The definitions about these order relation are in Section 4.

## 3 Proposed Algorithm

As previously stated, the proposed algorithm is an adaptation of the Ford-Moore-Bellman algorithm [15] for classic graphs. As Gondran and Minoux [21] stated, the Ford-Moore-Bellman algorithm analysis all the nodes per iteration and not only one. As the considered algorithm is an adaptation of the Ford-Moore-Bellman algorithm, the structure is maintained.

This is an iterative algorithm, possessing as stop criterion the number of iterations or the non-alteration of the costs of the paths between the previous iteration and the current one. If the number of iterations is equal the number of nodes and the costs of the paths between the related iteration and the previous one are different then, there is a negative circuit and the cost of paths are reducing in each iteration, therefore the algorithm has an infinite loop and then it is finished.

The algorithm outperforms the following steps: in Step 0, similar to the classic Ford-Moore-Bellman algorithm, initial labels and lengths (costs) are attributed to the paths, that is, is attributed a path with value $\infty$ between the nodes 1 and $i$ $(i > 1)$. In Step 1, all the paths between 1 and $i$ are found, the order relation chosen is applied and are eliminated the dominated paths. In Step 2 the stop criterion is verified: if it is satisfied, the paths are built in the Step 3, otherwise it is necessary to return to Step 1.

### 3.1 Description of the Algorithm

Some of the notations used for the algorithm are presented:

$(m+\beta)^i$: right spread $(m+\beta)$ of cost of arc $(i,j)$;

$it$: iteration counter;

$\tilde{c}_{ji}$: cost (length) of edge $(j,i)$;

$\tilde{c}^{it}_{(i,k)}$: length of path between node 1 and $i$ with label $k$ in the iteration $it$;

$M$: a large number, substituting $\infty$ in Ford-Moore-Bellman's algorithm;

$\Gamma_i^{-1}$: set of predecessor nodes of $i$;

$f(\tilde{c}^{it}_{(i,k)})$: order relation applied in the length of path with label $k$, between nodes 1 and $i$, in the iteration $it$.

---

**Algorithm**

---

**Step 0:** [Initialization]

1. $\tilde{c}^{0}_{(1,1)} = (0,0,0)$
2. $\tilde{c}^{0}_{(j,1)} = (M+2,1,1)$, $j = 2,3,...,r$
   - where:
     - $r$ is a number of nodes;
     - $M = \sum_{i=1}^{E} \left|(m+\beta)^i\right|$;
     - $E$ : number of nodes;
3. $it \leftarrow 1$.

**Step 1:** [Determination of the paths and verifications of the order relation]

1. $\tilde{c}^{it}_{(1,1)} = (0,0,0)$;
2. $\forall j \in \Gamma_i^{-1}$, $i = 2,3,\dots r$, do:
   - $\tilde{c}^{it}_{(i,k1)} = \tilde{c}^{it-1}_{(j,k2)} \oplus \tilde{c}_{ji}$
3. Label scan and dominance check
   Between all the labels of the node $i$ do:
   - If $f(\tilde{c}^{it}_{(i,k1)}) > f(\tilde{c}^{it}_{(i,k2)}) \Rightarrow$ delete the label $k1th$ label;

- If $f(\tilde{c}^{it}_{(i,k1)}) < f(\tilde{c}^{it}_{(i,k2)}) \Rightarrow$ delete the label $k2th$ label;

**Step 2:** [Stop criterion]

1. If ($\tilde{c}^{it}_{(i,k1)} = \tilde{c}^{it-1}_{(i,k1)}$ or $it = r$) ($N$ : set of nodes)
   - If $it = r$ and ($\tilde{c}^{it}_{(i,k1)} \neq \tilde{c}^{it-1}_{(i,k1)} \Rightarrow$ Step 4 (negative circuit $\Rightarrow$ infinite loop);
   - Otherwise go to Step 3.
2. Otherwise $it \leftarrow it + 1 \Rightarrow$ return to Step 1.

**Step 3:** [Shortest paths composition]

- Find the shortest paths from 1 to $i$ ($i = 2, 3, ..., r$).

**Step 4:** [END]

---

It is important to quote that although we considered in this algorithm the parameters as triangular fuzzy numbers, for trapezoidal fuzzy numbers the generic order relation is the same and the complexity is not different.

### 3.2 Computational Complexity

Since Ford-Moore-Bellman's algorithm converges in the case of non-existence of a negative circuit, in a maximum of $r - 1$ iterations (where $r$ is the number of nodes); then the proposed algorithm will also converge in the maximum of $r - 1$ iterations.

Considering that Okada and Soper's order relation is the most expansive, the proposed algorithm complexity will be applied on it: in the Step 1 has a maximum of $rV_{max}$ additions to calculate the cost for each path, where $V_{max}$ is the maximum number of labels of all the nodes. In the Step 2 has in the maximum $rV^2_{max}$ comparisons of dominances. Then, the complexity for each iteration is $O(rV^2_{max})$. Therefore, the proposed algorithm has a complexity of $O((r-1)(rV^2_{max})) = O(r^2V_{max})$.

## 4 Computational Results

In this section we present the definitions of the implemented order relations and an illustrative example. This example presents some tourist points of San Salvador City (Brazil), where the objective is to find the shortest path between Centro (region which has some hotels) and other important tourist points. The uncertainties are in the time, where we have the time (in minute) to traverse between two tourist points (Figure 2). These uncertainties are formulated as triangular fuzzy numbers.

### 4.1 Implemented Order Relations

As said previously, the following order relations were implemented in this chapter: Yager's center of gravity method [16, 17, 18], García and Lamata order relation [19], Liou and Wang's index [20] and Okada and Soper's order relation [7].

These order relations have been chosen because of the following reasons: if the decision-maker wishes to use a defuzzification index that considers the center of gravity, (s)he could use Yager's index. If (s)he would like to use the upper spread of the fuzzy number, (s)he would choose the Liou and Wang index with $\lambda = 1$. In the case that decision-maker would like to use the lower spread, then (s)he could consider the García and Lamata index, with $\lambda = 0$ and $\delta = 0.01$. When decision maker were to be interested in the modal value, (s)he could choose García and Lamata's index, with $\lambda = 0$ and $\delta = 1$. Besides, Okada and Soper relation will be implemented because this relation finds a solution set of nondominated paths, and hence the decision-maker can get a set of path to choose. These order relations are defined in the following.

**a) Yager's first index** [16, 17, 18]
**Definition 6.** The Yager's first defuzzification index is the center of gravity method and it is defined as: for any $\tilde{a} \in S$, then

$$f(\tilde{a}) = \frac{\int_0^1 \alpha \tilde{a}_\alpha d\alpha}{\int_0^1 \tilde{a}_\alpha d\alpha}. \tag{4}$$

**b) Liou and Wang index** [20]
Liou and Wang considered an ordinance method of fuzzy numbers with integral values. This method is independent of the type of membership function and the functions are not normalized. This method can analyze more than two numbers simultaneously. They defined the ordinance index in accordance with the areas related to the spreading to the right and to the left of the fuzzy number in question.

**Definition 7.** For any fuzzy number $\tilde{a} = (m, \alpha, \beta) \in S$, the Liou and Wang index is defined as:

$$LW^\lambda(\tilde{a}) = \lambda S_D(\tilde{a}) + (1 - \lambda) S_I(\tilde{a}) \tag{5}$$

where,

$$S_D(\tilde{a}) = m + \int_m^{m+\beta} f_{\tilde{a}}^R(x)dx = \int_0^1 f_{\tilde{a}}^{R^{-1}}(y)dy \tag{6}$$

$$S_I(\tilde{a}) = (m - \alpha) + \int_{m-\alpha}^{m} f_{\tilde{a}}^L(x)dx = \int_0^1 f_{\tilde{a}}^{L^{-1}}(y)dy \tag{7}$$

are the left and right areas associated to the respective spreads, and $\lambda \in [0, 1]$ is a degree that might reflect the decision-maker's optimism/pessimism.

**c) García and Lamata index** [19]
As Liou and Wang's index is not capable of differentiating certain fuzzy numbers, García and Lamata [19] proposed an alternative solution. The proposed index improves the discrimination between fuzzy numbers, retaining an optimism degree ($\lambda$) and including the modality index $\delta \in [0, 1]$. This is called the *modality index* and represents the proximity of the choice of the decision-maker in relation to the modal one. In other words, $\delta$ give us the weighting of the central value and $(1 - \delta)$ the weighting of the extreme values.

**Definition 8.** Let $\tilde{a} = (m, \alpha, \beta)$ be the fuzzy number, the García and Lamata index is defined as:

$$I(\tilde{a}) = (1 - \delta)[\lambda S_D(\tilde{a}) + (1 - \lambda)S_I(\tilde{a})] + \delta m \quad (8)$$

where: $m$ is the modal value and $S_D$ and $S_I$ are the left and right areas.

**d) Okada and Soper relation** [7]
A generalized order relation for flat fuzzy numbers is given in Okada and Soper [7] and as a special case; the order relation for triangular fuzzy numbers can be developed as:

**Definition 9.** Let $\tilde{a} = (m_1, \alpha_1, \beta_1)$ and $\tilde{b} = (m_2, \alpha_2, \beta_2)$ be two triangular fuzzy numbers,

$$\tilde{a} \prec \tilde{b} \Leftrightarrow m_1 \leq m_2, \alpha_1 \leq \alpha_2, \beta_1 \leq \beta_2 \text{ and } \tilde{a} \neq \tilde{b} \quad (9)$$

### 4.2 Illustrative Example

The $21^{st}$ century shows a steady increase in tourism all over the world. The level of organization, methods of transport and the facilities available at destination points have enjoyed an accelerated pace of improvement.

We live in a world formed by networks: biological, social, technological and in the tourist case the networks of transportation. The study and characterization of such networks have boosted recently by the emergency of new ideas, increasing network database and available computational power to test models and link them to data.

In this context we present in the following a small tourist network in Salvador de Bahia, Brazil (Figure 2) to which we will apply the algorithm described in Section 3.1, here implemented in Matlab 7.0.

It is important to point out that decision-maker will be able to choose the order relation that best accomplishes his wishes, and hence depending on the order relation selected, an appropriate solution will be obtained.

In this network it is assumed that the uncertainties are in the parameters, more specifically with the time. The algorithms was executed considering two different situations: in the first, we have the time (in minutes) that one tourist can spend going by foot between two tourist points. The second situation is like the above one but computing now the time for handicaped people. As it may be patent, the time in this second context will be greater than in the former one.

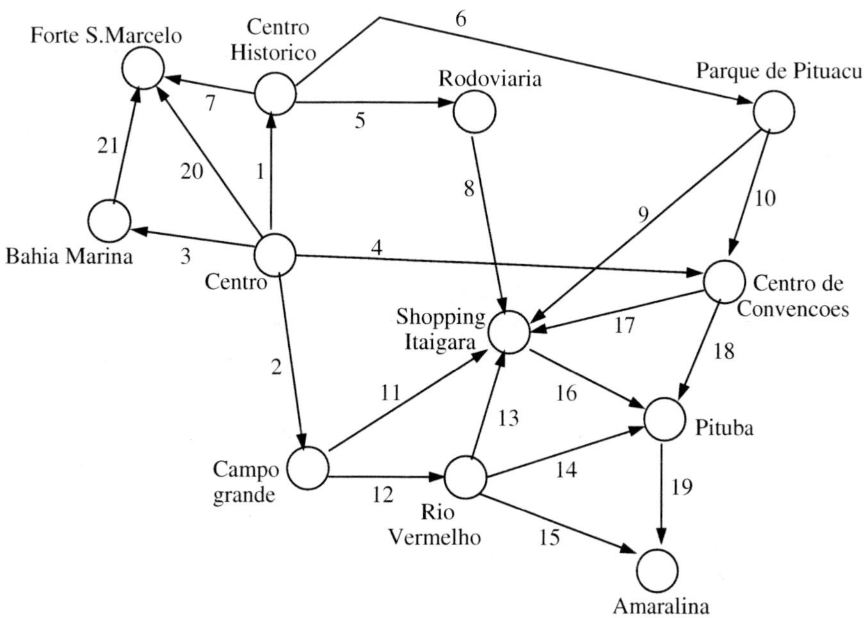


**Fig. 2.** Tourist points of San Salvador de Bahia

Considering that some hotels are near Centro and that some important tourist points are in Shopping Itaigara, Amaralina and Forte S. Marcelo, then the objectives of this problem are to find the shortest paths between Centro and these three tourist points.

Results are in the Table 2 and Table 3. In Table 2 one has the solution when the tourists do not use the handicaped way, and in Table 3 are the results when the tourists need to use the handicaped way.

Looking at Table 2, we are to conclude these different results are due to we have used different order relations,each characterized by its own properties and meanings (center of gravity, optimism/pessimism degrees, modality, etc.)

In short, from these results we can conclude that:

If the tourist wants to visit the city in a day that does not have traffic, is interesting that (s)he uses the García and Lamata order relation ($\lambda = 0$ and $\delta = 0.01$), therefore probably (s)he goes to traverse these places faster and will go to save time. The suggested paths are:

- Centro → Centro Histrico → Forte S. Marcelo;
- Centro → Campo Grande → Rio Vermelho → Shopping Itaigara;
- Centro → Centro de Convencoes → Pituba → Amaralina.

If the tourist visits the places in a weekday (normal traffic), (s)he would use the García and Lamata order relation ($\lambda = 0$ and $\delta = 1$), in this case we have three paths between Centro and Shopping Itaigara. The suggested paths are:

- Centro → Centro Historico → Forte S. Marcelo;
- Centro → Campo Grande → Rio Vermelho → Shopping Itaigara;

**Table 1.** Edge Information - Figure 2

| Arc | Source Node | Destination Node | Time[1,2] | Time[1,3] |
|---|---|---|---|---|
| 1 | Centro | Centro Historico | (6 1 2) | (15 6 4) |
| 2 | Centro | Campo Grande | (10 3 5) | (23 10 6) |
| 3 | Centro | Bahia Marina | (5 1 1) | (7 3 3) |
| 4 | Centro | Centro de Convencoes | (14 2 6) | (24 6 10) |
| 5 | Centro Historico | Rodoviaria | (8 1 1) | (18 3 2) |
| 6 | Centro Historico | Parque de Pituacu | (16 1 3) | (30 5 10) |
| 7 | Centro Historico | Forte S. Marcelo | (4 1 3) | (10 2 7) |
| 8 | Rodoviaria | Shopping Itaigara | (9 3 1) | (23 8 7) |
| 9 | Parque de Pituacu | Shopping Itaigara | (13 1 5) | (20 3 6) |
| 10 | Parque de Pituacu | Centro de Convencoes | (4 1 1) | (10 3 4) |
| 11 | Campo Grande | Shopping Itaigara | (9 3 1) | (23 8 7) |
| 12 | Campo Grande | Rio Vermelho | (6 1 1) | (8 2 2) |
| 13 | Rio Vermelho | Shopping Itaigara | (5 2 1) | (10 2 5) |
| 14 | Rio Vermelho | Pituba | (10 2 2) | (20 4 4) |
| 15 | Rio Vermelho | Amaralina | (10 2 2) | (14 4 3) |
| 16 | Shopping Itaigara | Pituba | (5 1 1) | (10 4 5) |
| 17 | Centro de Convencoes | Shopping Itaigara | (7 3 3) | (18 8 8) |
| 18 | Centro de Convencoes | Pituba | (6 1 2) | (13 6 6) |
| 19 | Pituba | Amaralina | (5 1 1) | (7 3 2) |
| 20 | Centro | Forte S. Marcelo | (11 1 0) | (20 4 4) |
| 21 | Bahia Marina | Forte S. Marcelo | (6 1 0) | (14 4 4) |

[1] Approximations.

[2] Path by foot (for not handicaped).

[3] Path by an special way (for handicaped).

**Table 2.** Results of Figure 2 (by foot)

| Destination Node | Arcs | Total Time | Order Relations |
|---|---|---|---|
| Forte S. Marcelo | 3 and 21 | (11 2 1) | Yager and OS |
| Forte S. Marcelo | 20 | (11 1 0) | LW(1) and OS |
| Forte S. Marcelo | 1 and 7 | (10 2 5) | GL(0,0.1), GL(0,1) and OS |
| Shopping Itaigara | 2, 12 and 13 | (21 6 7) | Yager, GL(0,0.1), GL(0,1) and OS |
| Shopping Itaigara | 2 and 11 | (21 5 6) | LW(1), GL(0,1) and OS |
| Shopping Itaigara | 4 and 17 | (21 5 9) | GL(0,1) |
| Amaralina | 4, 18 and 19 | (25 4 9) | Yager, GL(0,0.1), GL(0,1), LW(1) and OS |
| Amaralina | 2, 12 and 15 | (26 6 8) | OS |

- Centro → Campo Grande → Shopping Itaigara;
- Centro → Centro de Convencoes → Shopping Itaigara;
- Centro → Centro de Convencoes → Pituba → Amaralina.

If the tourist prefers to visit the places in days with traffic jam (holiday), (s)he could to choose the Liou and Wang index ($\lambda = 1$). The paths are:

- Centro → Campo Grande → Shopping Itaigara;
- Centro → Centro de Convencoes → Pituba → Amaralina;
- Centro → Forte S. Marcelo.

If the tourist wants to know the central value (center of gravity), the First Yager index is the ideal and the paths are:

- Centro → Campo Grande → Shopping Itaigara;
- Centro → Centro de Convencoes → Pituba → Amaralina;
- Centro → Bahia Marina → Forte S. Marcelo.

The Okada and Soper order relation presents a set solution of nondominated paths. If the tourist wants to know a paths set, this order relation is ideal. The paths are:

- Centro → Centro Historico → Forte S. Marcelo;
- Centro → Bahia Marina → Forte S. Marcelo;
- Centro → Forte S. Marcelo
- Centro → Campo Grande → Rio Vermelho → Shopping Itaigara;
- Centro → Campo Grande → Shopping Itaigara;
- Centro → Centro de Convencoes → Pituba → Amaralina;
- Centro → Campo Grande → Rio Vermelho → Amaralina.

In the Table 3 we have the results of the proposed algorithm for tourists that need the bus to move around.

**Table 3.** Results of Figure 2 (by an special way)

| Destination Node | Arcs | Total Time | Order Relations |
|---|---|---|---|
| Forte S. Marcelo | 3 and 21 | (21 7 7) | GL(0,0.1) and OS |
| Forte S. Marcelo | 20 | (20 4 4) | Yager, LW(1), GL(0,1) and OS |
| Shopping Itaigara | 2, 12 and 13 | (41 14 13) | GL(0,0.1), GL(0,1) and OS |
| Shopping Itaigara | 2 and 11 | (42 14 10) | Yager, LW(1) and OS |
| Amaralina | 4, 18 and 19 | (44 15 18) | GL(0,0.1), GL(0,1) and OS |
| Amaralina | 2, 12 and 15 | (45 16 11) | Yager, LW(1)and OS |

The conclusions are the similar of the Table 2. The results are the following: García and Lamata order relation ($\lambda = 0$ and $\delta = 0.01$):

- Centro → Bahia Marina → Forte S. Marcelo;
- Centro → Campo Grande → Rio Vermelho → Shopping Itaigara;
- Centro → Centro de Convencoes → Pituba → Amaralina.

García and Lamata order relation ($\lambda = 0$ and $\delta = 1$):

- Centro → Forte S. Marcelo;
- Centro → Campo Grande → Rio Vermelho → Shopping Itaigara;
- Centro → Centro de Convencoes → Pituba → Amaralina.

Liou and Wang index ($\lambda = 1$):

- Centro → Forte S. Marcelo;
- Centro → Campo Grande → Shopping Itaigara;
- Centro → Campo Grande → Rio Vermelho → Amaralina.

Yager index:

- Centro → Forte S. Marcelo;
- Centro → Campo Grande → Shopping Itaigara;
- Centro → Campo Grande → Rio Vermelho → Amaralina.

Okada and Soper order relation:

- Centro → Bahia Marina → Forte S. Marcelo;
- Centro → Forte S. Marcelo;
- Centro → Campo Grande → Rio Vermelho → Shopping Itaigara;
- Centro → Campo Grande → Shopping Itaigara;
- Centro → Centro de Convenes → Pituba → Amaralina;
- Centro → Campo Grande → Rio Vermelho → Amaralina.

## 5 Conclusions

In problems of graphs involving uncertainties, the shortest path problem is one of the most studied topics since it has a wide range of applications in different areas and therefore deserves special attention.

In this chapter was presented a generic algorithm for this problem. This algorithm can be implemented using the order relation chosen by the decision-maker, and can work with crisp numbers, using defuzzification indices, or with fuzzy numbers. Depending on the order relation used by the decision-maker, this algorithm can return a set of shortest paths or a single path as the solution. It is worth emphasizing that unlike algorithms considered in literature, this can be executed on a network with negative parameters and can detect the existence of a negative circuit.

This algorithm was executed in a tourist context and it was implemented using some order relations. But the algorithm can be adapted for different fuzzy numbers and the generic order relation can be used for different defuzzification index and fuzzy numbers.

In addition to these advantages, the computational structure of the algorithm is easy to implement, and its computational complexity $O(r^2 V_{max}^2)$ is better than, for example, Okada and Soper's algorithm $O(r^3 V_{max}^2)$.

In future works is intended to study other order relations, to execute this algorithm on networks of differing sizes and densities, to include of time windows, and the implementation of a decision support systems.

## Acknowledgements

Research carried out under projects CAPES (1249/05), TIN2005-02418, TIN2005-08404-C04-01, TIC-00129 (MINAS) and TIN2005-024790-E.

## References

1. Dubois, D., Prade, H.: Ranking fuzzy numbers in the setting of possibility theory. Information Sciences 30, 183–224 (1983)
2. Dubois, D., Prade, H.: Fuzzy sets and systems: Theory and applications. Academic Press, New York (1980)
3. Takahashi, M.T.: Contribuições ao estudo de grafos fuzzy: Teoria e aplicações (in Portuguese). Thesis, State University of Campinas, Campinas, Brazil (2004)
4. Klein, C.M.: Fuzzy shortest paths. Fuzzy Sets and Systems 39, 27–41 (1991)
5. Okada, S., Gen, M.: Order relation between intervals and its application to shortest path problem. In: Proceedings of the 15th Annual Conference on Computers and Industrial Engineering, vol. 25, pp. 147–150 (1993)
6. Okada, S., Gen, M.: Fuzzy shortest path problem. In: Proceedings of the 16th Annual Conference on Computers and Industrial Engineering, vol. 27, pp. 465–468 (1994)
7. Okada, S., Soper, T.: A shortest path problem on a network with fuzzy arc lengths. Fuzzy Sets and Systems 109, 129–140 (2000)
8. Blue, M., Bush, B., Puckett, J.: Unified approach to fuzzy graph problems. Fuzzy Sets and Systems 125, 355–368 (2002)
9. Eppstein, D.: Finding the k-shortest paths. In: Proceedings of the IEEE Symposium on Foundations of Computer Science, pp. 154–165 (1994)
10. Okada, S.: Fuzzy shortest path problems incorporating interactivity among paths. Fuzzy Sets and Systems 142(3), 335–357 (2004)
11. Nayeem, S.M.A., Pal, M.: Shortest path problem on a network with imprecise edge weight. Fuzzy Optimization and Decision Making 4, 293–312 (2005)
12. Ahuja, R.K., Magnanti, T.L., Orlin, J.B.: Network flows. Prentice Hall, Englewood Cliffs (1993)
13. Hernandes, F., Lamata, M.T., Verdegay, J.L., Yamakami, A.: A generic algorithm for the shortest path problem on a network with fuzzy parameters. In: Proceedings of the International Symposium on Fuzzy Rough Sets, Santa Clara, Cuba (2006)
14. Hernandes, F., Lamata, M.T., Verdegay, J.L., Yamakami, A.: The shortest Path Problem on Networks with Fuzzy Parameters. Fuzzy Sets and Systems 158, 1561–1570 (2007)
15. Bellman, R.E.: On a routing problem. Quarterly Applied Mathematics 16, 87–90 (1958)
16. Yager, R.R.: Ranking fuzzy subsets over the unit interval. In: Proceedings of the CDC, pp. 1435–1437 (1978)
17. Yager, R.R.: On choosing between fuzzy subsets. Kybernetes 9, 151–154 (1980)
18. Yager, R.R.: A procedure for ordering fuzzy subsets of the unit interval. Information Sciences 24, 143–161 (1981)
19. García, M.S., Lamata, M.T.: A modification of the index of Liou and Wang for ranking fuzzy numbers. International Journal of Uncertainty, Fuzziness and Knowledge-Based Systems (2007)

20. Liou, T.-S., Wang, M.-J.: Ranking fuzzy numbers with integral value. Fuzzy Sets and Systems 50, 247–255 (1992)
21. Gondran, M., Minoux, M.: Graphs and algorithms. John Wiley and Sons, New York (1984)

# PID Control with Fuzzy Adaptation of a Metallurgical Furnace

Mercedes Ramírez Mendoza[1] and Pedro Albertos[2]

[1] Department of Automatic Control, Faculty of Electrical Engineering
Universidad de Oriente, Cuba
mramirez@fie.uo.edu.cu

[2] Department of Systems Engineering and Control
Universidad Politécnica de Valencia, Spain
pedro@aii.upv.es

**Summary.** In this chapter a control strategy based on a combination of local PID con-trollers whose contribution is adjusted by means of fuzzy techniques is presented. The final control action is the result of a fuzzy interpolation of the control action computed by the local controllers. The proposed control structure is applied to control the temperature in a metallurgical furnace from which several local models have been experimentally obtained.

## 1 Introduction

Fuzzy logic control is one of the most fruitful research areas in fuzzy set theory and many practical applications to industrial process, as well as theoretical studies on itself, have been reported in many research works [1].

The control methodology based on the fuzzy logic is able to integrate in the same frame a great number of control problems. This is due to the intrinsic possibility of handling information (data, objectives, and models) expressed in an approximate way or with uncertainty and the capacity to implement, by means of fuzzy controllers, controllers designed by using other methodologies [2]. On the other hand, one of the characteristics that can be easily incorporated in a fuzzy system is learning capacity.

In this way, the fuzzy methodology can be considered as a framework for practical applications able to integrate different approaches in the several control levels acting in a plant, interacting with subsystems handling either numerical or logical information as well as with humans. Though the advances in theory require a formalism that escapes to the final user, perhaps the basic idea of compute with words should be maintained while it is possible, and this idea is the one that justifies their practical ap-plication instead of other techniques [3].

In this work, the control of non-linear processes is addressed. The typical control design options are: look for a specific non-linear controller, implement a robust controller able to operate in the full range of operation of the nonlinear system, or adapt the parameters of a single linear controller. The proposal here is to design a set of simple linear controllers based on a set of local models and to

R. Bello et al. (Eds.): Granular Computing, STUDFUZZ 224, pp. 321–332, 2008.
springerlink.com 

combine the control actions by using a pure gain scheduling strategy or a softer adaptation structure. In order to illustrate the advantages of this approach, the design control strategy is applied to an industrial metallurgical furnace.

For that purpose, a brief description of a metallurgical furnace, the control objectives and the motivation for the development of the proposed control system are first presented. Several experimental models for different operating conditions are obtained. The design and implementation of the appropriate controller for each zone, together with the use of different adaptation/weighting techniques to derive the final control actions, will complete the control design procedure. This control structure is illustrated with the control of the fundamental variable of the furnace: the temperature of the hottest hearth or fireplace. The control action is obtained by means of a fuzzy inference system that allows an appropriate weighting of the actions provided by the different controllers.

The essence of the scheme is that at every time instance, the controller evaluates the trend of the controlled process output to detect the possible deviation from a prescribed course. If a deviation is found, an appropriate control action according to the nature of the deviation will be generated instantaneously to correct it.

## 2 The Metallurgical Furnace

Multiple hearth furnaces are used for various gas-solid reaction processes in the extractive metallurgy industries [4]. In this case, they are dedicated to the selective reduction of lateritic ores. Selective reduction of nickel re-quires a narrow range of operating temperatures as well as a good control of the gaseous composition.

At the Nicaro plant, where our control system has been experimented, there are 11 identical furnaces in continuous operation with a total production capacity of about one million tons of reduced calcine (Ni 1.3%) per year. Each furnace consists of a metallic cylinder of 21.3 m high and 6.7 m diameter, recovered inside with fireproof material. They have 17 bricked superimposed circular hearths, numbered from up to down -from the H-0 to H-16, as shown in Fig. 1.

The mineral enters the furnace at the upper hearth (H-0) after being graded by a computerized weighing system. A central rotating shaft drives rabble arms spreading the material across the roaster hearths, turning over the concentrate charge and transferring it via drop holes to the next lower hearth. The gas flow takes place from lower to upper hearths, producing a counter current motion between gas and solid. In the first hearths, (H-0/H-4), the ore is heated and dehydrated. Then it goes to a transitional area (H-5/H-9) where a partial reduction and dissociation takes place. Finally, a strong reduction begins in H-10 [4] and reduced calcine, the final product, lies at the bottom.

### 2.1 The Influence of the Temperature in the Reduction Process

The temperature is a fundamental parameter in pyrometallurgical processes like this one, facilitating the weakening of crystalline structures of ore and, consequently, the development of the reduction reactions. In the furnace, the profile of

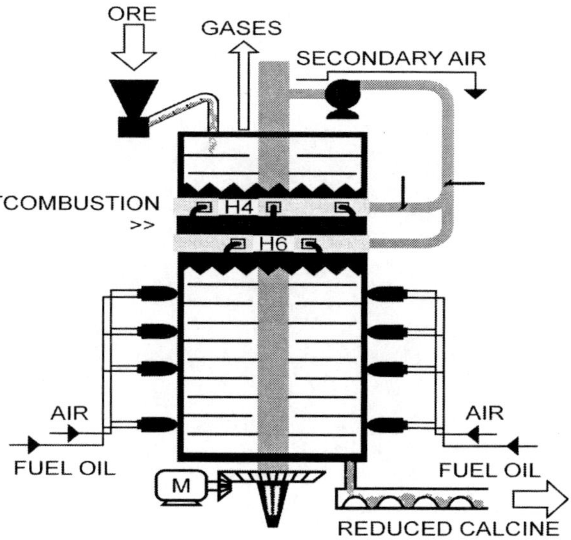


**Fig. 1.** Reduction furnace

temperature is regulated by acting on the fuel burners located in the combustion chambers coupled to the furnace, and acting on the so-called secondary air that is injected in hearths 4 and 6.

During the operation, a certain prescribed temperature profile should be maintained. The temperature should increase from the upper part to the central one in order to guarantee a gradual heating of the ore. In the hearth 4 the heating of the ore is more intense, since it is the place where the temperature reaches the highest value.

The temperature stability in hearth 4 is significantly important due to its influence over the temperature in the other hearths. Practice has shown that if a stable operation is obtained in H-4, the temperature in the other hearths is maintained constant without great difficulties. The correct reduction process demands a stable and well defined temperature profile in the furnace.

### 2.2 Description of the Combustion Process with Secondary Air in Hearth 4

In H-4 secondary air is supplied with a double purpose: to burn the reducers elements that have not intervened in the reduction reactions taking advantage of the heat removed for the heating of ore and to reduce the danger of explosions if the gassy concentrations exceed the permissible values (in particular, $CO$: 3,5% and $H_2$: 2,5%).

In this process a severe non-linearity is present due to the relationship between the temperature of the hearth 4 and the opening of the air regulation valve at the stationary state, which is in bell form and the slope of the curve depends on several factors. One of the crucial factors is the flow of mineral [5]. Several linear models can be used to represent this nonlinear behavior for different ranges of furnace load.

The existence of temperature values below certain given bounds may cause a shift of the thermal zones of the furnace, resulting in a decrease in the yield of nickel and cobalt. On the other hand, the control of temperature in the mentioned hearths contributes to a decrease in environmental contamination that takes place due to the release of polluting compounds $CO$ and $H_2$ out to the ecosystem.

## 3 The Process Model

The most important objective in controlling this process is to keep the temperature at the hearth H4 ($T_4$) at the required reference value, by acting on the airflow which passes through the valve $f_4$. For this purpose, and having the furnace in operation, several experiments have been designed to model its dynamics under different operating conditions.

Under normal operating conditions, the fuel flow has been adjusted to have a reference temperature ($T_4 = 780C$). Then, pseudo-random binary sequences have been applied on the valve opening for different load conditions in the furnace. The transfer functions which have been obtained, as well as the static gain and the dominant "time constant" of the responses are shown in Table 1. The experimental details that led to these results can be consulted in [6]. The strong difference in the linear models points out the nonlinear behavior of the process. The same happens if different output/input relationships are considered and models like these ones can be obtained for all transfer functions between any pair output/input.

To get a unique nonlinear model and design an appropriate nonlinear controller is rather difficult. Also, this behavior is easily changing due to aging or some other environmental conditions. In the same way, the variations in the characterizing parameters of the response being very large the robust controller designed to offer similar behavior under all load conditions is too conservative [7].

**Table 1.** Linear models of $T_4/f_4$

| Load (t/h) | t.f. : $T_4/f_4$ | G | T |
|---|---|---|---|
| 22 | $g_1 = \dfrac{-0.09(s-47.4)}{(s+1.5)(s+1.14)}$ | 2.5 | 1.5 |
| 23 | $g_2 = \dfrac{0.0486(s^2-8s+114)}{(s+4.626)(s+0.2885)}$ | 4.15 | 2.5 |
| 24 | $g_3 = \dfrac{0.00736(s-40)(s-9.336)}{(s+0.9846)(s+0.8526)}$ | 3.28 | 1.4 |
| 25 | $g_4 = \dfrac{0.0532(s+16.16)}{(s+13.33)(s+0.1326)}$ | 4.86 | 5.2 |

T: Response time for 63% (min).

**Table 2.** PID controllers

| $K_i$/Load (t/h) | Controller parameters | | |
|---|---|---|---|
| | $k_c$ | $T_d$ (min) | $T_i$ (min) |
| $K_1/22$ | 17.598 | 0.0688 | 1/0.275 |
| $K_2/23$ | 7.58 | 0.1175 | 1/0.47 |
| $K_3/24$ | 3.036 | 0.21 | 1/0.84 |
| $K_4/25$ | 150 | 0.4 | 1/0.1 |

### 3.1 Linear Local Controllers

Thus, different proportional-integral-derivative controllers (PID) have been designed for each local model, trying to obtain a similar behavior, although they are only valid in a limited load range. The PID controller is the type of controller most frequently used in engineering, being found in approximately 90% of the control applications [1]. There are many different industrial implementations and it is the main component in all distributed systems for process control.

The PID controller generates a control $u(t)$ based on the closed-loop error $e(t)$. It has the following standard form:

$$u(t) = k_c \left[ e(t) + T_d \frac{de(t)}{dt} + \frac{1}{T_i} \int e(t)\, dt \right] \tag{1}$$

where $k_c$, $T_d$, $T_i$ are the proportional gain, the derivative time and the integral time of the controller, respectively.

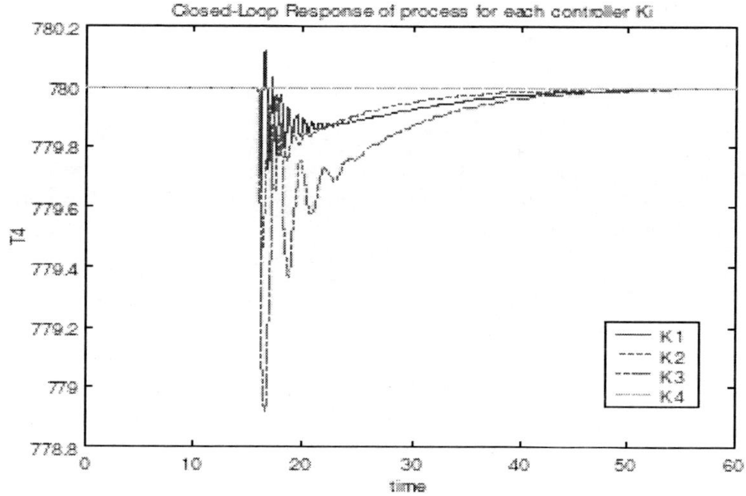


**Fig. 2.** The load disturbance responses at each zone of operation

In this case the controllers have been adjusted using the Ziegler-Nichols tuning formula, which is a classical and pioneer method being also known as the closed-loop or on-line tuning method. In this method the dynamic characteristics of the process are represented by the ultimate gain of a proportional controller and the ultimate period of oscillation of the loop [8]. Based on these two measurements, the PID parameters are selected.

In Table 2 the PID controllers' parameters are shown for each one of the transfer functions of the process in each zone. In Fig. 2, it can be seen that the temperature is appropriately controlled by each controller in its respective zone.

The set point $T_{4ref}$ is 780C and a static load disturbance of 0.5 $t/h$ is introduced into the process at $t = 10s$.

## 4 Adaptive Control

Local controllers operate satisfactorily as far as the process is in their do-main of operation (design). If the operating conditions are changing, the control should be switched from one controller to another. This results in a bumping transient and even in instability of the global process. For that purpose, the adaptation of a controller can provide robustness in the case of processes with time varying or non-linear dynamics. Another important feature is that it provides techniques for the self-tuning of controllers when the design method is based on approximate models of the process.

A common point of view in literature [9] is that the adaptive control is a special case of non-linear feedback control, where the states of the process can be separated in two categories. The states with slow dynamics are considered as parameters. This induces the idea of two time scales: a quick scale for the common feedback control and a slow one to updating the regulator parameters.

A revision of the specialized literature on adaptive fuzzy control points out the existence of a considerable quantity and variety of realizations about adaptive fuzzy control, among them the self-organizing controllers, the fuzzy adaptation of classic controllers and adaptive controllers with multiple criteria (see, e.g. [10] for the design of this type of controllers). There are also many papers describing successful applications with fuzzy tuning of PID controllers [11].

In this work the adaptation is first carried out by using a gain scheduling strategy. The weighted combination of the control actions delivered by the local PID controllers is applied to the process. Later on, and in order to avoid a continuous change in the gains, the weights are adapted following a fuzzy reasoning.

Coming back to the furnace application, a robust control able to stabilize and to provide appropriate performance to the furnace under any operating conditions is not possible. As previously mentioned, the variability of the f.d.t. T4 / f4 for different material flow (fm) is too strong (Table 1). Thus, it is possible to design local controllers and then obtain the control action from interpolation of the actions calculated by each controller.

There are diverse options to design and to adapt the controllers. In [12], this problem is analyzed from the perspective of the robust adaptive control: for

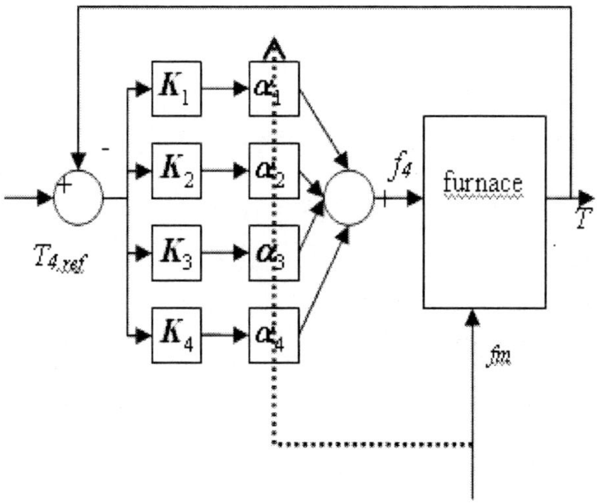


**Fig. 3.** Multi-model control of the temperature $T_4$

each operation zone a linear model is obtained, with bounded uncertainty, and a local robust controller is designed. By means of an appropriate weighing of the actions provided by the different controllers, the control action to be applied to the process is obtained.

A block diagram, valid for non linear processes with uncertainty, is illustrated in Fig. 3 for the control of temperature $T_4$ under variable load conditions.

Each one of the controllers $K_i (i = 1 \ldots 4)$ satisfies the control requirements for the corresponding transfer function $g_i$ of the Table 1. As it is well known, switching from one controller to another one, as far as the system is operating in one or another range of operation, does not guarantee the stability of the overall system, even if the controllers are suitable for a steady operation in their respective zone. Thus, the contribution of each controller in the control action is weighted by a single parameter $\alpha_i$ so that the weights add up to 1. The weights are determined by considering the correspondence with the load conditions. This compensation can be purely proportional. For example,

$$\alpha_3 = \begin{cases} 1 - |f_m - 24| & 23 < f_m < 25 \\ 0 & \text{otherwise} \end{cases} \tag{2}$$

or it can be calculated by means of any other interpolation technique, such as fuzzy logic. A simple fuzzy map is formed in such a way that it updates $\alpha_i$ in accordance to the current regulation error and error rate.

The fuzzy inference system designed in this case is of Mamdani-type, probably the most commonly used approach. Mamdani's method was used in the first control systems built using fuzzy set theory. It has two input variables: the error $e$ and the time derivative of error $\frac{de}{dt}$ of temperature $T_4$ . The error signal $e(t)$ is defined by

$$e(t) = T_{4ref} - T_4 \tag{3}$$

where $T_{4ref}$ is the reference temperature and $T_4$ is the current temperature in the hearth H4. These variables were fuzzified into the fuzzy variables $E$ and $DE$, as defined later on. The controller output for each zone of operation contributes with a corresponding weighing coefficient $\alpha_i$.

### 4.1 Definition of Partitions for Different Variables

Fuzzy parameters include the shape and position of membership functions. For input variables trapezoidal membership functions were used while for the output variable singletons were chosen.

For each input variable five fuzzy sets were considered:

$$\begin{aligned} E &= \{NL, NS, Z, PS, PL\} \\ DE &= \{NL, NS, Z, PS, PL\} \end{aligned} \tag{4}$$

where, as usual, the meanings of the acronyms in (4) are PL for positive large, PS for positive small, Z for zero, NS for negative small and NL for negative large, respectively.

The inference system output variable has been denoted by $X$, an auxiliary variable that, once defuzzified will become $x(t) = \Delta\alpha(t)$. For this variable, seven singleton subsets have been considered:

$$X = \{-3, -2, -1, 0, 1, 2, 3\} \tag{5}$$

Based on the load conditions, the variation in the weight for each controller is being apportioned in a similar way to (2).

The rule base is composed of 15 rules relating the linguistic input variables Eq. (4) to the singleton outputs (5).

### 4.2 Fuzzy Operators

For fuzzy reasoning the compositional operator "sup-product" of Kaufmann was used. That is, the product was taken to calculate the Cartesian product (logic connective "and") and the maximum was used to evaluate the union (logic connective "also"). In fuzzy logic controller (FLC) applications, the sup-min and sup-product compositional operators are the most frequently used. The reason is obvious if the computational aspects of an FLC are considered. The inferred results employing in the last compositional operator are better than those obtained by using the sup-min operator.

For the defuzzification process the centroid method was used. In all adaptive schemas the evaluation of the behavior of the system is required for the adaptation. The index used to evaluate the behavior of the system is the pseudo-damping rate $d_r$ [13] given as

$$d_r = \frac{r(t_2)}{r(t_1)} \tag{6}$$

$$r(t) = e^2(t) + p^* \left(\frac{de}{dt}\right)^2 \tag{7}$$

where $p > 0$ is a weighing constant and $t_1$ and $t_2$ are two consecutive sampling points with $t_2 > t_1$. As $r(t)$ is a function of both $e(t)$ and $\frac{de}{dt}$, it describes both the present and the future behavior of the system. A small value of $r(t)$ means that both $e(t)$ and $\frac{de}{dt}$ are small in absolute value and, thus, the temperature is close to its desired position, and large changes will not be undergone in the near future. On the other hand, a large value of $r(t)$ means that either the present state of the system differs greatly from the desired state or it will change greatly in the near future. It is expected that when $d_r(t) < 1$ for all $t$, $r(t)$ will decrease monotonically, which indicates that the system exhibits good performance. When $d_r(t) > 1$, $r(t)$ will increase monotonically, and the system will be unstable or, at least, the controller will need to be adjusted. From the evaluation of this index, the algorithm decides if the adaptation takes place. If this is the case, the $\alpha$ parameters are updated, the change being determined by the modification in the material flow.

## 5 Simulation Results

To carry out the simulation experiments, the package MATLAB-SIMULINK was used as tool. The block diagram in Fig. 3 was implemented, while the fuzzy inference system was implemented by means of the fuzzy logic toolbox according to the structure described in the previous section.

The first experiment consists on the application of step change of two tons in the disturbance (ore flow) at ten minutes and then returning to the original

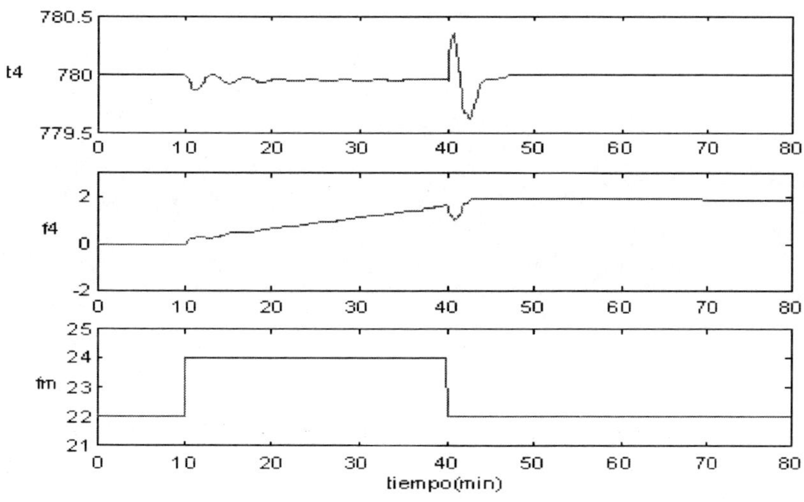


**Fig. 4.** Response to change in the ore flow from 22 to 24 $t/h$

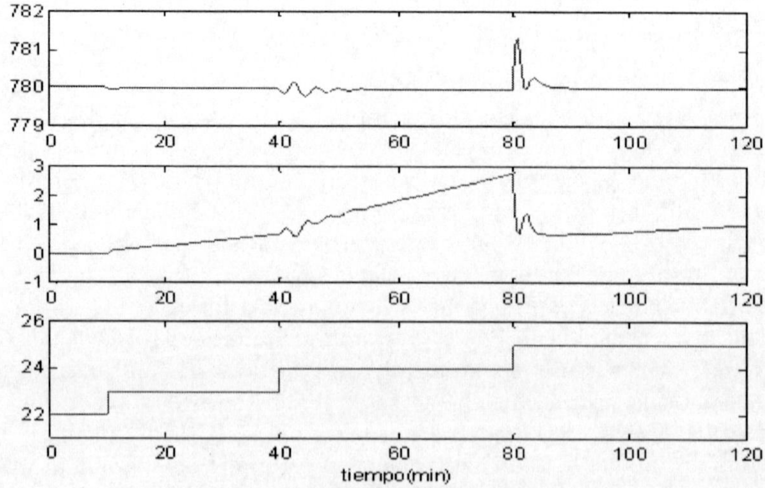


**Fig. 5.** Response to continuous load increase by 1 $t/h$

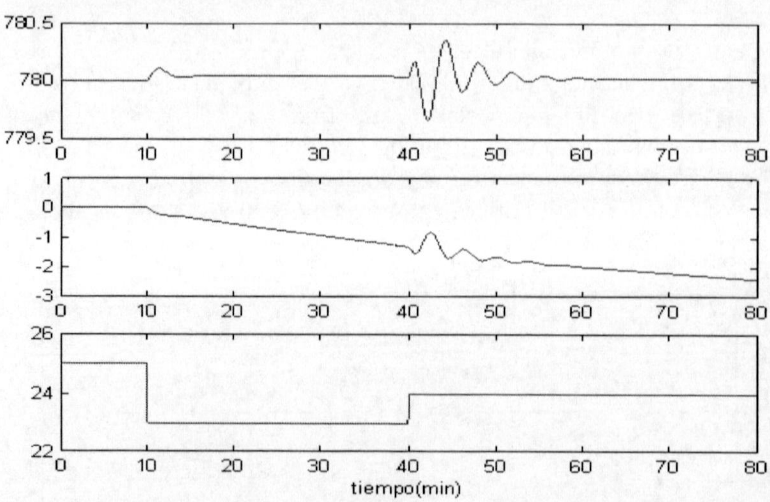


**Fig. 6.** Response to a decrease/increase in the load

value at 40 min. The results of the experiment can be appreciated in Fig. 4. In this case, first, the local controller 3 overcomes the local con-troller 1 and later on $\alpha_1$ recovers its maximum which allows the action of the local controller 1 to be applied.

Another experiment consisted on repeatedly increasing the mineral flow at an average rate of 1 $ton/h$. The results can be observed in Fig. 5. The system response only has small deviations off the reference ($T_{4ref} = 780C$) after the

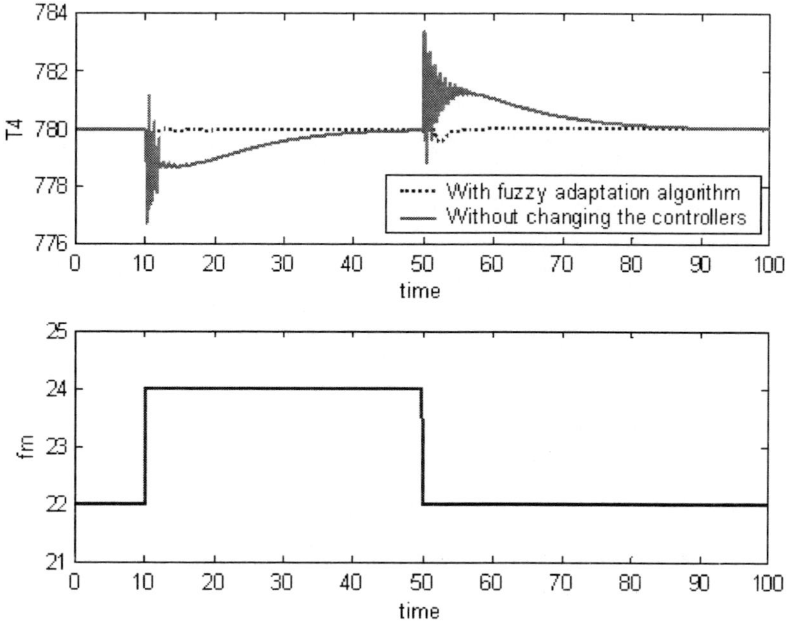


**Fig. 7.** Response to change in the ore flow from 22 to 24 $t/h$

appropriate selection of the more suitable local controllers for each value of the load.

The third experiment consisted on introducing an abrupt decrease and then an increase of the mineral flow. A good behavior of the system is also obtained, as depicted in Fig. 6.

Figure 7 clearly shows the remarkable disturbance response performance of the fuzzy adaptation algorithm over the actions without change in the controllers with shorter rise time, shorter settling time and less overshoot.

Observe that, unlike normal fuzzy controllers where lowering the over-shoot is often at the expense of slowing down considerably the rise time, this scheme seems to reconcile these two requirements. The reason is roughly that the adaptation allows the selection of different PID controllers for controlling the process. The algorithm is designed so that it picks up an appropriate PID controller combination at each case.

## 6 Conclusions

The implementation of PID controllers and their fuzzy combination improves the gain scheduling, and it allows the integration in a single system of a global control that includes the supervision and the operation with different operating environments. This corroborates that the fuzzy logic is an interesting alternative for the control of complex processes.

From the experiments carried out it is observed that the controlled variable in all cases responds with appropriate response time, the overshoot does not exceed the permissible limits values and it presents negligible stationary state error for different changes in ore flow, which is the main disturbance in this process. The simulations indicated that multi-model PID control strategy with fuzzy adaptation exhibits a good performance and robustness.

## References

1. Levine, W., et al.: The Control Handbook. CRC Press, USA (1996)
2. Lewis, F., Liu, K.: Automática 32(2), 167–181 (1996)
3. Albertos, P., Sala, A.: RIAI 1(2), 22–31 (2004)
4. Chang, A.: Minería y Geología 16(1), 76–82 (1999)
5. Angulo, M.: Identificación y control extremal de un horno de reducción. PhD Thesis, Czech Technical University, Prague (1982)
6. Ramírez, M.: Control borroso multivariable de la postcombustión en un horno de reducción de múltiples hogares. PhD Thesis, Universidad de Oriente, Santiago de Cuba, Cuba (2002)
7. Ramírez, M., Albertos, P.: Opciones de control en un horno metalúrgico. In: AMCA Congress, Cuernavaca, Mexico (2005)
8. Aström, K., Hägglund, T.: Automatic tuning of PID controllers. Instrument Society of America, USA (1988)
9. Astrõm, K.J.: Automatica 19, 471–486 (1983)
10. Wang, L.X.: Adaptive fuzzy systems and control. Prentice-Hall, Englewood Cliffs, NJ (1994)
11. He, S., Tan, S., Xu, F., Wang, P.: Fuzzy Sets and Systems 56, 37–46 (1993)
12. Athans, M., Fekri, S., Pascoal, A.: Issues on robust adaptive feedback control. In: IFAC World Congress, Prague (2005)
13. Raju, G., Zhou, J.: IEEE Transactions on Systems, Man and Cybernetics 23(4), 973–980 (1993)

# Index

# Author Index